T0332912

# Reproductive Biology of Angiosperms

This book is designed to introduce the basics of different aspects of the biology of reproduction in a concise and coherent manner. The book aims to equip students with the fundamentals of the biology of reproduction and also update them with the most recent advances in the field of reproduction. The book has been organized into 16 chapters that introduce and explain different aspects in a stimulating manner. Each chapter is supplemented with a summary and relevant illustrations. A glossary has been added to help the students to understand some important scientific terms.

The book offers comprehensive coverage of the important topics including:

- Flower structure and development
- Development and structure of male and female gametophytes
- Pollination biology, fertilization and self-incompatibility
- Endosperm, embryo and polyembryony
- Apomixis and seed biology

A separate topic on experimental plant reproductive biology (experimental embryology) has been provided, which includes basics of cell, tissue and organ culture, anther culture, pollen culture, flower, ovary, ovule culture, embryo culture, somatic embryogenesis, synthetic seeds, protoplast culture and other aspects of plant biotechnology.

The book aims to cater to the needs of the advanced undergraduate and postgraduate students in Botany, Forestry, Agriculture and related fields.

# Reproductive Biology of Angiosperms

Arun K. Pandey

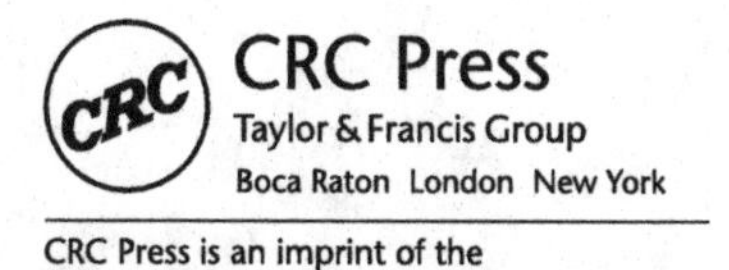


CRC Press is an imprint of the
Taylor & Francis Group, an **informa** business

First edition published 2023
by CRC Press
6000 Broken Sound Parkway NW, Suite 300, Boca Raton, FL 33487-2742

and by CRC Press
4 Park Square, Milton Park, Abingdon, Oxon, OX14 4RN

*CRC Press is an imprint of Taylor & Francis Group, LLC*



*Library of Congress Cataloging-in-Publication Data*

Names: Pandey, A. K. (Arun K.), author.
Title: Reproductive biology of angiosperms / Arun Pandey.
Description: First edition | Boca Raton, FL : CRC Press, 2023. | Includes bibliographical references and index.
Identifiers: LCCN 2022023858 | ISBN 9781032196206 (hardback) | ISBN 9781032196213 (paperback) | ISBN 9781003260097 (ebook)
Subjects: LCSH: Angiosperms--Reproduction.
Classification: LCC QK495.A1 P36 2023 | DDC 583--dc23/eng/20220707
LC record available at https://lccn.loc.gov/2022023858

ISBN: 9781032196206 (hbk)
ISBN: 9781032196213 (pbk)
ISBN: 9781003260097 (ebk)

DOI: 10.1201/9781003260097

Typeset in Times New Roman
by Deanta Global Publishing Services, Chennai, India

# Contents

# Preface

Plant reproductive biology is the study of mechanisms and processes of sexual and asexual reproduction in plants. Reproductive biology involves the study of phenology, floral biology, pollination biology (including pollen–pistil interaction and stigma receptivity) and breeding systems. Reproductive biology of flowering plants is important for determining barriers to seed and fruit set, for conservation, and for understanding pollination and breeding systems that regulate the genetic structure of populations. Studies on reproductive biology also help in developing strategies to preserve the genetic potential of rare and threatened species and are crucial for their restoration and reintroduction. There is growing concern regarding food and nutritional security for the increasing population and conservation of biological diversity. In view of this, the reproductive biology of flowering plants has acquired special significance. Any conservation approach has to be based on an in-depth study of the reproductive biology of plants.

The book is designed to meet the requirements of advanced undergraduate and post-graduate students of Plant Science. The book presents fundamental aspects of the subject in a simple and concise form. An attempt has been made to include up-to-date information collected from the latest journals and reference books. The selected references at the end of each chapter would be helpful to those students who wish to obtain further information on a topic.

In undertaking the task of preparing a textbook, it is necessary to use information from the publications of many researchers. I am grateful to all those whose information and interpretations have been used in this book.

I am deeply indebted to my revered teacher, the late Dr Ram Pratap Singh, CSIR-National Botanical Research Institute, Lucknow under whose guidance I learnt the basics of plant reproductive biology. I am thankful to Drs M. Anis, Ashok Sharma, C.R. Deb and Dinesh Agrawala for providing photographs. Help rendered by Dr Shruti Kasana, Department of Botany, University of Delhi is thankfully acknowledged.

I hope that students and teachers will find this book stimulating and useful. Comments and suggestions for improvement of the book are welcome.

**Arun K. Pandey**

# About the Author

**Arun K. Pandey**, former Professor of Botany and Dean of Colleges, University of Delhi, is presently Vice-Chancellor at Mansarovar Global University, India. After graduating from the University of Lucknow, India, he worked for a doctorate degree at CSIR-National Botanical Research Institute, Lucknow. He was a post-doctoral fellow at Ohio State University, USA. He has over four decades of teaching and research experience. Dr Pandey has supervised 32 PhD students; published over 180 research papers in peer-reviewed journals; and edited/authored 12 books including 2 edited volumes on Reproductive Biology of Angiosperms. He has described seven species new to science. He is Fellow of the National Academy of Sciences, India. He was INSA Visiting Fellow at the University of Munich, Germany, University of Vienna, Austria, and the Korea Research Institute of Bioscience & Biotechnology, and Bass Fellow at the Field Museum, Chicago. He was DAAD Visiting Fellow at the Technical University of Munich. He is the recipient of prestigious awards including P. Maheshwari and Y.S. Murty Medal of the Indian Botanical Society, V. Puri medal of the Indian Science Congress Association, V.V. Sivarajan medal of the Indian Association for Angiosperm Taxonomy, and Saligram Sinha Memorial award of the National Academy of Sciences, India. He is past President of the Indian Botanical Society, Indian Science Congress Association (Plant Science Section) and Indian Association for Angiosperm Taxonomy. Presently he is President of the East Himalayan Society for Spermatophyte Taxonomy.

# 1 Introduction

## HISTORICAL BACKGROUND

Arabs and Assyrians in third century B.C. used to perform the ritual of artificial pollination of date palm in which a man climbed up a male tree, brought down the inflorescence and handed it over to the high priest, who touched the female inflorescence with it, in order to get a good yield of dates. However, they were not aware of sexuality in plants.

Grew (1682) for the first time mentioned stamens as the male organs of the flower. He believed that the pollen grains, by merely falling upon the stigma, transmitted to the ovary a "vivifick effluvium" which prepared it for the production of the fruit. Camerarius (1694) noticed that a female mulberry tree, near which no male plant was growing, produced only abortive seeds. Inspired by this discovery, he experimented with some female plants of *Mercurialis annua* and kept them completely isolated from male plants. None of the fruits were found to contain fertile seeds. This encouraged him to make further observations in *Ricinus* and *Zea mays*. He concluded that anthers are the male sex organs and ovary, with its style, the female sex organ.

Kolreuter (1761) performed several experiments on the sexuality of plants and presented a detailed account of the role of insects in pollination. He made notable contributions to pollen morphology and was able to identify two distinct layers in the covering of pollen grains. He further noticed that if the stigma of a plant received its own pollen and that of another species at the same time, normally, only the pollen grains of the same species were effective in fertilization. Kolreuter also produced hybrids in *Dianthus*, *Hyoscyamus*, *Mathiola* and *Nicotiana*.

Credit for the discovery of the pollen tube goes to Amici (1824), an Italian mathematician, astronomer and meticulous microscope maker, who discovered the pollen tube in *Portulaca oleracea*. He observed that a pollen grain attached to the stigmatic hair suddenly split open and gave out a tube or "gut" which grew along the side of the hair and entered the tissues of the stigma. Based on his detailed examination of *Hibiscus syriacus*, Amici (1830) concluded that the pollen tube, after coming out of the pollen grain, elongates bit by bit and finally comes in contact with the ovules, one tube for each ovule.

Brogniart (1827) also examined several pollinated pistils. He observed the pollen tubes (spermatic tubules) but imagined that they burst and discharged the "spermatic granules". He also visualized the passing of spermatic granules through the style, and their reaching the placenta, as spermatozoids do in animals.

Schleiden (1837) confirmed Amici's statement that the pollen tubes enter from the stigma to ovule but believed that the tip of the pollen tube forms the

DOI: 10.1201/9781003260097-1

embryonal vesicle which after repeated divisions gives rise to the embryo. He further believed that the embryo sac served as an incubator within which the end of the pollen tube was nourished to form a new plantlet. Such a conclusion would obviously negate sexuality in plants. Nevertheless, Schleiden was able to muster support for his absurd idea; notable among his supporters was Schacht.

Amici (1842) contradicted Schleiden's idea and tried to prove that the embryo developed from a portion already existing in the ovule and was fertilized by the fluid in the tube. Schleiden (1845) vehemently challenged the observations of Amici and invited those who opposed his idea to see for themselves that his assertion was the real truth.

Amici (1847) collected evidence to demonstrate that in *Orchis* the "germinal vesicle" was already present inside the embryo sac before the arrival of the pollen tube and embryo developed from the vesicle, of course stimulated by the presence of the pollen tube. Amici's idea got support from several researchers.

Hofmeister (1849) studied 38 species belonging to 19 genera and published his findings in "Die Entstehung des Embryo der Phanerogams". He concluded that in all cases the embryo developed from a pre-existing cell in the embryo sac and not from the pollen tube. Hofmeister (1849) also studied the formation of tetrads during microsporogenesis in *Tradescantia*, development and organization of the female gametophyte and development of the cellular type of endosperm in *Monotropa hypopytis.*

Schleiden and Schacht continued to hold their ideas for some time, but soon evidence against them became so overwhelming that both had to withdraw their views. Radalkofer (1856), while accepting the findings of Hofmeister conclusively, proved that the embryo originated from a cell within the embryo sac and not from the pollen tube. With the publication of Radalkofer's review, controversy ended between Amici and Schleiden. Hanstein (1870) studied the development of embryos of *Capsella* and *Alisma*. He was the first to follow the sequence of early cell division in embryogeny. Famintzin (1879) confirmed the observations of Hanstein. Trueb (1879) traced the embryogeny in several orchids and reported the presence of suspensor haustoria.

Strasburger (1879) traced the development of adventive embryos from the nucellus in *Citrus* and some other plants. He also studied the development of the female gametophyte in *Polygonum divaricatum* and further pointed out the orientation and establishment of polarity at binucleated embryo sac followed by the organization of eight nucleate gametophyte. The most significant contribution of Strasburger is the discovery of syngamy in 1884. He observed actual fusion of the male gamete with the female gamete (egg) in *Monotropa hypopitys*. Guignard (1881) gave a detailed account of massive suspensors of the Leguminosae. Treub (1891) reported chalazogamy in *Casuarina*.

S.G. Nawaschin (1898) discovered double fertilization in *Fritillaria* and *Lilium*. He observed that one of the male gametes fused with the egg (syngamy) and other with the two polar nuclei (triple fusion). Double fertilization was soon demonstrated in other species of angiosperms, and later on it was found to be of universal occurrence in angiosperms.

By 1900, most of the details about the development of male and female gametophytes and embryos had been discovered. Coulter and Chamberlain (1903) summarized this data in their book *Morphology of Angiosperms*. Soueges (1913–1948) traced embryogeny in a large number of taxa belonging to both dicotyledons and monocotyledons. Karl Schnarf published his two books *Embryologie der Angiospermen* (1929) and *Vergleichende embryologie der Angiospermen* (1931) which still serve as valuable reference books. Johansen's *Plant Embryology* (1950) provided ample details and analyses of embryonic development in all groups of seed plants.

P. Maheshwari published his book *An Introduction to the Embryology of Angiosperms* (1950), which is still an excellent text in embryology. This was followed by an edited volume entitled *Recent Advances in the Embryology of Angiosperms* (Maheshwari, 1963).

G.L. Davis (1966), an Australian embryologist, brought out *Systematic Embryology of Angiosperms*. Recently three edited books, namely *Experimental Embryology of Vascular Plants* (Johri, 1982), *Embryology of Angiosperms* (Johri, 1984) and *Comparative Embryology of Angiosperms* (Johri et al., 1992) have been published.

From 1965 to 1975 considerable work was carried out on the organization of the female gametophyte and the ultrastructure of its constituent cells. W.A. Jensen and his colleagues at the University of California, Berkeley demonstrated that synergids are metabolically more active than the egg and play an important role in the process of fertilization. A pollen tube after entering the embryo sac discharges its contents in one of the synergids from where one of the sperms migrates to the egg and another sperm to the central cell. The occurrence of filiform apparatus in synergids has been correlated with this function of synergids.

Cass (1973) demonstrated that the two sperms produced by a male gametophyte remain together for a considerable time. Subsequent researchers reported the presence of a "male germ unit" in which a vegetative cell remains in contact with the sperm cells (see Chapter 4). Kranz et al. (1991) first achieved *in vitro* fertilization of maize using isolated sperm and egg cells and finally generated fertile plants (Kranz and Lörz, 1993).

With the accumulation of comparative data of a large number of angiosperms, it became possible to utilize the embryological characters in taxonomy, especially in cases where there were disputes regarding the systematic positions of taxa belonging to different orders and families. The first comparative account of comparative embryology of angiosperms is by Schnarf (1931), followed by Davis (1966) and, more recently, by Johri et al. (1992). That the embryological characters can be useful in taxonomy was indicated by both Hofmeister (1856) and Strasburger (1878, 1900) towards the end of the 19th century. Subsequently, the invention of rotary microtome and widespread use of paraffin sectioning resulted in the considerable accumulation of embryological data for systematic considerations. This prompted Schnarf (1937), Mauritzon (1939), Maheshwari (1950) and Kapil and Bhatnagar (1994) to pinpoint some conservative features which could have systematic significance. Dahlgren and Clifford (1982) designed a series of diagrams for comparative representation of several embryological characters in various families. Bhatnagar and Pandey (2021) worked out the embryology of the genus *Corokia* (Argophyllaceae) and highlighted the role of embryological characters in taxonomic considerations.

The beginning of experimental embryology dates back to the observations of Massart (1920) who treated the ovaries of certain plants with dead pollinia, aqueous extracts of pollen and spores of *Lycopersicon*, and obtained swelling of the ovary. Since then, experimental embryology has travelled a long way, and we now have a much deeper understanding of the development of anther, pollen, embryo sac, control of pollination, fertilization, endosperm, embryo and seed.

Laibach (1925, 1929) isolated embryos from non-viable seeds of the cross *Linum perenne* × *Linum austriacum* and reared them to maturity on a nutrient medium. Based on his observations, he demonstrated the practical application of embryo culture in plant breeding. LaRue (1942) was the first to attempt the culture of ovaries of a few angiosperms. Reinert (1958) and Stewards et al. (1958) discovered an embryo-like structure in tissue culture of *Daucus carota* which became an object of general biological interest. Since then, somatic embryogenesis has been observed in a large number of plants. Artificial seeds have been produced by encapsulating somatic embryos in a protective coating.

Over the last decades, the female gametophyte (FG) of flowering plants has attracted the attention of plant embryologists. It is clear that the female gametophyte serves as an attractive model system for studies of cell specification, cell–cell communication and programmed cell death (Tekleyohans et al., 2017).

Kanta et al. (1962) cultured excised ovules and pollen grains together on the same medium. They observed that the pollen grains germinated and fertilized the ovules. This technique was termed "test tube fertilization". Sachar and Guha (1962) cultured ovary of *Ranunculus sceleratus* and induced polyembryony. Guha and Maheshwari (1964, 1966) reported development of embryos in anther cultures of *Datura innoxia* and subsequently confirmed their origin from pollen grains (Guha and Maheshwari, 1966). Anther culture technique has been widely practiced and has been applied with varying degrees of success in obtaining haploids in many crop plants since 1964.

Bourgin and Nitsch (1967), using *Nicotiana* anthers, observed that the pollen embryos developed directly into haploid plants. Experimental embryology has been helpful in revealing the physiological, biochemical and genetical processes involved in the growth, development and differentiation of reproductive organs. Liu et al. (1993) succeeded in achieving complete embryogenesis in the cultures of excised 8-celled proembryos of *Brassica campestris* (mustard), by using a rich nutritive medium and a modified culture technique. Kranz and Lorz (1993) and Holm et al. (1994) achieved full fertile plants from *in vitro* and *in vivo* formed zygotes of maize and barley, respectively, by supplementing the culture medium with nurse cells.

Ploidy breeding, the modification of the number of chromosome sets in a plant genome, is frequently used in ornamental plant breeding to induce novel variations and create homogenous lines (Eechaut et al., 2018). A commonly used protocol for ploidy breeding is haploid induction, which refers to haploid regeneration via a single gamete cell under specific conditions, using either pollen (anther or microspore culture) or egg cells (ovule culture) (Kiełkowska, 2018).

## DEVELOPMENT AND SCOPE

The combined efforts of embryologists, geneticists and molecular biologists have resulted in the discovery of specific genes that control meiosis, egg cell development and early stages of embryogenesis. In recent years, several molecular studies on sperms and eggs have been published. Numerous sperm-expressed genes have been identified and their transcribed products have been shown to be essential for fertilization and normal embryogenesis (Frank and Johnson, 2009; Russel et al., 2012). Flores-Tornero et al. (2019) have provided a protocol for the manual isolation of egg apparatus cells from the basal angiosperm, *Amborella trichopoda.*

Our understanding of the cellular and molecular basis of induction of microspore embryogenesis has improved in recent years. However, the mechanism underlying the change in microspore cell fate is still largely unknown (Testillano, 2019). Recent years have witnessed a surge of information on identification and control of the haploid genome, which regulates activities during initial pollen germination and fertilization, characterization and structure of sperm cells, and structure and composition of pollen tubes. Artificial means have been developed for storing pollen for long periods. Pollen storage has important applications in both fundamental and applied areas of pollen biology. The main aim is to establish "pollen banks" so that the breeder can get the required pollen at any time of the year, from any part of the world. Currently, gaseous and cryobiological methods are used to prolong the longevity of pollen. The use of organic substances, such as hexane and ethyl alcohol ensures their successful transportation across continents.

The development of male and female gametophytes has drawn considerable attention because this knowledge has found application in generating male sterility. The sterile male plants can be cross-pollinated and are widely used in commercial hybrid seed production. Male sterility can also be induced experimentally: the basic strategy is to use tissue-specific and developmentally regulated expression of some of the genes that interfere with the normal growth and development of anthers and pollen grains. Using a tapetum-specific promoter, Rnase or glucanase have been engineered into some plants.

Extensive studies have been carried out on the cytology, genetics, physiology and biochemistry of self-incompatibility (SI). Apart from its importance in fundamental and applied aspects of reproductive biology, SI has provided an attractive model to study cellular recognition and communication in higher plants. Biochemical studies carried out since the 1960s have established that S-allele produces a glycoprotein which coseggregates with the S-gene. In recent years, considerable progress has been made in molecular characterization of S-alleles. Immunological principles are being used to explain the nature of the product of the sporophytic cells (stigma and style) involved in gametophytic and sporophytic incompatibility.

Concerted application of electron microscopy, autoradiography and cytochemistry has led to a deeper understanding of the functioning of reproductive structures. The embryo sac, though morphologically reduced, has been shown to be a highly specialized entity, and its constituent cells exhibit intricate metabolic interrelations. The study of cleared, unsectioned ovules and of squashed ovules with phase contrast or Nomarski interference optics (Herr, 1971, 1973) facilitates the study of archesporial differentiation, megasporogenesis and megagametogenesis by circumventing time-consuming sectioning and allows more quantitative research of these reproductive processes within or between closely related species (Bouman, 1984).

During the last decade or two, focus has shifted from a purely descriptive or comparative approach of embryo development to a more dynamic one. This has resulted in regarding the embryo as a reaction system interacting in a special environment. The role of polar auxin transport is now clearly established in embryo development and specification of the position of cotyledons.

The prospect of using plant somatic embryos produced in tissue cultures as "synthetic seeds" is a subject of increased interest since this technology is expected to have a significant impact on plant propagation in the future (Mora-Gutierrez et al., 2012; Javed et al., 2017).

During the last five years, the progress made in the area of freeze preservation of plant cell, tissue and organ cultures and regeneration of entire plants from such retrieved cultures has created an atmosphere of optimism about the feasibility of employing cryogenic methods as a meaningful tool for the long-term storage of germplasm.

Plants regenerated from cultured cells, tissues, organs and embryos have become models for studying mechanisms of differentiation and molecular genetic laws of morphogenesis of embryonal structures (Batygina, 2009). One of the most significant developments in experimental embryology has been the isolation, culture and fusion of protoplasts which has far-reaching implications in crop improvement (Ahmed et al., 2021). Isolated protoplasts from unrelated or distantly related species can be fused resulting in somatic hybrids. DNA uptake into protoplasts is now a routine and universally accepted procedure in plant biotechnology for introducing and evaluating both short-term (transient) and long-term (stable) expression of genes in cells and regenerated plants.

Over the last decade, plant reproduction research has been expanding rapidly. There is an unprecedented emphasis on crop improvement for increased and higher-quality food production. Consequently, as a necessary basis for crop improvement, plant reproduction research has attracted much attention, and several projects focused on plant fertility, fertilization, embryogenesis, endosperm development, and meiosis have been funded. All of the grants promoted plant reproduction research using unique technologies, e.g., live-cell imaging, microfluidics, chemical synthesis, structural biology and bioinformatics. As a consequence, plant reproduction research is one of the most active and leading fields in biological sciences. There is an urgent need to fully understand the fundamental molecular processes that regulate plant reproduction and lead to development and formation of seeds.

In recent years remarkable progress has been made in the field of plant reproduction, mainly due to the use of the latest advanced technologies such as next-generation sequencing, high throughput DNA sequencing, cutting-edge genetics, cell imaging, new cellular isolation technologies, and molecular and quantitative approaches (Pereira and Coimbra, 2019).

## REPRODUCTION AND REGENERATION IN PLANTS

Reproduction is defined as a biological process in which an organism gives rise to offspring similar to itself. Reproduction allows a continuity of species from one generation to the next: it ensures their perpetuation as a distinct species. Reproduction is the basis of survival and sustenance of populations and species in natural habitats.

Angiosperms represent 90% of all living plant species (Paton et al., 2008). Flowering plants are able to reproduce both sexually and asexually, leading to the formation of seeds that enable them to survive to a new generation.

## SEXUAL REPRODUCTION

Sexual reproduction starts with the landing of pollen grains onto the surface of a receptive stigma. It involves the fusion of male and female gametes (reproductive cells) during fertilization. Sexual reproduction is the major source of genetic variation by way of meiotic crossing over and segregation of genes during gamete formation as well as the recombination of genes during fertilization. Sexual reproduction allows plants to adapt to an ever-changing environment by generating genetic variability. In angiosperms, the flower is the unit of sexual reproduction. However, in a few species, seeds develop without fertilization. This phenomenon is known as apomixis. Sexual reproduction generates recombination and enables the species to adapt to changing environment. In most of our crop plants, seeds and/or fruits are the economic products, and are the result of sexual reproduction.

When two parents of opposite sex take part in the reproductive process resulting in the fusion of a male and a female gamete, it is sexual reproduction.

In the sexual process, the pollen grains produce male gametes. The egg and secondary (fused polars) nucleus are formed in the embryo sac. One male gamete fuses with the egg (syngamy), and the other with the secondary nucleus or the two polars, resulting in triple fusion. These processes (the syngamy and triple fusion) are referred to as "double fertilization". Fertilization takes place with the fusion of the male and female gametes each with only half the number of chromosomes which will result in the formation of a zygote, a cell with a full set of chromosomes. The zygote (fertilized egg) usually develops into an embryo, which is the progenitor of the next generation. The primary endosperm nucleus, usually a product of triple fusion (fertilized secondary nucleus), produces the endosperm. As the sexually produced offspring do not arise from a single ancestral cell, but from the fusion of the male and female gametes, the offspring are not identical to the parents or among themselves.

In flowering plants, the reproductive phase is easily observed when they begin to flower. Some plants flower annually (e.g., wheat, maize, rice, marigold, etc.) and are called annuals. Some plants, like carrots, are biennials, producing flowers and seeds in their second year of growth. Some plants live for a number of years, growing and flowering year after year and are known as perennials.

Some plants exhibit an unusual flowering phenomenon. For example, bamboo species flowering occurs only once in their lifetime, generally after 50–100 years. Such plants are called monocarpic. They produce a large number of flowers and fruits and finally die. In the self-compatible *Strobilanthes kunthiana* (Acanthaceae), in the Western Ghats of India, flowering occurs after a regular gap of 12 years.

The events in the process of reproduction follow a regular sequence. Sexual reproduction is characterized by fertilization, zygote formation and development of the embryo. These sequential events may also be grouped as pre-fertilization, fertilization and post-fertilization events.

The natural process of sexual reproduction has been a useful tool for plant breeders in developing novel plants with superior agronomic traits by hybridizing selected elite parental lines. Several improved cultivars of crop plants have been produced by employing this method (Borlaug, 1983).

## ASEXUAL/VEGETATIVE REPRODUCTION

A species may reproduce with the participation of one or two individuals. When an offspring is produced by a single parent with or without the involvement of gamete formation, it is asexual reproduction.

In plants, asexual reproduction occurs by vegetative propagation. Vegetative propagation occurs through rhizome, tuber, runner, sucker, bulbil, offset, etc., which can give rise to new individuals. These structures are called vegetative propagules. The process of their formation does not involve two parents; hence this process is asexual. The offspring formed by vegetative propagation can be called clones as they are genetically identical, having arisen from a single parent. New plants are also regenerated from a portion of the vegetative part of a plant. In some plants, the organs like root, stem, leaf and even node, are variously modified for vegetative propagation. Besides vegetative propagation, these modifications may also perform some special functions, e.g., storage of food and tolerance under unfavourable conditions.

During favourable conditions, new buds arise from the nodes of the underground stem modifications like rhizomes (e.g., ginger), tubers (e.g., potato), bulbs (e.g., onions) and corms (e.g., gladiolus). When the nodes come into contact with damp soil or water, these buds develop roots and grow into a new plant.

Some plants have aerial modifications, e.g., runners (*Centella*), stolons (*Colocasia*), offsets (*Pistia*, *Eichhornia*) and suckers (*Chrysanthemum*) which also serve as means of vegetative propagation. Some plants like sweet potato, asparagus, tapioca, and dahlia are propagated through modified roots. Gardeners propagate *Bryophyllum* and *Kalanchoe* through adventitious buds produced in the leaf notches. These buds develop into new plants. In garlic (*Allium sativum*), some of the lower flowers of the inflorescence are modified into small multicellular bodies called bulbils that serve as units of propagation. Some plants like the American aloe (*Agave*) are propagated through the reproductive buds or bulbils that often take the place of many flowers of the inflorescence. In other plants, like the wild yam (*Dioscorea*), bulbils are produced in the leaf axil, while in wood-sorrel (*Oxalis*) bulbils arise on the top of the swollen tuberous roots.

The aquatic plant water hyacinth (*Eichhornia*) is an invasive weed that multiplies rapidly by vegetative propagation. It uses most of the dissolved oxygen in stagnant water. Its vegetative propagules are offsets, which are like short runners, or stems that grow horizontally giving rise to new plants from the axillary or terminal buds.

Vegetative reproduction through propagules in many species, including crops, is an advantage that circumvents failure in sexual reproduction and helps them survive through clonal expansion.

## TERMINOLOGY

### Embryology

Plant embryology deals with the sexual cycle of the plant. It includes microsporogenesis, megasporogenesis, development of gametophytes, fertilization and development of endosperm, embryo and seed.

Embryological studies can be classified into three broad areas: (i) descriptive embryology, (ii) comparative embryology and (iii) experimental embryology. Descriptive embryology deals with the critical study of the various developmental processes that take place in a plant from the initiation of the sex organs to the maturation of the embryo. Comparative embryology is concerned with the taxonomic evaluation of embryological characters. Experimental embryology aims at controlling and inducing changes in the embryological processes.

### Reproductive Biology

Reproductive biology has emerged as a rapidly evolving discipline which includes embryology, carpology, genetics, physiology and plant breeding. Plant reproductive biology is the study of mechanisms and processes of sexual and asexual reproduction in plants. Reproductive biology involves the study of flower organogenesis, anthesis, pollination, pollination biology (including pollen–pistil interaction and stigma receptivity), fertilization, embryogenesis, seed maturation, dispersal and germination and propagation by seeds. The reproductive biology of flowering plants is important for determining barriers to seed and fruit set, for conservation and for understanding pollination and breeding systems that regulate the genetic structure of populations.

Reproductive biology is important especially for the introduction and repatriation of rare, threatened and endangered plants including wild relatives of agricultural crops. Any conservation approach has to be based on an in-depth study of the reproductive biology of flowering plants, in order to determine the barriers to fruit and seed set and for understanding pollination and breeding systems that regulate the genetic structure of populations. Studies on reproductive biology also help in developing strategies to preserve the genetic potential of rare and threatened species.

### Pollen Biology

Pollen biology encompasses pollen production, its transfer to the stigma, pollen germination, and details of pollen–pistil interaction leading to fertilization and seed set. Pollen grains at the time of shedding are either two-celled (a large vegetative cell and a generative cell) or three-celled (vegetative cell and two sperm cells). Pollen biology studies are a prerequisite for any program aimed at optimization and improvement of the yield of crop plants. Pollination ecology is also a part of pollen biology.

### Pollen Biotechnology

Pollen biotechnology refers to the manipulation of various aspects of pollen biology for production and improvement of crops. It is one of the most challenging areas of plant reproductive biology and plays an important role in crop improvement programs. Pollen biotechnology can be integrated to conventional breeding programs for improvement of crops and other related economic products. Pollen biotechnology not only decreases the time and cost involved, but also greatly increases the efficacy of conventional breeding methods.

## SUMMARY

- Plant reproductive biology is the study of mechanisms and processes of sexual and asexual reproduction in plants. Sexuality in plants was first discovered by Camerarius (1694).
- Reproductive biology involves the study of phenology, floral biology, pollination biology (including pollen–pistil interaction and stigma receptivity), and breeding systems.
- Syngamy was discovered by Strasburger (1884). S.G. Nawaschin (1898) discovered the phenomenon of double fertilization.
- The study of reproductive biology is important for determining barriers to seed and fruit set, for conservation, and for understanding pollination and breeding systems that regulate the genetic structure of populations.

- Sexual reproduction is the major source of genetic variation by way of meiotic crossing over and segregation of genes during gamete formation and the recombination of genes during fertilization.
- Vegetative propagation occurs through the formation of units like rhizomes, bulbs, tubers, runners and suckers that are capable of giving rise to new offspring.

## SUGGESTED READING


Ahmed, A.A.A., M. Miao, E.D. Pratsinakis, Zhang, H., Wang, W., Yuan, Y., Lyu, M., Ifftikha, J., Yousef, A.F., Madesis, P., and Wu, B. 2021. Protoplast isolation, fusion, culture and transformation in the woody plant *Jasminum* spp. *Agriculture* 11 (8): 699. http://www.doi.org/10.3390/agriculture11080699.

Amici, G.B. 1824. Observation microscopiques sur diverses species de plants. *Ann. des Sci. Nat. Bot.* 2: 41–70, 211–248.

Amici, G.B. 1830. Note sur le mode d'action du pollen sur le stigmata. Extrait de'une lettere de'Amaici a Mirbel. *Ann. des Sci. Nat. Bot.* 21: 329–332.

Amici, G.B. 1844. Quatrieme reunion des naturalalistes italians. Padua, 1843. *Flora* 1:359.

Amici, G.B. 1847. Sur la fecundation des Orchidees. *Ann. des Sci. Nat. Bot.* 7/8: 193–205.

Batygina, T.B. 2009. *Embryology of Flowering Plants. Terminology and Concepts. Vol. 3. Reproductive Systems.* Enfield, NH: Science Publishers.

Bhatnagar, A.K. and A.K. Pandey. 2021. Embryology and systematic position of *Corokia* A. Cunn. (Argophyllaceae, Asterales). *Taiwania.* http://www.doi.org/10.6165/tai.2021.66.

Bhojwani, S.S. and W.Y. Soh (eds.). 2001. *Current Trends in the Embryology of Angiosperms.* Dordrecht: Kluwer Academic Publishers.

Borlaug, N.E. 1983. Contributions of conventional plant breeding to food production. *Science* 219: 689–693.

Bouman, F. 1984. The ovule. In *Embryology of Angiosperms*, ed. Johri, B.M. Berlin and Heidelberg: Springer. https://doi.org/10.1007/978-3-642-69302-1_3.

Bourgin, J.P. and J.P. Nitsch. 1967. Obtention de Nicotiana haploides à partir d'étamines cultivées in vitro. *Ann. Physiol. vég.* 9: 377–382.

Brongniart, A. 1827. Memoire sur la generation et le developpement del'embryon dans les vegetaux phanerogamiques. *Ann. des Sci. Nat. Bot.* 12: 14–53, 145–172, 225–298.

Camerarius, R.J. 1694. *De sexu plantarum epistola.* Leipzig: Wilhelm Engelmann.

Cass, D.D. 1973. An ultrastructural and Nomarski-interference study of the sperms of barley. *Can. J. Bot.* 51: 601–605.

Chauhan, Y.S. and A.K. Pandey. 1995. *Advances in Plant Reproductive Biology, Vol. I.* Delhi: Narendra Publishing House.

Coulter, J. and C.J. Chamberlain. 1903. *Morphology of Angiosperms.* New York: D. Appleton.

Cresti, M. and A. Tiezzi. 1992. *Sexual Plant Reproduction.* Berlin: Springer-Verlag.

Dahlgren, R.M.T. and H.T. Clifford. 1982. *The Monocotyledons. A Comparative Study.* London: Academic Press.

Davis, G.L. 1966. *Systematic Embryology of Angiosperms.* New York: John Wiley.

Eeckhaut, T., Van der V, J., Dhooghe, E., Leus, L., Van Laere, K., and Van Huylenbroeck, J.2018. Polidy breeding in ornamentals. In *Ornamental Crops*, ed. Johan Van Huylenbroeck, 145–173. New York: Springer International Publishing

Famintzin, A. 1879. Embryologische Studien. *Mem. Acad. Imp. des Sci. St.Petersburg VII*, 26 (10): 1–19.

Flores-Tornero, M., S. Proost, M. Mutiwil, C.P. Scutt, T. Dresselhaus and T. Sprunks. 2019. Transcriptomics of manually isolated *Amborella trichopoda* egg apparatus cells. *Plant Reprod.* 32: 15–28.
Frank, C.A. and M.A Johnson. 2009. Expressing the Diphtheria toxin A subunit from the HAP2 (GCS1) promoter blocks sperm maturation and produces single sperm-like cells capable of fertilization. *Plant Physiol.* 151 (3): 1390–1400.
Grew, N. 1682. *The Anatomy of Plants.* London: W. Rawlins.
Guha, S. and S.C. Maheshwari. 1964. In vitro production of embryos from anthers of Datura. *Nature (London)* 204: 497.
Guha, S. and S.C. Maheshwari. 1966. Cell division and differentiation of embryos in the pollen grains of Datura in vitro. *Nature* 212: 97–98.
Guignard, L. 1881. Reserches d'embryogenie vegetale compare. I. Ligumineuses. *Ann. des Sci. Nat. Bot.* 12: 5–166.
Hanstein, J. 1870. Die Entwickelung des Keimes der Monocotylen and Dicotylen. *Bot. Abhandl. Bonn.* 1: 1–112.
Herr, J.M., Jr. 1971. A new clearing-squash technique for the study of ovule development in angiosperms. *Am. J. Bot.* 58: 785–790.
Herr, J.M., Jr. 1973. The use of Nomarski-interference microscopy for the study of structural features in cleared ovules. *Acta Bot. Ind.* 1: 35–40.
Hofmeister, W. 1849. *Di Entstehungen des Embryo der Phanerogamen.* Leipzig.
Hofmeister, W. 1859. Neue Beitrage zur Kenntnis der Embryobildung der Phaneorogamen. I. Dikotyledonen mit urs prunglich einzelligem, nur durch Zelltheilung wachsendem Endosperm. *Abh. Konigl. Sachs. Gesell. Wiss.* 1859: 535–672.
Holm, P.B., S. Knudsen, P. Mouritzen, D. Negri, F.L. Olsen and C. Roue. 1994. Regeneration of fertile barley plants from mechanically isolated protoplasts of the fertilized egg cell. *Plant Cell* 6: 531–543.
Javed, S.B., A.A. Alatar, M. Anis and M. Faisal. 2017. Synthetic seeds production and germination studies, for short term storage and long distance transport of *Erythrina variegata* L.: A multipurpose tree legume. *Ind. Crops Products* 105: 41–46.
Johri, B.M. 1982. *Experimental Embryology of Vascular Plants.* Heidelberg: Springer Berlin.
Johri, B.M. 1984. *Embryology of Angiosperms.* Berlin: Springer-Verlag.
Johri, B.M., K.B. Ambegaokar and P.S. Srivastava (eds.). 1992. *Comparative Embryology of Angiosperms*, Vols. 1 and 2. Heidelberg: Springer Berlin.
Kanta, K., N.S. Rangaswamy and P. Maheshwari. 1962. Test-tube fertilization in a flowering plant. *Nature* 194: 1214–1217.
Kapil, R.N. and A.K. Bhatnagar. 1994. The contribution of embryology to the systematics of the Euphorbiaceae. *Ann. Miss. Bot. Gard.* 81 (2): 145–159.
Kiełkowska, A., A. Adamus and R. Baranski. 2018. Methods in Molecular Biology. In *Plant Cell Culture Protocols*, eds. Víctor M. Loyola-Vargas and Neftalí Ochoa-Alejo, 301–315. New York: Springer.
Kolreuter, J.C. 1761. *Vorlaufige Nachricht von einigen das Geschlecht der Pflanzen betreffenden Versuchen und Beobachtungen.* Leipzig: In der-Gleditschischen Handlung.
Kranz, E. and H. Lörz. 1993. In vitro fertilization with isolated, single gametes results in zygotic embryogenesis and fertile maize plants. *Plant Cell* 5: 739–746.
Kranz, E., Bautor, J. and H. Lörz. 1991. In vitro fertilization of single, isolated gametes of maize mediated by electrofusion. *Sex. Plant Reprod.* 4: 12–16.
Laibach, F. 1925. Das Taubwerden der Bastardsamen und die kunstliche Aufzucht fruh absterbender Bastardembryonen. *Ztschr. F. Bot.* 17: 417–459.
Laibach, F. 1929. Ectogenesis in plants: Methods and genetic possibilities of propagating embryos otherwise dying in the seed. *J. Hered.* 20: 201–208.

La Rue, C.D. 1936. The growth of plant embryos in culture. *Bul. Torrey Bot. Club* 63: 365–382.

Liu, C.M., Z. Xu and N.H. Chua. 1993. Auxin polar transport is essential for the establishment of bilateral symmetry during early plant embryogenesis. *Plant Cell* 5(6): 621–630.

Maheshwari, P. 1950. *An Introduction to Embryology of Angiosperms.* New York: McGraw Hill Book Co., Inc.

Maheshwari, P. (ed.). 1963. *Recent Advances in the Embryology of Angiosperms.* Delhi: International Society of Plant Morphologists.

Maheswari Devi, H., B.M. Johri, M.A. Rau, D. Singh, A.S.R. Dathan and R.K. Bhanwra. 1995. Embryology of angiosperms. In *Botany in India-History and Progress*, ed. B.M. Johri, Vol. II, 59–146. New Delhi: Oxford & IBH Pub. Co.

Massart, J. 1902. Sur la pollination sans fecundation. *Bull. Jrd. Bot. Brussels de P'Etat* 1: 85–95.

Mauritzon, J. 1939. Die Bedeutung der embryologischen Forschung fiir das naturliche System der Pflanzen. *Lunds Univ. Arsskr. N.F. II* 35 (15): 1–70.

Mora-Gutierrez, A., A.G. Gonzale-Gutierrez, B. Rodriguez-Garay, A. Ascencio-Cabral and L. Li-Wei. 2012. Plant somatic embryogenesis: Some useful considerations. In *Embryogenesis*, ed. Ken-ichi Sato, pp. 229–248. Croatia: Intech.

Nair, P.K.K. 1960. A modification in the method of pollen preparation. *J. Sci. Industr. Res.* 19C: 26–27.

Nawaschin, S.G. 1898. Resultate einer Revision der Befuchtungsvorgange bei *Lilium martagon* und *Fritillaria tenella. Bull. Acad. Imp. des Sci. St. Petersburg* 8: 345–357.

Pandey, A.K. and M.R. Dhakal. 1995. *Advances in Plant Reproductive Biology, Vol. II.* Delhi: Narendra Publishing House.

Pareira, A.M. and S. Coimbra. 2019. Advances in plant reproduction: From gametes to seeds. *J. Exp. Bot.* 70 (11): 2933–2936.

Paton, A.J., N. Brummitt, R. Govaerts, K. Harman, S. Hinchcliffe, B. Allkin and E. Nic Lughadha. 2008. Towards target 1 of the global strategy for plant conservation: A working list of all known plant species-progress and prospects. *Taxon* 57 (2): 602–611.

Radalkofer, L. 1856. *Die Befruchtung der Phanerogamen. EinBeitrag zur Entscheidung des daruber bestehenden Streites.* Leipzig: E. Engelmann.

Raghavan, V. 1986. *Embryogenesis andAngiosperms: A Developmental and Experimental Study.* Cambridge: Cambridge University Press.

Russel, S.D., X.G. Chui, W.X. Wang, T.Yuan, X. Wei, P.L. Bhalla and M. Singh. 2012. Genomic profiling of rice sperm cell transcripts reveals conserved and distinct elements in the flowering plant male germ lineage. https://doi.org/10.1111/j.1469-8137.2012.04199.x.

Sachar, J.A. and S. Guha. 1962. In vitro growth of achenes of *Ranunculus sceleratus* L. In *Plant Embryology - A Symposium*, 244–253. New Delhi: Council of Scientific and Industrial Research.

Schleiden, M.J. 1837. Einige Blicke auf die Entwicklungsgeschichte des vegetablischen Organismus bei den Phanerogamen. *Ac. Bwl. Natugeschichte, III*, 1: 289–320.

Schleiden, M.J. 1845. Uber Amicis letzten Beitrag zur Lehre von der Befruchtung der Pflanzen. *Flora*, 593–600.

Schnarf, K. 1929. *Embryologie der Angiospermen.* Berlin: Gebruder Borntraeger.

Schnarf, K. 1931. *Vergleichende Embryologie der Angiospermen.* Berlin: Gebruder Borntraeger.

Shivanna, K.R. and V.K. Sawhney. 1977. *Pollen Biotechnology for Crop Production and Improvement.* Cambridge: Cambridge University Press.

Steward, F.C., M.O. Mapes and K. Mears. 1958. Growth and organized development of cultured cells. II. Organization in cultures grown from freely suspended cells. *Am. J. Bot.* 45: 705–708.

Strasburger, E. 1878. Uber Polyembryonie. *Jenaische Ztschr. F. Naturw.* 12: 647–670.

Strasburger, E. 1879. *Die Angiospermen und Gymnospermen*. Jena.

Strasburger, E. 1884. *Neue Untersuchungen uber den Befruchtungsvorgang bei den Phanerogamen*. Jena.

Strasburger, E. 1900. Einige Bemerkungen zur Frage nach der "doppelten Befruchtung" bei Angiospermen. *Bot. Ztg. II* 58: 293–316.

Sun, M.X., W.C. Yang and T. Higashiyama. 2018. Special issue on plant reproduction research in Asia. *Plant Reprod.* 31: 1–2. https://doi.org/10.1007/s))497-018-0330-9.

Tandon, R., K.R. Shivanna and M. Koul. 2020. *Reproductive Ecology of Flowering Plants: Patterns and Processes*. Singapore: Springer.

Tekleyohans, D.G., N. Thomas and R. Groß-Hardt. 2017. Patterning the female gametophyte of flowering plants. *Plant Physiol.* 173 (1): 122–129.

Testillano, P.S. 2019. Microspore embryogenesis: Targeting the determinant factors of stress-induced cell reprogramming for crop improvement. *J. Exp Bot.* 70 (11): 2965–2978.

Treub, M. 1891. Sur les Casuarinees et leur place dans le systeme natural. *Ann. Jard. Bot. Buitenzorg*. 10: 145–231.

# 2 Flower

The outstanding characteristic of angiosperms is the flower. The name angiosperm is derived from the Greek *angeion*, for vessel or receptacle, and *sperma*, for seed. Probably the most distinctive structure of the flower is the carpel, the vessel or container enclosing the ovules, which develop after fertilization into the seeds.

## FLOWER AS A MODIFIED DETERMINATE SHOOT

A flower is a shoot of determinate growth in which the internodes are highly reduced, and the leaves function as different floral parts. The flowering phase in plants, despite its morphological and physiological complexities, represents a transitory stage between vegetative and reproductive parts and is of great importance to seed and fruit maturation.

At the time of floral initiation, the shoot apex transforms into a floral apex. The axis that bears the floral organs is called the receptacle. This consists of several shortened nodes which are closely brought together by the suppression of internodes. Both fertile and sterile appendages are borne on the receptacle (Figure 2.1). The shortening and suppression of internodes bring the appendages close together, either in spirals or whorls.

Flowers show a great deal of variation in shape, size, color and insertion of different floral whorls. Among flowering plants, duckweed (*Wolffia microscopica*) possesses the smallest flowers (about 0.1 mm in diameter) whereas *Refflesia*, a root parasite found in the forests of Malaysia, bears the largest flowers (about 1 meter in diameter).

Flowers are generally borne on special branches collectively known as inflorescences. Flowers borne singly are termed as solitary flowers. These flowers may be terminal or axillary. Typical flowers have a long or a short stalk at the base. It is called a pedicel and is the first internode of the branch below the receptacle. Such flowers are called pedicellate, whereas flowers that lack pedicels are termed as sessile.

The leaflike structures in the axil of which flowers develop are called bracts. Similar foliar structures present at the base of the pedicel and subtending the remaining parts of the flower are called bracteoles.

Both sterile and fertile appendages are borne on the floral receptacle in distinct whorls consisting of calyx, corolla, androecium and gynoecium. A flower that bears all four organs (sepals, petals, stamens and pistil) is complete (e.g., cotton, *Gossypium*), and if any one of these parts are missing, the flowers are termed as incomplete (e.g., wheat, *Triticum*). The latter type (incomplete) may be perfect

DOI: 10.1201/9781003260097-2

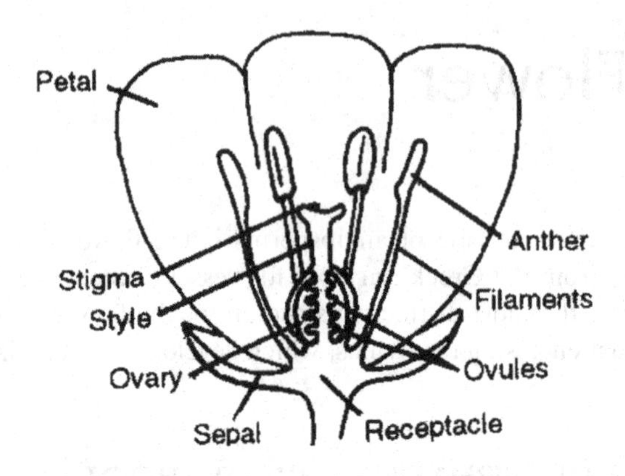


**FIGURE 2.1** The main parts of a flower.

– possessing both male and female sex organs – or imperfect – lacking one or the other sex organs.

Flowers with only male sex organs are known as staminate and those with only female sex organs are described as pistillate (or carpellate). In *Zea mays* (maize), the staminate and pistillate flowers are present on the same plant. If a flower possesses both the stamens and the carpels, it is designated as bisexual or hermaphrodite, and if found separately, the flower is called unisexual. A flower is said to be neuter or sterile if it has no functional stamen or carpel. If the male and female flowers occur on the same plant, it is monoecious (e.g., *Ricinus*) but if formed on separate plants, it is dioecious (e.g., *Cannabis*). Some plants bear both unisexual and bisexual flowers (e.g., *Mangifera*) and are called polygamous.

In some flowers, the corolla is made up of petals of similar shape that radiate from the center of the flower and are equidistant from each other. Such flowers are said to be actinomorphic, regular or radially symmetrical. In other flowers, one or more members of at least one whorl are of different form from the other members of the same whorl. These flowers are said to be zygomorphic, irregular or bilaterally symmetrical.

## CALYX

The outermost whorl of the flower is known as the calyx which is made up of individual sterile appendages, the sepals. Morphologically, sepals are modified leaves. Generally, sepals are small, green and leaflike, enclosing and protecting the young buds. When the sepals are free, the condition is polysepalous and when fused, it is gamosepalous. Sepals may be persistent or they may fall soon after anthesis. In some species sepals are conspicuous and serve as organs of advertisement to the pollinators (e.g., petaloid sepals in Proteaceae). In some cases, they may become petaloid (e.g., Proteaceae) and function to attract the insects for pollination. Where sepals are greatly reduced, they may be in the form of minute teeth, scales or bristles. All stages of calyx reduction are observed in *Cornus*. In some species sepals are persistent (remain green until fruit maturity). In gooseberry (*Physalis*) the sepals are persistent and protect the developing fruit.

## COROLLA

The corolla occupies a position between sepals and stamens and consists of petals (Figure. 2.1). Petals may be free or fused. Flowers with free petals are known as polypetalous (e.g., *Ranunculus*, *Brassica*), whereas those with fused petals are called gamopetalous (*Solanum*, *Helianthus*). Flowers which lack petals are called apetalous. Petals are variously colored and may or may not be fragrant. In insect pollinated flowers, nectaries may be present, often occurring at the base of the carpels. These nectaries (glands) secrete nectar to which insects are attracted. In some flowers, there may be modifications of the petals. Nectaries may also occur at the base of each petal (e.g., *Ranunculus sceleratus*). The corolla may show elongation of one petal, forming a spur (e.g., *Delphinium*). Sepals and petals commonly differ in form, size and other characteristics. In some families, they may be closely alike, as in most of the Liliaceae. In others, transitional forms occur, as in the Magnoliaceae. When sepals and petals are not distinguishable, they together constitute the perianth. The individual member of the perianth is called a tepal. The corolla and perianth protect the young reproductive organs. They also help in pollination through their attractive colours and curious features.

## ANDROECIUM

The androecium is formed by the male reproductive parts of the flower (stamens). A typical stamen consists of a filament and an anther. The pollen grains develop inside the anther. When mature, the pollen grains are released, often in large numbers, through narrow slits or pores in the anther. In between the filament and the anther, a continuation of the filament forms a strip of tissue that lies between the two anther lobes and is known as the connective. The structure and development of anthers are described in Chapter 3.

## GYNOECIUM

The gynoecium is the innermost whorl of the flower. It consists of one or more carpels. The carpels are borne laterally on the receptacle. Depending upon the position of the gynoecium in relation to other organs of the flower, the condition may be superior, inferior or semi-inferior. A typical carpel shows three distinct parts: the basal swollen portion called the ovary and the terminal pollen receptive part called the stigma. The middle-elongated portion is the style. The ovary bears the ovules in its locule(s). The female gametophyte develops inside the ovule. The number of ovules within an ovary may vary from one (sunflower) to several hundred (tobacco).

If the gynoecium comprises only one carpel, the condition is known as monocarpellary (e.g., Fabaceae), and if the gynoecium is composed of many carpels, the condition may be bicarpellary (e.g., Asteraceae, Brassicaceae), tricarpellary (e.g., Cyperaceae, Liliaceae) or multicarpellary (e.g., Malvaceae). A sterile carpel which does not bear ovules and becomes variously modified is called pistillode. A pistillode may be present in male flowers (e.g., Cucurbitaceae).

The portion of the ovary to which the ovules are attached is called the placenta. The arrangement of the placentae and, consequently, the ovules varies among different flowers. In some flowers, the placentation is parietal, i.e., the ovules are borne on the ovary wall or on extensions of it (*Brassica*). In other flowers, the ovules are borne on a central column of tissue in a partitioned ovary with as many locules as there are carpels. This is axile placentation (*Solanum*). In still others, the placentation is free central, where the ovules are borne upon a central column of tissue not connected by partitions with the ovary wall (*Dianthus*). And finally, in some flowers, a single ovule occurs at the very base of a unilocular ovary. This is basal placentation (*Helianthus*).

## ORGANOGENY OF FLORAL PARTS

Floral whorls develop from the swollen tip (thalamus) of the floral axis (pedicel). During the development of floral parts, the vegetative shoot apices are transformed into floral apices. The early stages of floral appendage development are like those of leaves. All organs develop laterally on the apical meristem, initiated by periclinal divisions below the surface layer. In niger (*Guizotia abyssinica*) floral apices are dome-shaped. It is on these dome-shaped structures that floral primordia and bracts start differentiating as small protuberances (Figure 2.2A), the former being placed in the axil of the latter.

The development of the floral primordia takes place in acropetal succession. In the beginning, the floral primordia are mound-shaped with a convex outline and are composed of a homogeneous mass of parenchymatous cells rich in cytoplasm. Very soon the primordia become concave owing to the initiation of petals. The carpel primordia develop after the initiation of the stamens (Figure 2.2B). Distinct sepal primordia are apparently absent during the organogeny of the floral parts, but epidermal cells at the basal level of the corolla tube (during further development of the flower) elongate and develop into uniseriately arranged many-celled hairs. The two carpel primordia grow in the form of a crescent and provide coverage at the apex enclosing the ovary cavity (Figure 2.2C, D).

## FLOWER DEVELOPMENT: GENETIC AND MOLECULAR ASPECTS

Flower development is under strict genetic control. It can be divided into several major steps, including floral induction, floral meristem formation and floral organ development. The control of different steps of flower development is achieved

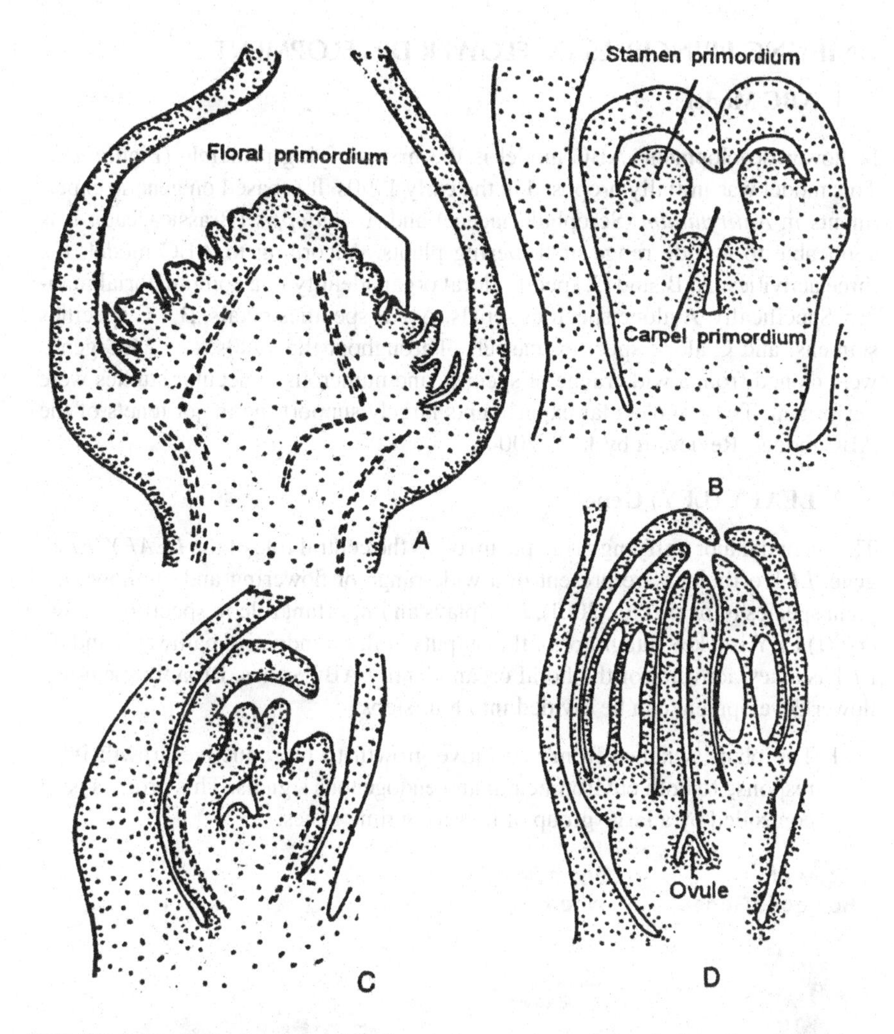


**FIGURE 2.2** *Guizotia abyssinica.* L.s. inflorescence and developing flower showing organogeny of various floral parts (after Chopra and Singh, 1974).

by a gene regulatory network (GRN) that is composed of interacting genes and their protein products. Flowering time genes mediate the switch from vegetative to reproductive development by activating meristem identity genes. Meristem identity genes control the transition from vegetative to floral meristems and act as upstream regulators of floral organ identity genes.

Flowering time control in plants is essential for their reproductive success. Plants have adapted several mechanisms to synchronize flowering so that they can maximize seed yields by carrying out fertilization and seed development at the optimal time (Purugganan and Fuller, 2009). Plants have evolved intricate mechanisms that measure fluctuations in day length to accurately time the onset of flowering throughout seasonal progression.

## UNIFYING PRINCIPLES OF FLOWER DEVELOPMENT

### 1. ABC Model

In flower development, ABC model is the first unifying principle (Figure 2.3). This model was initially proposed in the early 1990s. It is based on genetic experiments in *Antirrhinum* (Scrophulariaceae) and *Arabidopsis* (Brassicaceae). It is applicable to a wide range of flowering plants. According to ABC model, the three activities, A, B, and C, specify floral organ identity in a combinatorial manner. Specifically, A alone specifies sepals, A + B specifies petals, B + C specifies stamens, and C alone specifies carpels. Throughout the 1990s, the ABC genes were cloned from a wide range of species, and numerous molecular studies were performed. These molecular experiments largely support the major tenets of the ABC model (Reviewed by Jack, 2004).

### 2. LEAFY (LFY) Gene

The second major unifying principle involves the central role of the *LEAFY* (*LFY*) gene. *LFY* orthologs are present in a wide range of flowering and nonflowering plant species (Gocal et al., 2001). *LFY* plays an important role in specifying flowers: (1) LFY is a key integrator of the outputs of floral inductive pathways, and (2) *LFY* is a key activator of the floral organ identity ABC genes. Broadly speaking, flower development can be divided into four steps:

1. The plant switches from vegetative growth to reproductive growth in response to both environmental and endogenous signals. This process is controlled by a large group of flowering time genes.

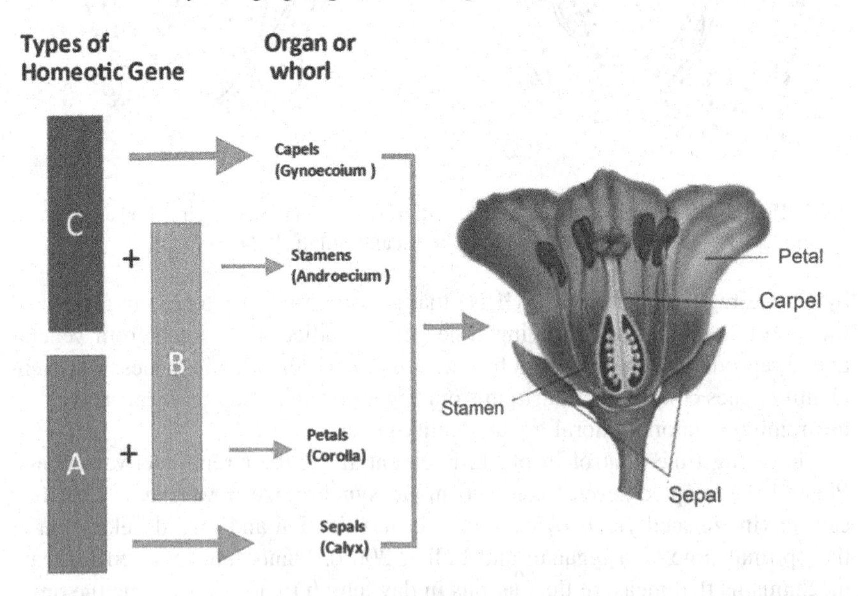


**FIGURE 2.3** ABC Model of flower development.

2. The signals from the various flowering time pathways are integrated. This leads to the activation of a small group of meristem identity genes that specify floral identity.
3. The floral organ identity genes are activated by the meristem identity genes in discrete regions of the flower.
4. The floral organ identity genes activate downstream "organ-building" genes. It finally specifies the various cell types and tissues that constitute the four floral organs.

## FLOWERING TIME GENES

The flowering time genes function on four major promotion pathways: long-day photoperiod, gibberellin (GA), autonomous and vernalization.

1. Mutants in the long-day photoperiod promotion pathway are late flowering when grown in long-day photoperiods. Many long-day pathway genes encode proteins involved in light perception. The light and clock components ultimately lead to the activation of *CONSTANS* (*CO*). Overexpression of *CO* results in very early flowering (Onouchi et al., 2000).
2. A second flowering time pathway involves the promotion of flowering by GA. Mutants defective in the biosynthesis of GA, such as *ga1*, exhibit dramatic delays in the timing of flowering when grown in short days but not long days, suggesting that GA is an important stimulator of flowering in the absence of long-day promotion (Moon et al., 2003).
3. Genes on the third pathway, the autonomous pathway, function to control flowering in a photoperiod-independent manner. As a facultative long-day plant, *Arabidopsis* flowers more rapidly when grown in long days, but it does eventually flower when grown in noninductive short-day photoperiods. Autonomous pathway components play a role in this promotion.
4. The fourth major pathway is the vernalization pathway. An extended cold treatment (vernalization) that mimics overwintering stimulates flowering in many *Arabidopsis* accessions.

The flowering time genes function to control the activity of a much smaller group of meristem identity genes. The meristem identity genes can be divided into two subclasses: the shoot meristem identity genes and the floral meristem identity genes.

**Shoot meristem identity genes:** Shoot meristem identity genes (e.g., *TERMINAL FLOWER1*, *TFL1*) specify the inflorescence shoot apical meristem as indeterminate and nonfloral. Ectopic expression of *TFL1* (e.g., 35S:TFL1) converts the normally floral lateral meristems that arise on the flanks of the shoot apical meristem into shoots.

**Floral meristem identity genes:** *APETALA1* (*AP1*) and *LEAFY* (*LFY*) are the major floral meristem identity genes. These genes specify lateral meristems to develop into flowers rather than leaves or shoots. In *Arabidopsis*, *LFY* and *AP1* specify the lateral primordia to develop as flowers rather than shoots. Ectopic expression of *LFY* or *AP1* converts the inflorescence shoot apical meristem to a flower. Although *AP1* and *LFY* are necessary to specify floral meristem identity, they do not function independently of one another.

## FLORAL ORGAN IDENTITY GENES

One of the important functions of the floral meristem identity genes is to activate the ABC floral organ identity genes.

**A class genes:** A class genes specify the identity of sepals and petals that develop in whorls 1 and 2, respectively. In *Arabidopsis*, both A class genes, *AP1* and *AP2*, are present.

**B class genes:** The B class genes *AP3* and *PISTILLATA* (*PI*) are needed to specify the identity of petals in whorl 2 and stamens in whorl 3.

**C class genes:** The C class gene *AGAMOUS* (*AG*) is necessary to specify the identity of whorl 3 stamens and whorl 4 carpels. The second major function of the C class is to repress A class activity in whorls 3 and 4.

## A NEW ADDITION TO THE ABC MODEL: SEPALLATA (SEP)

A fourth set of genes, the *SEPALLATA* (*SEP*) genes, are necessary for proper floral organ identity (Pelaz et al., 2001). The *SEP* genes also are called E class genes. The revised "ABCE" model postulates that sepals are specified by A activity alone, petals by A + B + E, stamens by B + C + E, and carpels by C + E (Figure 2.4).

Recent studies have demonstrated detailed molecular mechanisms by which different signals are integrated into FT expression in leaves. Flowering time control via vernalization has been demonstrated (Bloomer and Dean, 2017; Xu and Chong, 2018). There is still less information available for the signal integration in the shoot apical meristem (SAM) to reorganize its identity upon the arrival of FT protein. Future studies will elucidate such mechanisms more precisely and will deepen our knowledge on developmental plasticity (Kinoshita and Richter, 2020).

## SUMMARY

- A flower is a modified shoot in which internodes are highly reduced and leaves function as different floral parts.
- A typical flower consists of calyx, corolla, androecium and gynoecium. Flowers with only male sex organs are described as staminate and those with only female sex organs are called pistillate or carpellate.
- Duckweed (*Wolffia microscopica*) bears the smallest flowers (about 0.1 mm in diameter).
- *Refflesia*, a root parasite, bears the largest flowers (about 1 meter in diameter).

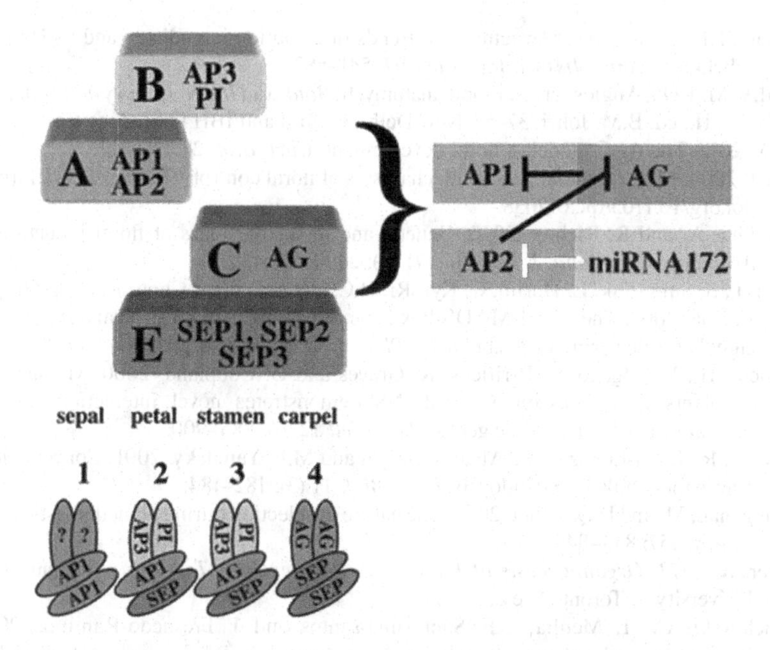


**FIGURE 2.4** A revised ABC Model of flower development. A, B, C, and E are four activities that are present in floral whorls. These four activities are postulated to function in combination to specify the identity of sepals, petals, stamens and carpels (after Jack, 2004).

- Flower development is under strict genetic control. *APETALA1* (*AP1*) and *LEAFY* (*LFY*) are the major floral meristem identity genes.
- Flowering time genes are responsible for the switch from vegetative to reproductive phase by activating meristem identity genes. *SEPALLATA* (*SEP*), also referred to as E class genes, are necessary for proper floral organ identity.
- Meristem identity genes control the transition from vegetative to inflorescence, and floral meristems act as upstream regulators of floral organ identity genes.

## SUGGESTED READING

Barnier, G. 1986. The flowering process as an example of plastic development. *Soc. Expt. Biol.* 40: 257–286.

Bloomer, R.H. and C. Dean. 2017. Fine-tuning timing: natural variation informs the mechanistic basis of the switch to flowering in Arabidopsis thaliana. *J. Exp. Bot.* 68 (20): 5439–5452.

Chopra, S. and R.P. Singh. 1974. Effect of gamma rays and 2,4-D on the development of flower and pollen grains in *Guizotia abyssinica*. *Phytomorphology* 24: 305–313.

Eames, A.J. 1961. *Morphology of the Angiosperms*. New York: Mc Graw Hill Book Co., Inc.

Endress, P.K. 2010. Flower structure and trends of evolution in eudicots and their major subclades1. *Ann. Missuri Bot. Gard.* 97: 541–583.

Govil, C.M. 1995. Angiosperms: Floral anatomy. In *Botany in India- History and Progress*. Vol. II., ed. B.M. Johri, 37–57. New Delhi: Oxford and IBH Publishing Co.

Iris, V. 2017. The ABC model of floral development. *Curt. Biol.* 27: 887–890.

Jack, T. 2004. Molecular and genetic mechanisms of floral control. *The Plant Cell*. https://doi.org/10.1105/tpc.017038.

Kinoshita, A. and R. Richter. 2020. Genetic and molecular basis of floral induction in *Arabidopsis thaliana*. *J. Exp. Bot.* 71 (9): 2490–2504.

Moon, J., S. Sung-Suk, L. Horim, C. Kyu-Ri, H.C. Bong, P. Nam-Chon, K. Sang-Gu and L. Ilha. 2003. The SOC1 MADS-box gene integrates vernalization and gibberellin signals for flowering in Arabidopsis. *Plant J.* 35 (5): 613–623.

Onouchi, H., M.I. Igeno, C. Perilleus, K. Graves and G. Coupland. 2000. Mutagenesis of plants overexpressing CONSTANS demonstrates novel interactions among Arabidopsis flowering-time genes. *Plant Cell* 12 (6): 885–900.

Pelaz, S., R. Tapia-Lopaz, E.R. Alvarez-Buylla and M.F. Yanofsky. 2001. Conversion of leaves intopetals in Arabidopsis. *Curr. Biol.* 11 (3): 182–184.

Purugganan, M. and D.Q. Fuller. 2009. The nature of selection during plant domestication. *Nature* 457: 843–848.

Sattler, R. 1973. *Organogenesis of Flowers: A Photographic Text Atlas*. Toronto, ON: University of Toronto Press.

Schuchovski, C., T. Meulia, B.F. Sant'Anna-santos and J. Fresnedo-Ramirez. 2021. Inflorescence development and floral organogenesis in *Taraxacum kok-saghyz*. *Plants* 9 (10): 1258. https://doi.org/10.3390/plants 9101258.

Schwabe, W.W. 1971. Physiology of vegetative reproduction and flowering. In *Plant Physiology*, ed. F.C. Steward, pp. 233–411. New York: Academic Press.

Wang, S.-L, K.K. Viswanath, C. Tong, H.R. An, S. Jang and F.C. Chen. 2019. Floral induction and flower development of orchids. *Front. Plant Sci.* https://doi.org/10.3389/fpls.2019.01258.

Xu, S. and K. Chong. 2018. Remembering winter through vernalization. *Nat. Plants* 4 (12) https://doi.org/10.1038/s41477-018-0301-z.

Zik, M. and V.F. Irish. 2003. Flower development: Initiation, differentiation and diversification. *Ann. Rev. Cell. Dev. Biol.* 19: 119–140.

# 3 Microsporangium

The anther is the male reproductive organ in seed plants. Its main function is to produce and disperse pollen. In the majority of angiosperms, a typical anther consists of four elongated microsporangia (Figures 3.1A, B) but it may be bisporangiate (Adoxaceae, Malvaceae), 8-sporangiate (Bixaceae) or multisporangiate (Loranthaceae, Rhizophoraceae). In some cases, there may be only one sporangium per anther (e.g., *Arceuthobium*).

At maturity, the two sporangia of each side become confluent owing to the breakdown of the partition wall between them (Figures 3.1C, D, Figure 3.2). Usually, each anther lobe consists of two microsporangia (dithecous) but in some taxa (e.g., *Wolffia*, *Moringa*) each anther lobe has only one microsporangium (monothecous).

## ANTHER WALL: STRUCTURE AND FUNCTIONS

In angiosperms, the anther wall consists of four layers, namely, the epidermis, endothecium, middle layer(s) and tapetum. Anther wall layers carry out different functions for pollen formation and release.

## ONTOGENY

A cross section of a young anther is almost oval in outline and is composed of richly cytoplasmic cells surrounded by an epidermis (Figure 3.3A). The outline soon becomes more or less rectangular, and the archesporial cells differentiate in the hypodermal region at four corners (Figure 3.3B). The anther soon becomes four-lobed. The archesporial cells are distinguishable by their large size, dense cytoplasm and prominent nuclei. The archesporial cells soon undergo periclinal divisions (Figures 3.3C, D, Figure 3.4) giving rise to the primary parietal cells on the outer side and primary sporogenous cells on the inner side. The primary parietal cells undergo further periclinal division giving rise to endothecium, middle layer(s) and tapetum (Figure 3.5). The primary sporogenous cells either directly or after a few mitotic divisions function as microspore mother cells. A schematic representation of the ontogeny of the anther wall layers is given in Figure 3.6.

## ANTHER WALL DEVELOPMENT

Based on the behaviour of the secondary parietal layer, anther wall developments have been divided into four types:

1. **Basic Type:** The outer and inner secondary parietal layers divide periclinally and form the endothecium, two middle layers and tapetum.

DOI: 10.1201/9781003260097-3

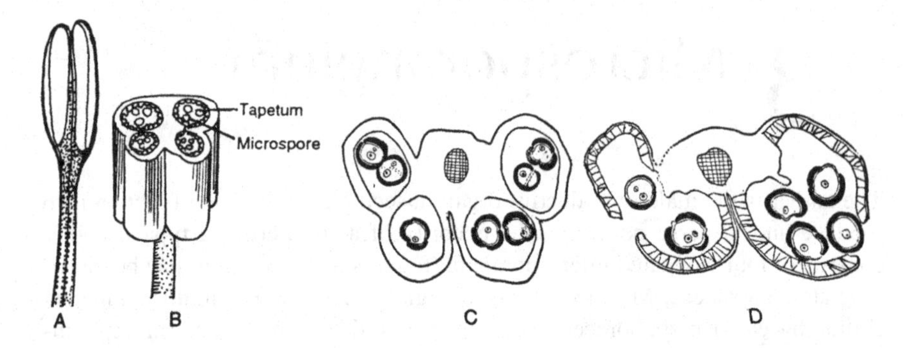


**FIGURE 3.1** A. Stamen; B. Anther cut transversely to show sporangia; C. Cross section of mature anther.

2. **Dicotyledonous Type:** The outer secondary parietal layers divide, giving rise to the endothecium and middle layer while the inner layer directly functions as tapetum.
3. **Monocotyledonous Type:** The inner secondary parietal layer divides and produces the middle layer and tapetum, whereas the outer one forms the endothecium.
4. **Reduced Type:** The outer and inner secondary parietal layers are transformed into endothecium and tapetum respectively, and a middle layer is absent.

## ANTHER WALL LAYERS

The mature anther wall consists of the epidermis, endothecium, middle layer(s) and tapetum.

## EPIDERMIS

During the development of anther wall layers, the epidermal cells undergo division only in the anticlinal plane for some time, and later on, they get stretched just to keep pace with the developing anther. In a mature anther, though continuity of epidermal cells gets disrupted, they persist as greatly stretched cells. However, in certain taxa, epidermis remains as a well-defined layer (e.g., Amaryllidaceae, Magnoliaceae, Liliaceae). In *Arceuthobium*, the epidermis develops fibrous bands. In *Zeuxine longilabris* (Orchidaceae), the epidermal cells simulate tapetum. The epidermis performs a protective role in anther development.

## ENDOTHECIUM

The endothecium is single layered in the majority of plants. However, multilayered endothecium has also been reported in many taxa viz., *Capsicum annuum*, *Datura alba* and *Sesamum indicum*. The endothecium originates from the parietal layer and is confined to the protuberant part of the anther. The cells of the

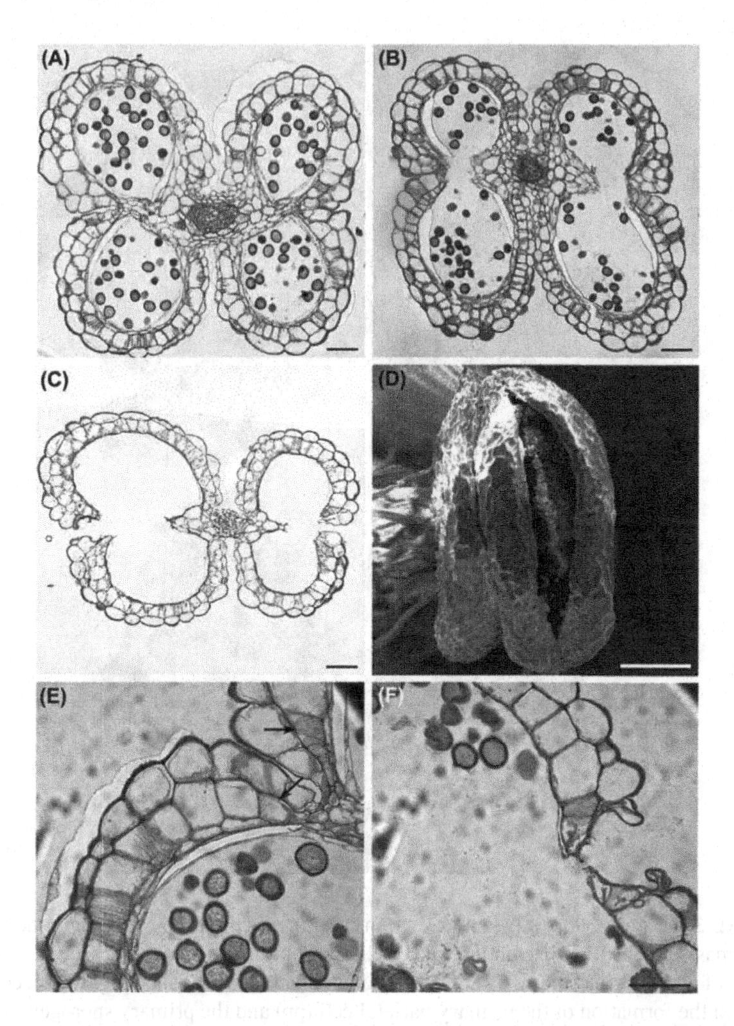


**FIGURE 3.2** Anther anatomy (*Acer oblongum*). A. Transverse section of bithecous anther showing the degenerating tapetum and middle layers; B. Partially dehisced anther, where two sporangia on either side of an anther lobe become joined due to breakdown of partition wall; C. Completely dehisced anther. D. SEM image of a completely dehisced anther showing empty anther lobe; E. Portion of a transverse section of anther near stomium showing endothecium cells with fibrous endothecial thickenings. Arrow indicates the endothecial thickening in the cells near the stomium; F. portion of anther showing opening of stomium valves at time of dehiscence (after Yadav et al., 2017).

endothecium are radially elongated and attain their maximum development at the time of anther dehiscence. Fibrous thickenings (Figure 3.7) develop from the inner tangential wall of each cell (e.g., *Brassica, Alectra*). However, fibrous thickenings are absent in many members of Hydrocharitaceae, cleistogamous forms and plants exhibiting anther dehiscence by pores (e.g., *Erica*). Reticulate thickenings have been reported in *Magnolia stellata* (Kapil and Bhandari, 1964).

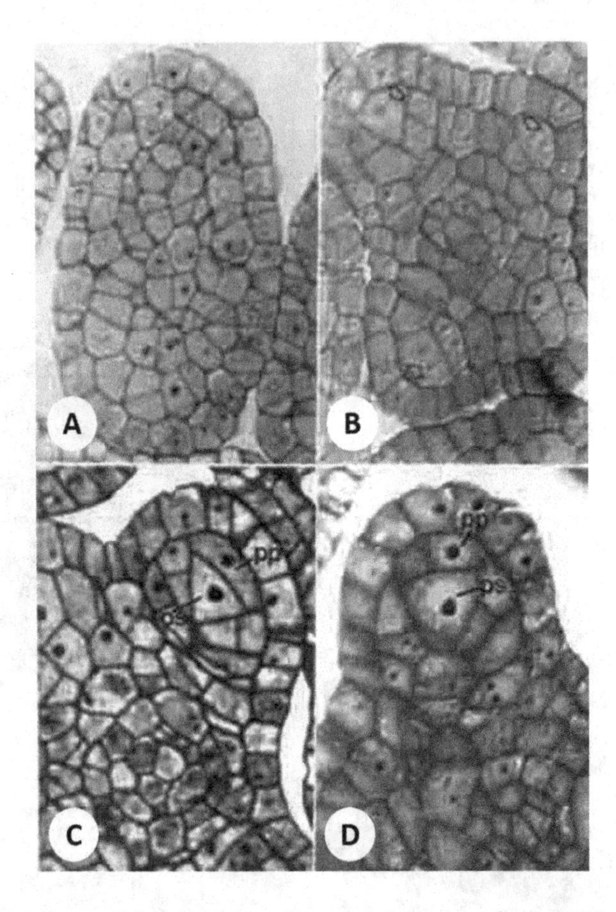


**FIGURE 3.3** *Oryza sativa.* A cross-section of anther primordium showing a homogeneous mass of cells; B. Differentiation of archesporial initials (arrows) as seen in a cross-section of anther primordium; C, D. Cross-section of anther lobes showing two successive stages in the formation of the primary parietal cell (pp) and the primary sporogenous cell (ps) (after Raghavan, 1988).

In *Senecio*, the endothecial cells are tabular, rectangular in periclinal view, and the thickenings are principally restricted in each cell to the longitudinal wall nearest the connective. Frequently, the transverse walls are also thickened. In periclinal view, these ribs appear as thickenings down the edge of the inner anticlinal wall. When viewed in a periclinal orientation, the ribs appear slightly convex and span the width of the cell between the periclinal walls. Some of these ribs possess a median transverse bar (Figure 3.8). The endothecium is responsible for anther dehiscence to disperse pollen when they are mature (van der Linde and Walbot, 2019). In anthers that open by longitudinal slits, the endothecial cells around the junction of the two sporangia lack these thickenings.

de Fossard (1969) reported that the endothecial thickenings contain a high proportion of alpha-cellulose. This has been further confirmed by Bhandari and

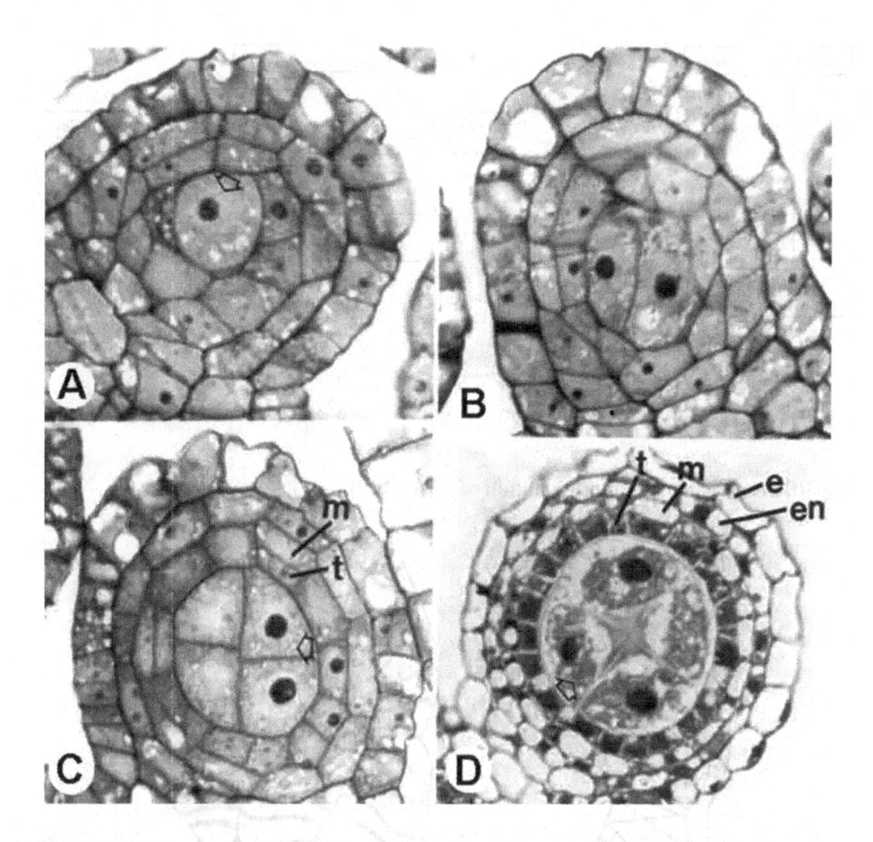


**FIGURE 3.4** *Oryza sativa.* Differentiation of wall layers and formation of microsporocytes. A. Cross-section of an anther lobe showing the primary sporogenous cell (arrow) surrounded by a three-layered wall; B. First division of the primary sporogenous cell; C. Cross-section of anther lobe showing differentiation of the middle layer (m) and tapetum (t) by division of cells of the innermost wall layer. The primary sporogenous cell has divided cross-wise to form four microsporocytes (arrow). D. Cross-section of an anther lobe showing the anther wall constituted of an epidermis (e), endothecium (en), middle layer (m) and tapetum (t) ( after Raghavan, 1988).

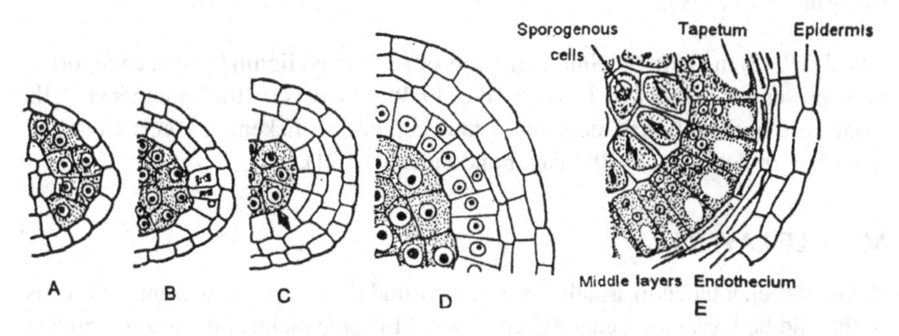


**FIGURE 3.5** Transverse section of a portion of anther showing development of anther wall layers

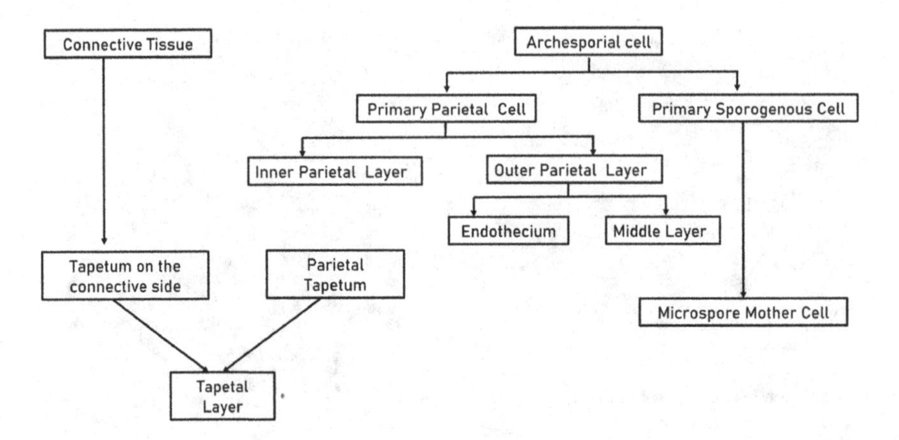


**FIGURE 3.6** Schematic representation of anther wall.

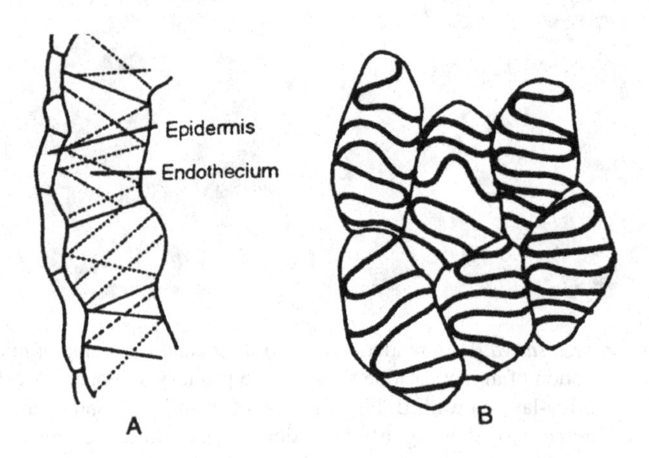


**FIGURE 3.7** A. *Cucumis maderaspatanus* (syn. *Melothria maderaspatana*). Longitudinal section of a part of mature anther wall showing epidermis and endothecium. B. *Momordica charantia*. Endothecium in surface view showing thickenings (after Deshpande et al., 1986).

Khosla (1982) in *Triticale*. Small amounts of pectin and lignin have been reported to be present in *Pisum* and *Lens* (Biddle, 1979). In *Corokia*, the endothecial cells elongate tangentially and develop fibrous cellulosic thickenings on their radial walls before dehiscence (Bhatnagar and Pandey, 2021).

## MIDDLE LAYERS

Below the endothecium usually one or two middle layers are present. The cells of the middle layers are generally ephemeral. In some plants, one or more middle layers may persist (*Lilium*). In *Nigella damascena* and *Linum* 2–5 middle layers persist until anther dehiscence. Sometimes middle layers present just below the

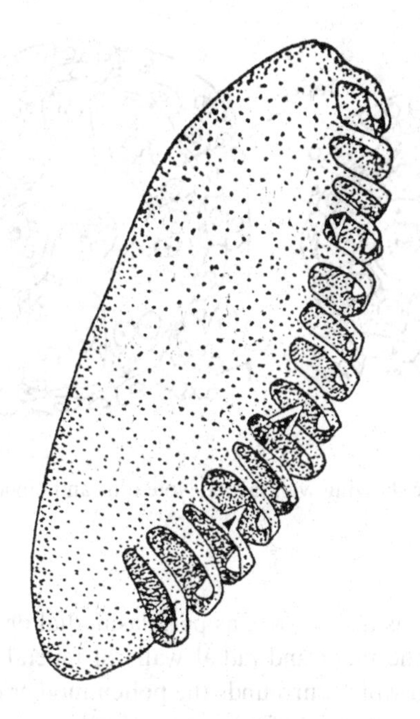

**FIGURE 3.8** *Senecio speciosus.* Diagram of a single endothecial cell illustrating convex ribs spanning the width of the inner anticlinal wall of the cell. Some ribs have a median transverse bar (arrow) (after Vincent and Getliffe, 1988).

endothecium may also develop fibrous thickenings as in *Agave*, *Argemone* and *Costus*. In many species the cells of the middle layers store starch and other reserve materials which get mobilized during later stages of pollen development. In a study of the EXCESS MICROSPOROCYTES$_1$ gene in *Arabidopsis*, the middle layer was observed to originate from outer secondary parietal (OSP) cells (Chang et al., 2011).

## TAPETUM

The layer of cells lying beneath the middle layers is the tapetum. The cells of the tapetum are densely cytoplasmic and surround the sporogenous cells completely (Figure 3.8). Generally, tapetum is single layered but biseriate tapetum has been reported in *Pyrostegia* and *Tecoma*. A multiseriate condition (Figure 3.9) is reported in *Oxystelma* (Maheswari Devi and Lakshminarayana, 1977).

The tapetum is usually derived from the primary parietal layer (Figure 3.5) and is composed of homogenous layer of cells, but in *Alectra thomsonii* and *Antigonon leptopus*, dimorphic tapetum is found which develops from the primary parietal layer and connective. In *Antirrhinum majus* and *Triticale*, the tapetum develops from the parietal layer and is homogeneous. The tapetum is of considerable physiological importance because all the food material to the sporogenous tissue must pass through it. In angiosperms, the tapetum is of two types, namely, amoeboid and secretory.

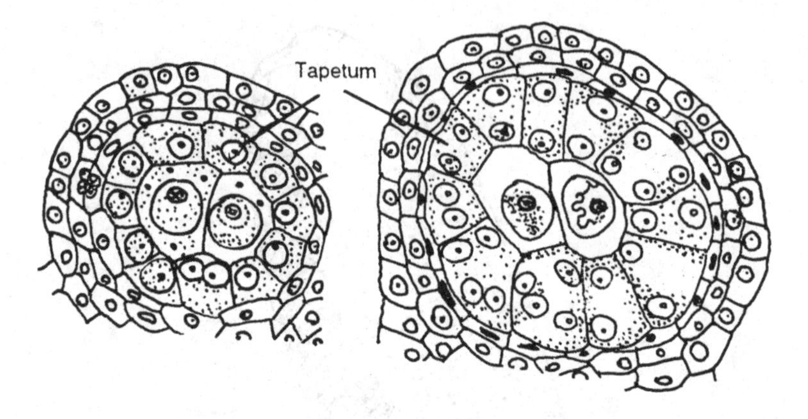


**FIGURE 3.9** T.s. anther showing wall layers. Note uni- and binucleate tapetum

## Amoeboid tapetum

The amoeboid tapetum is also known as periplasmodial or invasive tapetum. In this type of tapetum, the inner and radial walls of tapetal cells break down to form a periplasmodium which surrounds the pollen mother cells or microspores. This type of tapetum is commonly found in monocots.

In a detailed ultrastructural study of amoeboid tapetum in *Arum italicum*, Pacini and Juniper (1983) have shown that the anther wall comprises epidermis, fibrous endothecium, three middle layers and a two-layered tapetum. At the pre-meiotic phase, tapetal cells are interconnected by plasmodesmata. These interconnections are absent in the tapetal walls facing the middle layer and the microspore mother cells (Figure 3.10A). During the meiotic divisions of mother cells, the innermost middle layer flattens, and certain walls of the tapetal cells begin to dissolve. At leptotene (Figure 3.10B), the plasmodesmata in the tapetal cells widen to form cytomictic channels, and the tangential walls disappear at pachytene (Figure 3.10C). By anaphase, most of the radial walls between the tapetal cells also disappear and microtubules are seen towards the inner tangential wall. By the time tetrads are formed, the inner tangential wall of tapetal cells disappears and the amoeboid tapetum begins to envelop the tetrad (Figures 3.10D–F). Microtubules surround the tetrad facing the callose wall and the microspores. The tapetal plasmalemma facing the microspores retracts from the exine surface leaving roughly cone-shaped spaces (Figure 3.10H). With the release of microspores from the tetrads (early uninucleate stage) the outer wall of the tapetal cells also disappears. At this stage, tapetal contents are distinguishable as two distinct zones. The outer zone contains a few small vacuoles, nuclei, ribosomes, polysomes, mitochondria and plastids. The inner zone, on the other hand, contains polyribosomes, microtubules arcing parallel to the surface of the microspores, vesicles and a few dilated ER cisternae. The cytoplasm of the inner zone adheres to the microspore surface (Figure 3.10G). By the localized retraction of the plasma membrane, the spaces left on the microspore surface and tapetal periphery become filled with a material forming "spines" (Figure 3.10I).

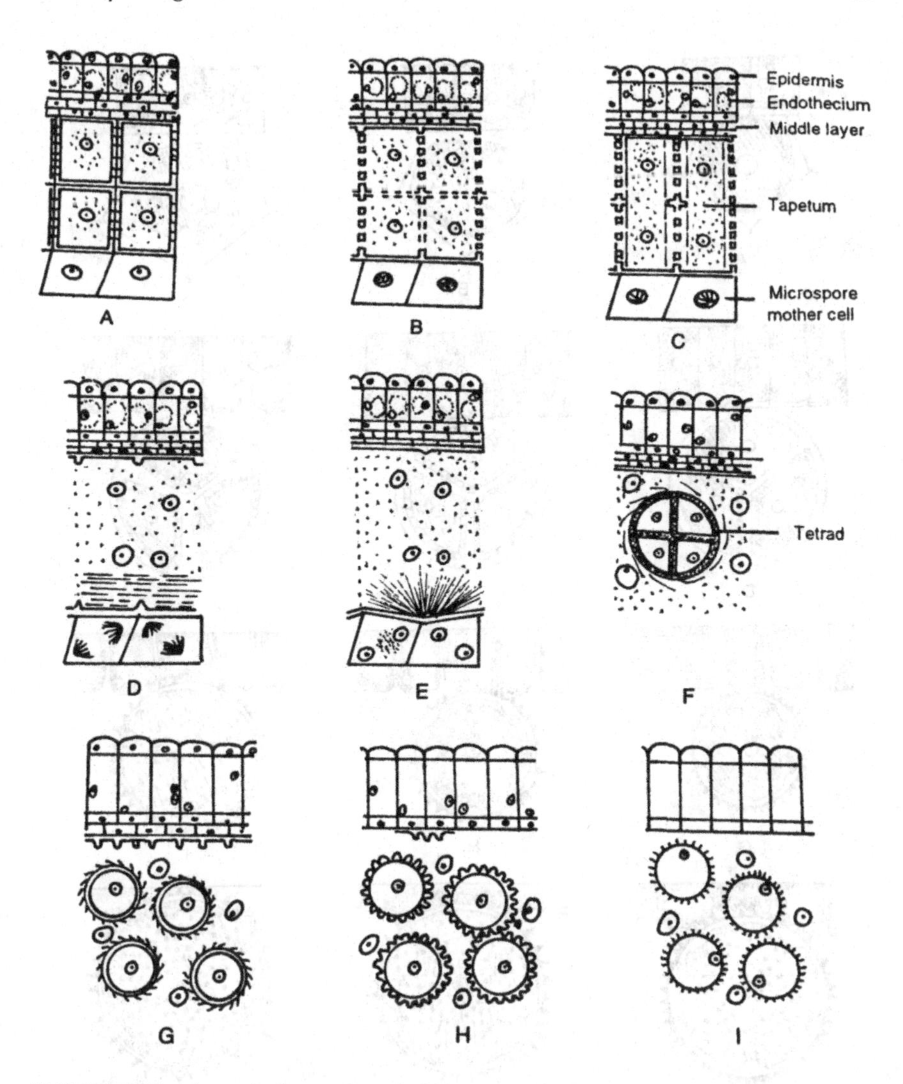


**FIGURE 3.10** *Arum italicum.* Amoeboid tapetum. A. Premeiotic phase; B. Leptotene; C. Pachytene; D. Meiosis I, anaphase; E. Meiosis I, telophase; F. Tetrad; G. Microspores surrounded by microtubules; H. Microspore later stage; I. Uninucleate pollen grain with intine and exine (after Pacini and Juniper, 1983).

### Secretory tapetum

The secretory tapetum is also known as parietal or glandular tapetum and is of common occurrence in the angiosperms. The cells of the tapetum remain intact throughout the development of the microspores, and their breakdown occurs when pollen grains reach maturity.

Ciampolini et al. (1993) reported 12 stages of tapetum–pollen development in *Cucurbita pepo* (Figure 3.11). In the late prophase stage, the callose wall is

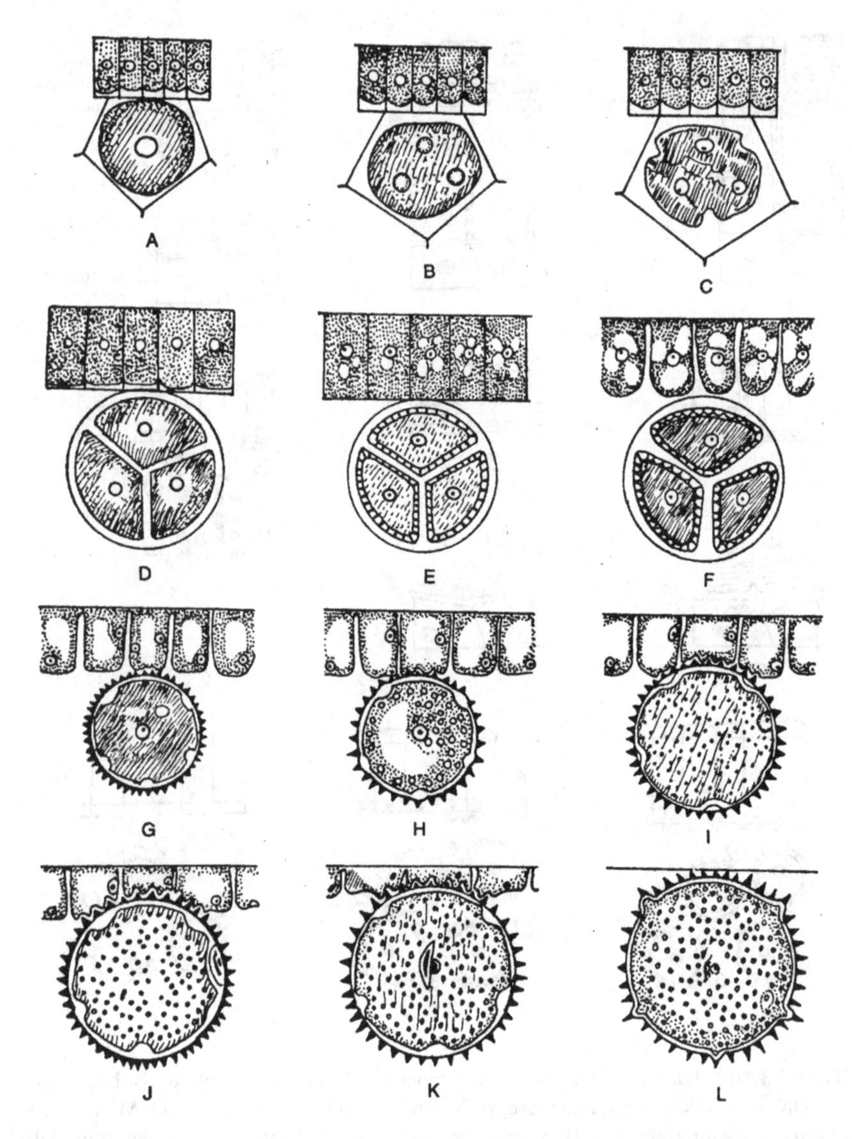


**FIGURE 3.11** Tapetum and pollen development in *Cucurbita pepo*. A. Late prophase; B. Second meiotic division; C. Meiotic cleavage; D. Early tetrad stage; E. Middle tetrad stage; F. Late tetrad stage; G. Early microspores stage; H. Middle microspore stage; I. Late microspore stage; J. Early bicellular stage; K. Middle bicellular stage; L. Late bicellular stage (after Ciampolini et al., 1993).

formed around the meiocytes. Plasmodesmata occur only in the radial walls, and plastids are undifferentiated. At telophase of second meiotic division, the callosic wall is more uniform. The thickness of the inner tangential wall of the tapetum increases. Tapetal cells remain uninucleate until degeneration. At the time of meiotic cleavage, the callose wall of the young microspore is formed centripetally and is complete by the early tetrad stage. At middle tetrad stage, small vacuoles of tapetal cytoplasm fuse and plastids continue to divide. The unswollen walls of tapetal cells are unchanged, but the swollen ones become multilayered with electron opaque inclusions. Active dictyosomes, producing numerous vesicles, are found parallel to the plasma membrane near dilated cisternae of the endoplasmic reticulum. At late tetrad stage, the inner tangential and the radial walls of the tapetal cells completely disappear. The tapetal plastids continue to divide. The tapetal cells reach their maximum size at early microspore stage, and their organelles remain unchanged, but each cell contains one big vacuole. At late microspore stage, tapetal plastids show electron opaque and electron translucent inclusions. Lipid bodies are observed in the tapetal cytoplasm. At early bicellular stage, degenerating plastids full of lipid inclusions are observed in the tapetal cytoplasm together with strongly electron opaque lipid droplets. By the time pollen grains reach middle bicellular stage, the content of the tapetal cytoplasm is reduced to a lipid mass becoming pollenkitt.

## Tapetal Membrane

Bhandari and Kishori (1969, 1973) reported that in *Nigella damascena* the original tapetal cell walls disintegrate, and a new membrane resistant to acetolysis is formed along the thecal face of tapetal cells. The Ubisch bodies get attached to the tapetal membrane.

In some taxa, an extra tapetal membrane is formed towards the middle layers. This membrane is called peritapetal membrane or extra-tapetal membrane. In Asteraceae, where the tapetum is of the amoeboid type, the extra-tapetal membrane is organized (Heslop-Harrison, 1969). In *Lilium* (Reznickova and Willemse, 1980), which show glandular tapetum, both tapetal and peritapetal membranes are organized. These membranes are resistant to acetolysis.

Banerjee (1967) reported that in grasses the tapetal membrane comprises three layers – fenestrated, reticular and orbicular. The fenestrated layer is present just outside the tapetal protoplast followed by a reticular layer which interconnects the Ubisch granules. The orbicular layer is constituted by the Ubisch granules themselves.

The precise function of tapetal membranes is not clear. According to Heslop-Harrison (1969), they probably act as a "culture sac". Tapetal membranes may also help in preventing a quick loss of water from within the anther locule.

## Behaviour of the Nucleus in Tapetal Cells

The cells of the tapetum show interesting behaviour, and the total DNA content may increase by a number of ways, including normal mitosis, endomitosis, formation of restitution nuclei and polyteny. Tapetum may be uninucleate, or become binucleate or multinucleate. Polyploidy up to 16n has been recorded in *Nigella damascena* and *Podolepis jaceoides.*

1. **Normal mitosis:** When division of the nucleus is not followed by cytokinesis, a multinucleate condition arises (e.g., *Guizotia abyssinica, Zea mays*). Thus, depending upon the number of nuclear divisions, a tapetal cell may have two (Figure 3.12), four, eight or more nuclei. Sometimes nuclei of multinucleate tapetum fuse. In the parietal tapetum, the nuclear divisions are rarely synchronous, whereas in the periplasmodial tapetum, they are highly synchronous.
2. **Endomitosis:** In this type of mitosis, the chromosomes split longitudinally within the same nucleus without the formation of the spindle, resulting in the formation of a polyploid nucleus (Figure 3.13). Endomitosis is reported in several taxa (e.g., *Spinacea, Cucurbita*). In the uninucleate tapetum of *Cucurbita pepo*, nuclei of four sizes corresponding to their degree of ploidy (2, 4, 8 or 16n) are formed as a result of endomitosis.
3. **Formation of restitution nuclei:** Here, normal mitosis occurs up to early anaphase stage. The two sets of chromosomes remain intact in a common nuclear membrane forming a restitution nucleus (Figure 3.14).
4. **Polyteny:** In polyteny, the number of the chromonemata per chromosome increases. This mode of DNA increase does not change the number of chromosomes per nucleus.

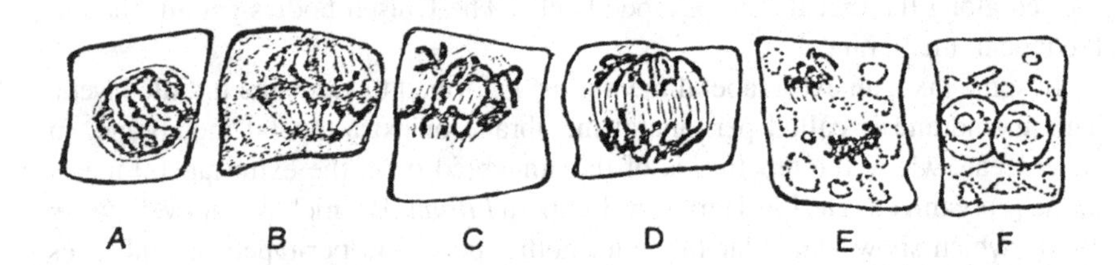


FIGURE 3.12 Normal mitosis in tapetal cells of *Zea mays.*

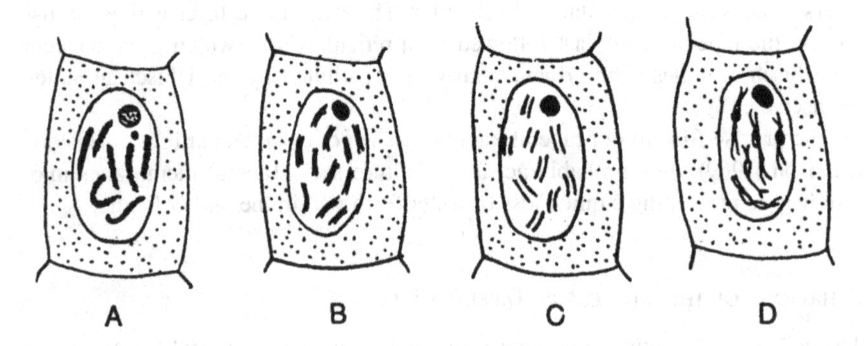


FIGURE 3.13 Endomitosis in tapetal cells. A. Endoprophase; B. Endometaphase; C. Endoanaphase; D. Endotelophase.

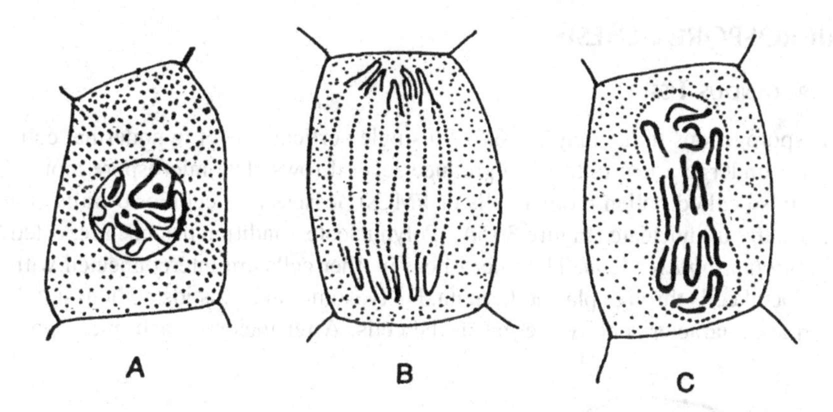


**FIGURE 3.14** Formation of restitution nucleus in tapetal cells.

### Functions of the Tapetum

The tapetum is a highly active layer of cells investing the sporogenous tissue and provides essential nutrients and materials for pollen development. It plays an important role in the formation of the pollen wall, as it supplies the major part of the sporopollenin of the exine. The contribution of sporopollenin through Ubisch bodies has been described in Chapter 4. Tapetum is involved in the synthesis and release of the materials which take part in the deposition of tryphine and the pollenkitt.

The precocious degeneration during premeiotic and meiotic stages and its cellular persistence for unusually long periods result in pollen sterility. Tapetum initiates the process of abortion of pollen in cytoplasmic male sterile lines (Heslop- Harrison, 1971; Laser and Lersten, 1972). Evidence from *Petunia* indicates that low levels of DNA synthesis in the tapetum are the first indicators of the CMS phenotype, followed by disruption of the young microspores.

In *Sorghum* (Raj, 1969), tapetal cells enlarge and squeeze the sporogenous tissue. Malfunctioning of tapetum causes arrested development of microspores. In such cases, microspores are represented by brownish narrow stripes. A number of tapetally expressed genes have been characterized. One class of tapetal genes, exemplified by the pLHM9B clone from *Lilium* and the A9 clone from *Brassica*, are highly tapetum-specific and encode a polypeptide with homology to seed proteins and enzymic inhibitors. Several tapetum genes involved in pollen wall formation have been cloned from male sterile lines of rice (Han et al., 2021).

In *Tradescantia bracteata* (Mepham and Lane, 1969), the plasmodial cytoplasm derived from the tapetum shows callase activity. Stieglitz and Stern (1973) measured callase activity in anther wall, and the microsporocytes starting from early meiosis to tetrad dissociation. They opined that sporophytic tissue, presumably tapetum is involved in the synthesis of callase enzyme for the release of microspores in a tetrad by degrading the callose wall.

## MICROSPOROGENESIS

### Sporogenous Tissue

The sporogenous tissue may consist of a single vertical row of hypodermal cells or may undergo a few mitoses to produce several rows. The microspore mother cells (also called pollen mother cells or PMCs) possess thin cellulosic walls and are usually uninucleate (Figure 3.15A). Polynucleate condition has been reported in *Santalum obtusifolium*. The microspore mother cells are interconnected with the tapetal cells through plasmodesmata. Dictyosomes and plastids without starch grain are characteristically present in the cells. After meiosis, each microspore

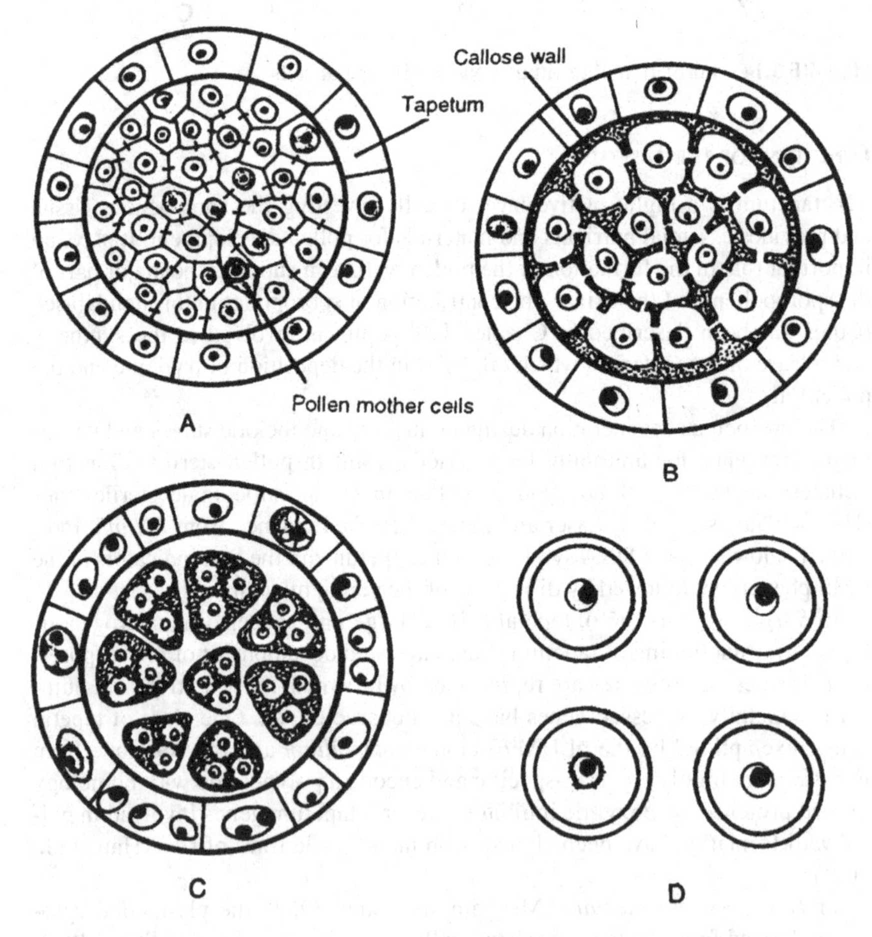


**FIGURE 3.15** A. Preleptotene stage; Note plasmodesmata connections between PMCs and between tapetal cells and PMCs; B. Meiotic prophase I. PMCs are enclosed in thick callose wall. Note cytoplasmic channels between PMCs; C. Tetrad stage; Cellular connections are cut off and microspores are independent cells; D. Microspores soon after their release from the tetrad.

mother cell gives rise to four haploid microspores. Aggregates of four microspores are referred to as microspore tetrads.

Which factors precisely trigger the mitotic to meiotic changeover is not clearly understood. Clutter and Sussex (1965) have shown that the stimulus for meiosis arises in the vegetative shoots of plants. This stimulus is highly specific in the sense that it affects only the spore mother cells, and not even the tapetal cells through which it must pass.

With the entry of PMCs into meiosis the connections between the tapetal cells and the PMCs are broken and the walls of the PMCs become thicker due to the deposition of callose (beta-1,3 glucan). During prophase I the callose deposition is initiated between the plasmalemma and the original cellulosic wall, particularly along the corners of the PMCs, and extends laterally until a continuous layer is laid down. In *Oryza sativa* (Raghavan, 1988), a dense fibrillar material fills each locule shortly before the first signs of callose deposition are seen on the side of microcytes facing the locule. Subsequent deposition of callose occurs along the radial walls of microsporocytes and along their sides facing the tapetum.

Concurrently with the deposition of callose, the plasmodesmatal connections between the PMCs are replaced by massive cytoplasmic channels. These channels provide passage for the movement of cytoplasmic contents from one cell to the other (Figure 3.15B).

After the completion of meiosis (Figure 3.15C), the callose wall is rapidly broken down and the microspores (Figure 3.15D) are released into the anther locule. According to Mepham and Lane (1969) the dissolution of callose is tapetum-mediated.

The callose gives mechanical isolation to the developing microspores. The callose layer also functions as a kind of chemical isolation, establishes a selective barrier between genetically different haploid cells, which must pass through their developmental stages unexposed to the influence of their sister spore, or of the adjoining spores and somatic tissue. Callose also protects the developing sporocytes from the harmful hormonal and nutritional influence of the adjoining somatic cells. Callose wall also plays an effective role in the establishment of the very first pattern of exine.

## CYTOKINESIS

The four microspores formed after meiosis are separated from each other due to the formation of cell wall. Two basic types of cytokinesis, successive and simultaneous, have been recognized.

1. **Successive:** In the successive type, two cell plates are laid down in a centrifugal manner immediately after the first and second meiotic divisions. A cell plate is formed in the centre and extends laterally. This is followed by the deposition of callose on either side of the plate. The two cells of the dyad undergo the second meiotic division forming a tetrad. This type

of cytokinesis is common in monocots, e.g., *Mayaca* (Figure 3.16), *Zea mays, Lilium, Canna* and *Senseviera.*

2. **Simultaneous:** In this type of cytokinesis, the first meiotic division is not followed by wall formation (Figure 3.17). A binucleate cell is formed after meiosis I, and there is no dyad stage. The two haploid nuclei undergo the second meiotic division simultaneously. The callose walls are formed after the second meiotic division, resulting in the formation of a tetrad. In this type, the isolation of the microspores occurs by concurrent centripetal furrows. This type of cytokinesis is common in monocots, e.g., *Melothria Nicotiana, Magnolia, Helleborus* and *Alectra.*

Sampson (1969) recognized three main categories of the process of cytokinesis:

1. Successive centrifugal cell plates at the end of meiosis I and II, e.g., *Zea mays.*
2. Simultaneous centripetal constriction furrows at the end of meiosis II, e.g., *Nicotiana.*
3. Simultaneous centrifugal cell plate formation at the end of meiosis II, e.g., *Helleborus.*

According to Davis (1966) simultaneous cytokinesis is derived from the successive type. The successive method predominates in the monocots (60%), and the simultaneous method in the dicots (93.6%).

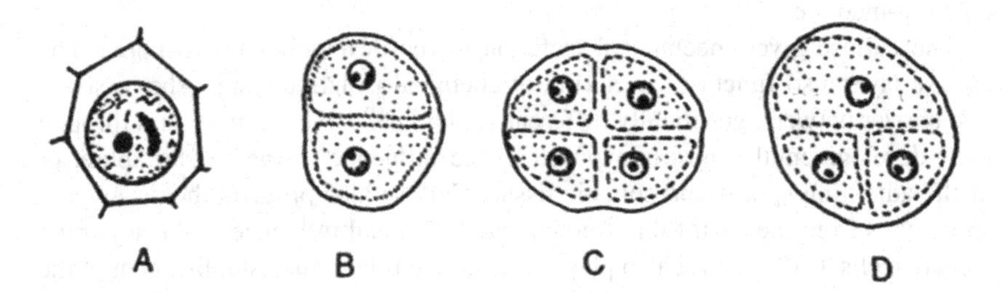


**FIGURE 3.16** A. Microspore mother cell; B. Dyad; C. Isobilateral tetrad; D. Decussate tetrads.

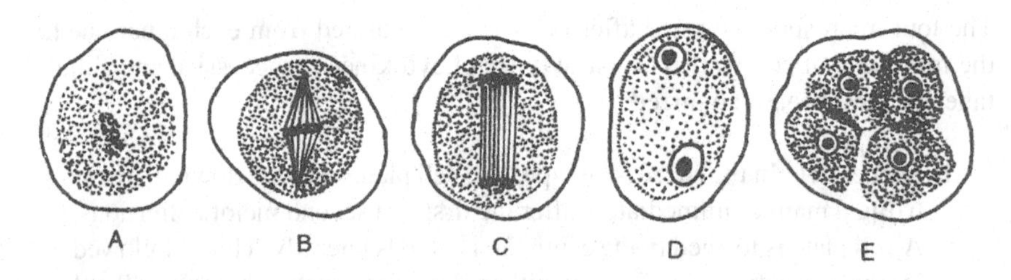


**FIGURE 3.17** Meiosis in PMCs. Simultaneous cytokinesis.

## MICROSPORE TETRADS

The aggregate of four microspores is referred to as microspore tetrads. Generally, the arrangement of microspores in a tetrad is tetrahedral or isobilateral (Figures 3.18A, B). The arrangement may be linear, T-shaped and decussate also (Figures 3.18C, D, E). In *Aristolochia elegans* and *Sparganium erectum*, all five types of tetrads have been observed. The position of the dividing microspore mother cell in the sporogenous mass and the impact of physical factors operating within the sporangium appear largely to control the variation in tetrad formation.

Individual microspore separates out from the tetrad by the sudden dissolution of callose wall. The deposition of the sporopollenin layer begins immediately after the release of microspores.

## MASSULAE AND POLLINARIA

Occurrence of more than four spores in a tetrad is called polyspory. In the majority of cases, the pollen grains of each tetrad separate from one another and lie free in the pollen sac. But many variations also occur, e.g., in *Philydrum lanuginosum*, the pollen grains remain in tetrads even when mature (Figure 3.19).

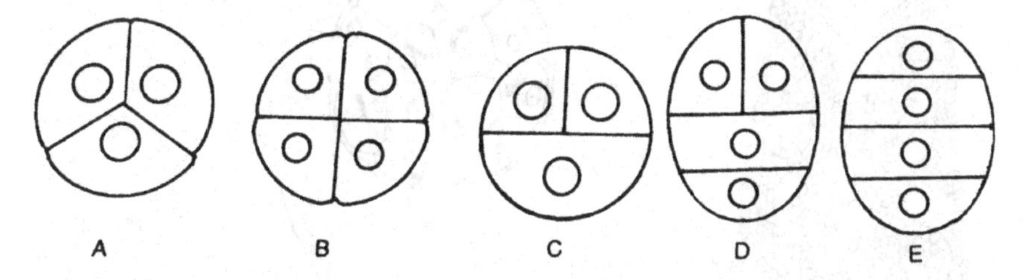


**FIGURE 3.18** Diagram showing different types of microspore tetrads. A. Tetrahedral; B. Isobilateral; C. Decussate; D. T-shaped; E. Linear.

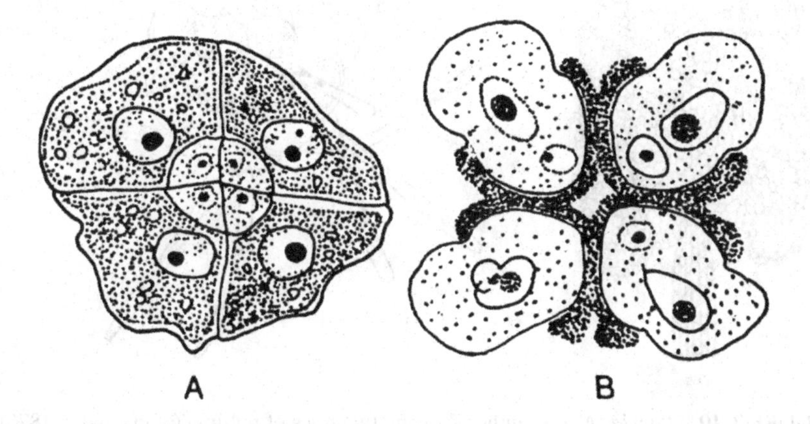


**FIGURE 3.19** *Philydrum lanuginosum*. Four binucleate pollen grains in permanent tetrads (after Kapil and Walia, 1965).

In *Allophylus serratus* (Sapindaceae), the pollen grains are produced as monads. The monads are radially symmetrical with triangular amb (Basith et al., 2019).

In Acacias the tetrads are stuck together in groups which may contain as many as 64 pollen grains, and these groups are found in separate chambers formed in the pollen sac. In *Parkia* (Mimosaceae) several sterile transverse septa appear, and eventually each compartment will contain only one pollinium (Figure 3.20A) nourished by its own tapetum. In *Mimosa pudica*, the tetrads form pollinia, but in *Mimosa hamosa*, a single tetrad as well as twin tetrads form pollinia. In *Albizzia lebbeck*, *Acacia baileyana* and *Calliandra*, all the microspores (in a sporangium) together form a single pollinium. In *Acacia*, pollen grains shed simultaneously in masses of 8 or 16 pollen grains (Figures 3.20B, C).

In Asclepiadoideae (family Apocynaceae), all the pollen grains of a sac are united in a single compact mass called a pollinium (Figure 3.21). Two pollinia along with retinaculum and corpusculum are known as pollinaria (e.g., *Oxystelma esculentum*). In *Pergularia daemia* (Asclepiadoideae), where pollinia are formed, no callose wall develops around microspore mother cells during or after meiosis (Vijayaraghavan and Shukla, 1977).

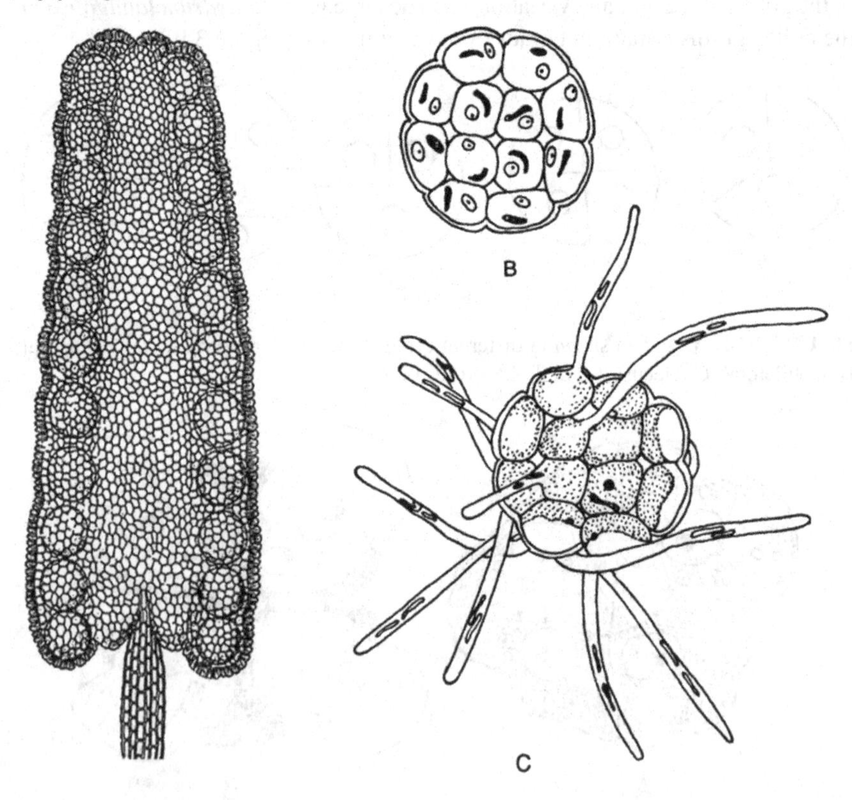


**FIGURE 3.20** *Parkia*. A. L.s. anther showing two rows of pollinia (after Engler, 1876). B., C. *Acacia dealbata*. Pollinium with two-celled pollen grains (after Poddubnaya-Arnoldi, 1954).

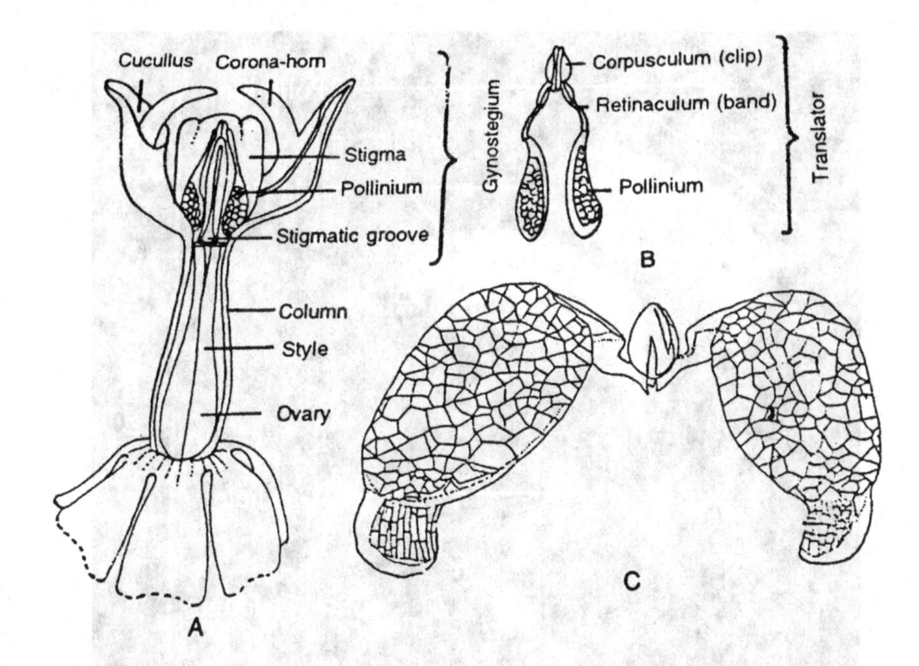


**FIGURE 3.21** *Asclepias curassavicas.* A. Corolla, column and corona. B. Transfer and pair of pollinaria. C. *Leptadenia reticulata.* Pair of pollinaria (after Pant et al., 1982).

In Orchidaceae, pollinaria are formed (Figure 3.22), but in certain cases, the pollen grains are loosely arranged (known as massulae) by means of viscin threads (Figure 3.23). These threads are PAS-positive and probably contain elastin but are not autofluorescent. In *Dendrobium*, the pollen grains are organized into distinct pollinia. The pollen grains are held together quite tightly. In *Oreorchis foliosa* (Orchidaceae), microspore mother cells undergo meiotic divisions and produce decussate, tetrahedral and isobilateral microspore tetrads which remain coherent to form pollinia.

**Pollinarium (pl. pollinaria):** Pollinaria are dispersal units of two pollinia, including a sterile interconnecting appendage.
**Massula (pl. massulae):** Unit of more than four pollen grains but less than the locular content of a theca, e.g., Orchidaceae.

## SUMMARY

- Anther is the male reproductive organ in seed plants. Its main function is to produce and disperse pollen.
- Anther wall development is of four types: basic, dicotyledonous, monocotyledonous and reduced.

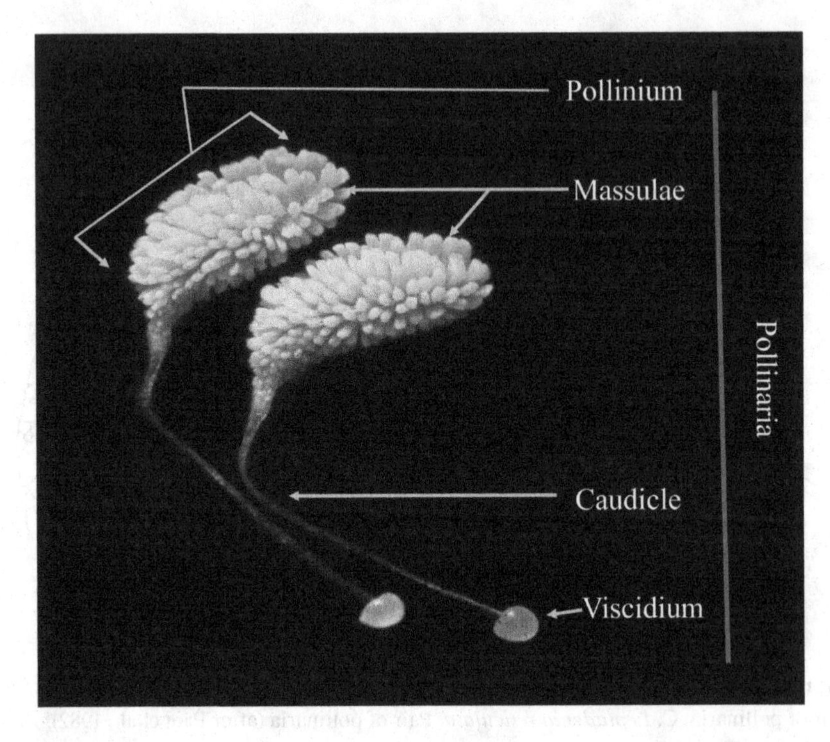


**FIGURE 3.22** Orchid pollinaria. *Habenaria arietina* (courtesy: Dr Dinesh Agrawala).

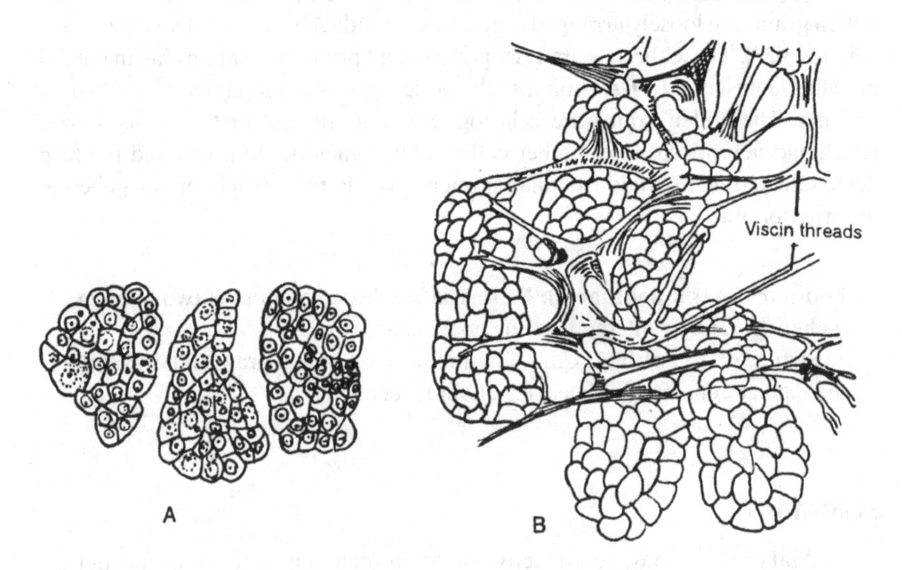


**FIGURE 3.23** A. Massulae; B. A group of massulae from a pollinium of an Orchidaceae member showing the connecting viscin threads.

- Anther wall consists of epidermis, endothecium, middle layer and tapetum.
- Tapetum is of two types: amoeboid (periplasmodial) and secretory (parietal or glandular).
- Tapetum is a highly active layer of cells investing the sporogenous tissue and provides essential nutrients and materials for pollen development.
- After meiosis, each microspore mother cell (pollen mother cell) gives rise to four haploid microspores.
- Cytokinesis is of two types, successive and simultaneous.
- Polyspory is the occurrence of more than four spores in a tetrad.
- In Asclepiadoideae, all the pollen grains of a sac are united in a single compact mass called pollinaria.

## SUGGESTED READING

Banerjee, U.C. 1967. Ultrastructure of tapetal membrane of grasses. *Grana Palynol.* 7: 365–377.

Basith, N., A.M. Abraham and P.S.S. Richard. 2019. Floral biology and floral visitors of *Allophylus serratus* (Hiren) Kurz. *Int. J. Plant Repro. Biol.* 11 (2): 159–166.

Bhandari, N.N. and R. Kishori. 1973. Development of tapetal membrane and Ubisch granules in *Nigella damascena*. A histochemical approach. *Beit. Biol. Pflanzen.* 49: 59–72.

Bhandari, N.N. 1984. The microsporangium. In *Embryology of Angiosperms*, ed. B.M. Johri, 53–121. Berlin: Springer Verlag.

Bhandari, N.N. and R. Khosla. 1982. Development and histochemistry of anther in *Triticum*. Some new aspects in early ontogeny. *Phytomorphology* 32: 18–27.

Bhandari, N.N. and R. Kishori. 1973. Development of tapetal membrane and Ubisch granules in *Nigella damascena*. *Betr. Biol. Pflanzen* 49: 59–72.

Bhatnagar, A.K. and A.K. Pandey. 2021. Embryology and systematic position of Corkia Cunn. (*Argophyllaceae, Asterales*). *Taiwania* 66 (2): 141–159.

Biddle, J.A. 1979. Anther and pollen development in garden pea and cultivated lentil. *Can. J. Bot.* 57 (18). https://doi.org/10.1139/b79-23.

Chang, F., Y. Wang, S. Wang and H. Ma. 2011. Molecular control of microsporogenesis in *Arabidopsis*. *Curr. Opin. Plant Biol.* 14 (1): 66–73.

Ciampolini, F., M. Nepi and E. Pacini. 1993. Tapetum development in *Cucurbita pepo* (Cucurbitaceae). *Pl. Syst. Evol. (Suppl.)* 7: 13–22.

Clutter, M.E. and I.M. Sussex. 1965. Meiosis and sporogenesis in excised fern leaves grown in sterile culture. *Bot. Gaz.* 126: 72–78.

Davis, G.L. 1966. *Systematic Embryology of Angiosperms*. New York: John Willey and Sons.

Deshpande, P.K., S.M. Bhuskute and V.H. Makde. 1986. Microsporogenesis and male gametophyte in some Cucurbitaceae. *Phytomorphology* 36: 145–150.

De Fossard, R.A. 1969. Development and histochemistry of the endothecium in the anthers of *in vitro* grown *Chenopodium rubrumo*. *Bot. Oaz.* 130: 10–22.

Dickinson, H.G. 1992. Microspore derived embryogenesis. In *Sexual Plant Reproduction*, eds. M. Cresti and A. Tiezzi, 1–15. Berlin: Springer-Verlag.

Galati, B.G. 1996. Tapetum development in *Sagittaria montevidensis* Cham. Et Schlech. (*Alismataceae*). *Phytomorphology* 46: 109–116.

Han, Y., S.D. Zhou, J.J. Fan, L. Zhou, Q-S. Shi, Y-F. Zhang, X-L. Liu, X. Chen, J. Zhu and Z-N. Yang 2021. OsMS188 is a key regulator of tapetum development and sporopollenin synthesis in rice. *Rice* 14 (4). https://doi.org/10.1186/s12284-020-00451-y.

Heslop-Harrison, J. 1969. An acetolysis-resistant membrane investing tapetum and sporogenous tissue in the anthers of certain Compositae. *Can. J. Bot.* 47: 541–542.

Heslop-Harrison, J. (ed.). 1971. *Pollen Development and Physiology*. London: Butterworths.

Heslop-Harrison, J. 1987. Pollen germination and pollen tube growth. *Int. Rev. Cytol.* 107: 1–78.

John, B. 1990. *Meiosis, Development and Cell Biology Series 22*. Cambridge: Cambridge University Press, p. 396.

Kamelena, O.P. 1980. *Sranitel'naya embriologiya semeistv Dipsacaceae I Morinaceae*. Leningrad: Nauka (in Russian).

Kapil, R.N. and N.N. Bhandari. 1964. Morphology of Magnolia Linn. *Proc. Natl. Inst. Sci. India B* 30: 245–262.

Kapil, R.N. and K. Walia. 1965. The embryology of *Philydrum lanuginosum* Banks ex Gaertn. and the systematic position of the Philydraceae. *Betr. Biol. Pfl.* 41: 381–404.

Laser, K.D. and N.R. Lersten. 1972. Anatomy and cytology of microsporogeneses in cytoplasmic male sterile angiosperms. *Bot. Rev.* 38: 425–454.

Lei, X. and B. Liu. 2020. Tapetum dependent male meiosis progression in plants: Increasing evidence emerges. *Front. Plant Sci.* https://doi.org/10.3389/fpls.2019.01667.

Maheswari Devi, H. and K. Lakshminarayana. 1977. Embryology of *Oxystelma esculentum*. *Phytomorphology* 27: 59–67.

Mepham, R.H. and G.R. Lane. 1969. Formation of development of tapetal periplasmodium in *Tradescantia bracteata*. *Protoplasma* 68: 175–192.

Pacini, E. and B.E. Juniper. 1983. The ultrastructure of the formation and development of the amoeboid tapetum in *Arum italicum* Miller. *Protoplasma* 117: 116–129.

Pant, D.D., D.D. Nautiyal and S.K. Chaturvedi. 1982. Pollination ecology of some Indian Asclepiads. *Phytomorphology* 32: 302–313.

Poddubnaya-Arnoldi, V.A. (ed.). 1976. *Cytoembryology of Angiosperms: Principles and Perspectives*. Moscow: Nauka (in Russian).

Raj, A.Y. 1969. Histologieal studies in male-sterile and male-fertrile *Sorghum*. *Indian J. Genet. Plant Breed.* 28: 335–341.

Raghavan, V. 1988. Anther and pollen development in rice (*Oryza sativa*). *Am. J. Bot.* 75 (2): 183–196.

Reznickova, S.A., A.C. van Aelst and M.T.M. Willemse. 1980. Investigation of exine and orbicule formationin the *Lilium* anther by scanning electron microscopy. *Acta Bot. Neer.* 129: 157–164.

Sampson, F.B. 1969. Cytokinesis in pollen mother eells of angiosperms, with emphasis on *Laurelia novae-zelandiae* ur (Monimiaeeae). *Cytologia* 34: 511–634.

Stieglitz, H. and H. Stern. 1973. Regulation of J-I,3 glueanase aetivity in developing anthers of *Lilium* mierosporoeytes. *Dev. Biol.* 134: 169–173.

van der Linde and Walbot, 2019

Venturelli, M. and F. Bouman. 1986. Embryology and seed development in *Mayaca fluviatilis* (Mayacaceae). *Act. Bot. Neerlandica* 35: 497–516.

Vijayaraghavan, M.R. and A.K. Shukla. 1977. Absence of callose around the microspore tetrad and poorly developed exine in *Pergularia daemia*. *Ann. Bot.* 41 (175): 923–926.

Vincent, P.L.D. and F.M. Getliffe. 1988. The endothecium in *Senecio* (Asteraceae). *Bot. J. Linn. Soc.* 91 (1): 63–71.

Xue, J.-S., C. Yao, Q.-L. Xu, C-X. Sui, X-L. Jia, W-J. Hu, Y-L. Lv, Y-F. Feng, Y-J. Peng, S-Y. Shen, N-Y. Yang, Y-X. Lou and Z-N. Yang 2021. Development of the middle layer in the anther of *Arabidopsis*. *Front. Plant Sci.* https://doi.org/10.3389/fpls.2021.634114.

Yadav, N., A.K. Pandey and A.K. Bhatnagar. 2017. Comparative anther and pistil anatomy of three flowering morphs of *Acer oblongum* Wall. ex DC. (Sapindaceae) and its adaptive significance. *Nord. J. Bot.* https://doi.org/10.1111/njb.01572.

Zaki, M.A. and H.G. Dickinson. 1992. Gene expression during microsporogenesis. In *Sexual Plant Reproduction*, eds. M. Cresti and A. Tiezzi, 17–29. Berlin: Springer-Verlag.

# 4 Male Gametophyte

The microspore is the first cell of the gametophytic generation. After meiosis, the tetrad of microspores is formed within the special callose wall. The life of a microspore is terminated by the first mitotic division forming the vegetative and generative cells. Older microspores, after their release from the tetrads, are referred to as pollen grains or microgametophytes.

The newly formed microspore is richly cytoplasmic with a centrally located nucleus (Figure 4.1A). The microspore cytoplasm is non-vacuolate when the spores are first released from the cellulose special wall at the end of the tetrad period (spore release period). Soon after the release from the tetrad, the volume of the microspore increases (Fig. 4.1B). The increase in volume of microspores and pollen grains as they develop is accommodated by the exine and intine elasticity. The rate of increase varies between species (Pandolfi et al., 1993). The initial expansion of the pollen grain is without an apparent vacuolation, but later vacuoles appear, and the cytoplasm comes to form a thin film lining the wall (Fig 4.1C).

In most tropical plants, the microspore nucleus begins to divide almost immediately. For example, in *Tradescantia reflexa* the resting period of the microspore is about four days or less. In plants belonging to colder regions, there is often a resting stage lasting from a few days to several weeks. In *Betula odorata*, the microspores pass the winter in the uninucleate stage.

## FORMATION OF VEGETATIVE AND GENERATIVE CELLS

The first division of the microspore gives rise to a small generative cell and a large vegetative cell (Figure 4.1C, D). Frequently during cytokinesis, the organelles of the microspore are positioned in such a way that most of them are transferred directly to the vegetative cell. The direct cause of the asymmetry has been attributed to a difference in the time of development of the two spindle poles, the wall ward or generative pole developing more slowly than the vegetative, presumably because of the smaller amount of cytoplasm associated with the former.

The mitotic spindle is either symmetrical or asymmetrical. It may be conical or flattened. A curved cell plate separates the two nuclei and extends up to intine. The curved wall grows between the intine and plasmalemma of the generative cell (Figure 4.1C). The generative cell gets separated from the wall of the pollen grain and comes to lie in the cytoplasm of the vegetative cell (Figure 4.1E, F).

When the first mitotic division occurs in a microspore within a tetrad, the position of the generative cell is usually predetermined due to certain genetic factors. The generative cell may be cut off towards the centre (proximal pole) or the outer wall side (distal pole). The polarity can be ascertained only when the pollen grains are attached in a tetrad (Figure 4.2).

DOI: 10.1201/9781003260097-4

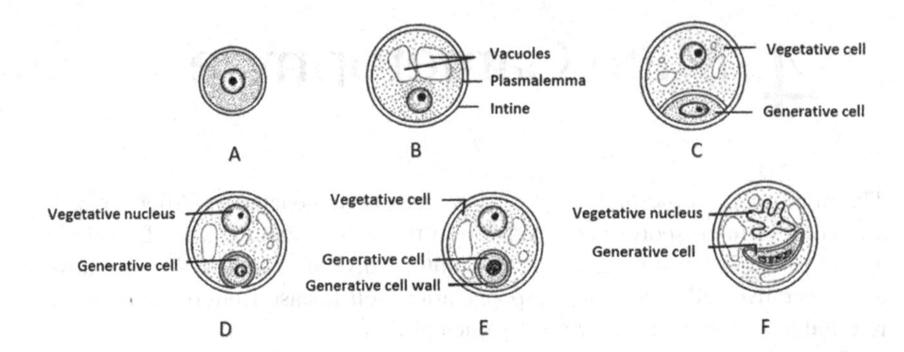


**FIGURE 4.1** Development of male gametophyte in angiosperms.

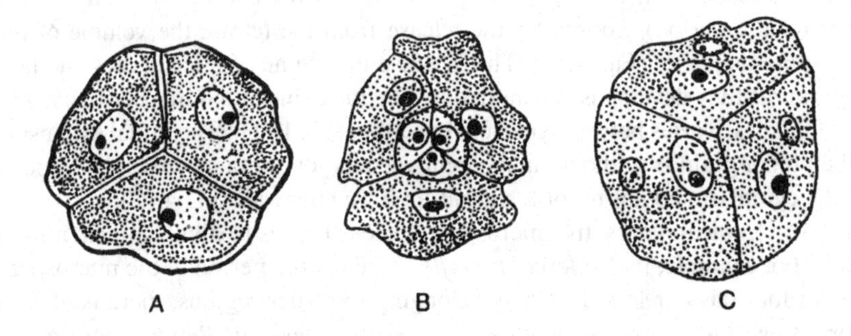


**FIGURE 4.2** *Juncus bufonius.* A. Single-celled pollen grains in a tetrahedral tetrad; B. Two-celled condition. The generative cell is cut off towards the centre; C. Two-celled pollen grains in a tetrahedral tetrad. Generative cell is cut off towards the periphery (after Shah, 1963).

A relationship has been demonstrated in many pollen types between the sitting of the aperture and the future position of the generative cell. According to Huynh (1976), the generative cell is cut off at a point away from the aperture. He considers that this siting might ensure that the vegetative nucleus would enter the pollen tube first.

Vinthanage and Knox (1980) recognised eight periods in the development of grass pollen grains (Table 4.1). Pandolfi et al. (1993) studied pollen development in *Pterostylis plumosa* and also recognised eight developmental stages (Figure 4.3).

## VEGETATIVE CELL

After the microspore mitosis, the vegetative cell continues to grow. The cell organelles increase in number as well as in size. The vacuole gradually disappears, and the nuclear envelope becomes highly convoluted. The nucleus of the vegetative cell is spherical or irregular in outline with thin chromatin material (Figure 4.1E, F). The nucleus may undergo endoduplication and become highly polyploid. The

**TABLE 4.1**
**Stages in Grass Pollen Development**

| Period | Features |
|---|---|
| Tetrad period | Four microspores within callose wall |
| Spore-release period | Microspores are released from callose wall with thin exine |
| Pre-vacuolation period | Grain spherical, with exine and pore clearly visible; dense non-vacuolate cytoplasm |
| Early vacuolate period | Spherical vacuole present, with diameter up to half that of the pollen grain |
| Mid-vacuolate period | Grain spherical with the vacuole filling the grain, giving a "signet ring" appearance |
| Late-vacuolate period | Grain distinctly ovoid in shape, showing buildup of cytoplasm so that vacuole is reduced to one-half to one-third of grain volume |
| Early maturation period | First pollen mitosis has occurred, cytoplasm fills the grain, further reducing the size of vacuole |
| Late maturation period | Second pollen mitosis has occurred, grains are tricellular, cytoplasm with starch grains completely fills the grain |

cytoplasm of the vegetative cell contains ribosomes, rough endoplasmic reticulum, active dictyosomes, plastids, mitochondria, fat and starch grains. RNA and proteins are also present in sufficient amounts. The final composition of the organelle population in the mature vegetative cell shows great diversity among species. In *Haemanthus katherinae* (Sanger and Jackson, 1971), the large nucleus begins to lose its spherical shape and is replaced by a small nucleolus. Despite this diversity, the vegetative cell may be regarded as a storage cell, equipped for the future formation of pollen tube and transmission of the male gametes.

## GENERATIVE CELL

After its formation, the generative cell separates from the intine and moves to a position where it is completely enclosed by the vegetative cell (Figure 4.1F). During this process, its shape changes. The generative cell is spherical in the beginning, but later on it becomes spindle-shaped (Figure 4.1F). The newly formed generative cell has a normal cell wall. The spindle-shaped generative cell is surrounded simply by a thin layer separating the plasma membranes of the two cells. The elongation of the generative cell has been attributed to the microtubules which are oriented longitudinally, parallel to the long axis of the cell.

The cytoplasm of the generative cell is enclosed in two plasma membranes; its own and the detached invagination of the plasmalemma of the vegetative cell. There is no cytoplasmic connection between the generative cell and vegetative cell. The cytoplasm of the generative cell contains mitochondria, microtubules, endoplasmic reticulum, dictyosomes and abundant free ribosomes. Normally generative cells do not contain plastids (*Beta*, *Petunia*), but the presence of plastids

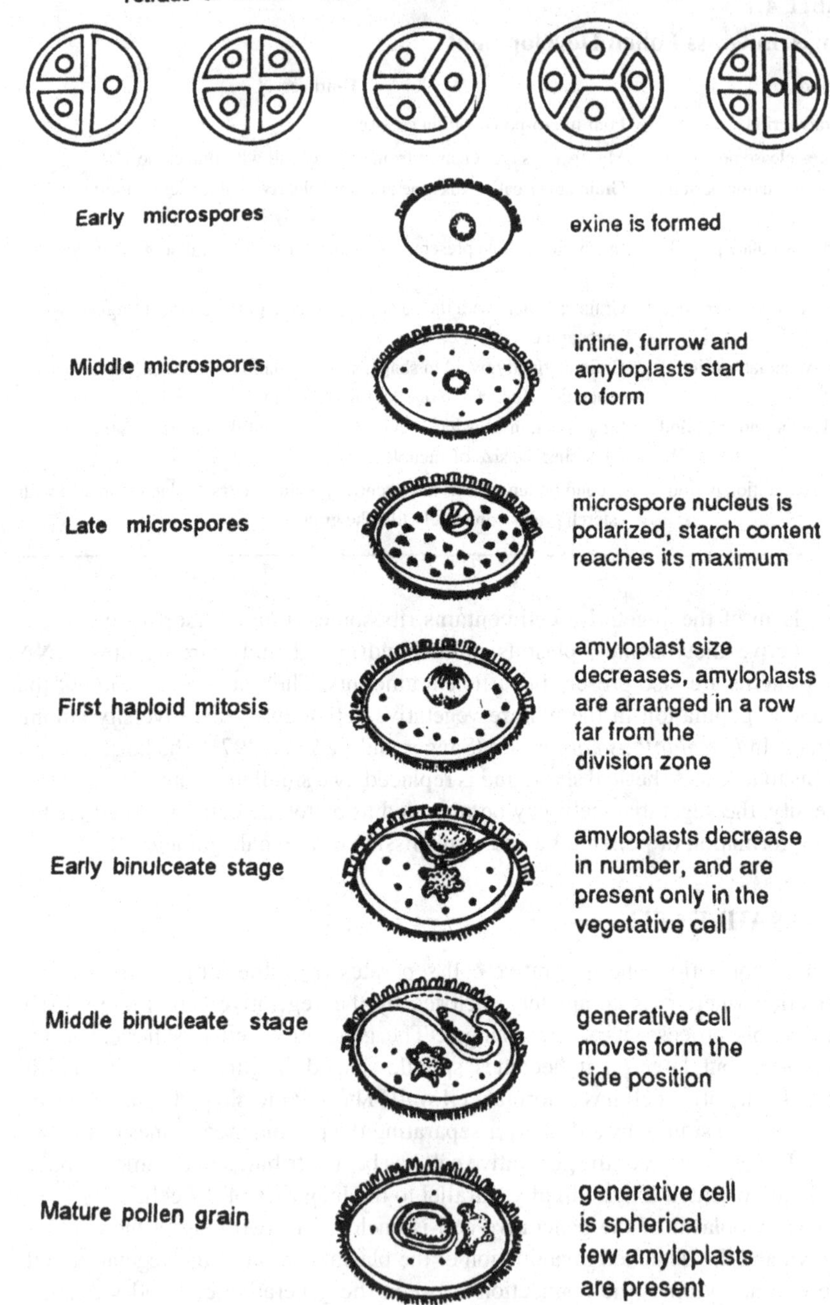


**FIGURE 4.3** Semischematic representation showing pollen grain development in *Pterostylis plumosa* (after Pandolfi et al., 1993).

has been reported in some other species (*Medicago sativa, Oenothera hookeri*). When present, the plastids do not contain starch.

## FORMATION OF SPERM CELLS

The sperm cells are formed by a mitotic division of the generative cell. The generative cell may divide before or after the dehiscence of the anther. In several species, the generative cell divides when the pollen grains are still within the anther locule, and the two sperm cells are formed. In such cases, shedding of the pollen grains takes place at the 3-celled stage. The sperm cells are oval or fusiform in shape. The fusiform shape of the cells permits them to move into the pollen tube.

In cases where pollen grains are shed at the 2-celled stage, the generative cell may divide inside the pollen grain after the deposition of the pollen tube before it reaches the embryo sac (Figure 4.4). The two sperm cells frequently remain connected to each other and become located near the vegetative nucleus, forming the male germ unit. The generative cell and sperm cell show relatively simple ultrastructure with only few organelles.

Chu and Hu (1981) observed the following four developmental periods in the tricellular pollen of wheat (*Triticum aestivum*).

1. **Naked cell stage:** After mitosis, the sperm cell is enveloped in a discontinuous plasma membrane.
2. **Walled cell stage:** The cell is enclosed within a plasma membrane, surrounded by a callose wall and the plasma membrane of the host vegetative cell.

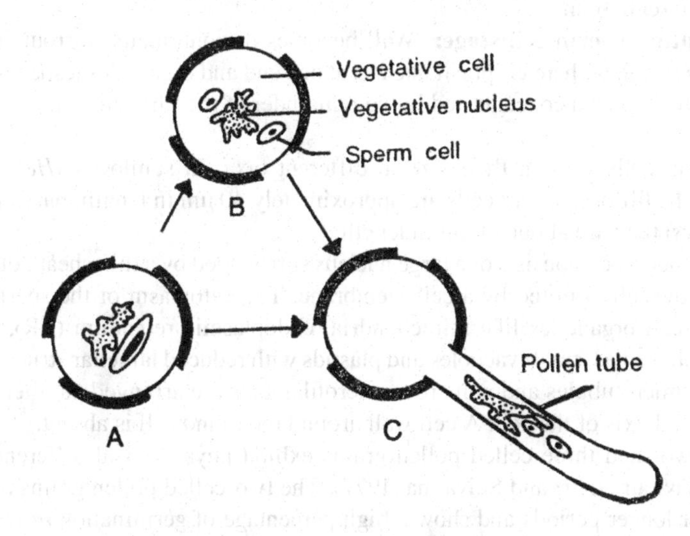


**FIGURE 4.4** A. Mature bicellular pollen grain; B. Tricellular pollen grain; C. Bicellular pollen becomes tricellular after germination (adapted from Dexheimer, 1970).

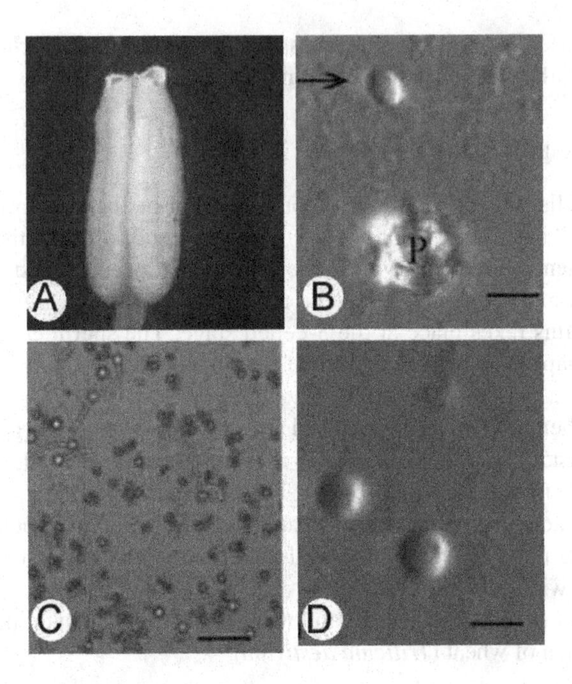


**FIGURE 4.5** Isolation of generative cells and sperm cells of *Solanum verbascifolium*. A. Anther at the time of anthesis; B. A released generative cell (arrow) from a pollen grain; C. Pollen grains with germinating pollen tubes; D. A pair of sperm cells (after Yang et al., 2015).

3. **Cytoplasm-increasing stage:** Period of organelle and cytoplasmic differentiation.
4. **Mature sperm cell stage:** Wall becomes discontinuous, surrounding the sperm with an elliptical nucleus at one end and organelles (especially mitochondria) concentrated in the remainder of the filiform cell.

The sperm cells vary in their size in different taxa. In sunflower (*Helianthus annuus*) the filiform sperm cells are approximately 40 µm in length, but in maize (*Zea mays*) they are about 14 µm in length.

Each sperm cell consists of a large nucleus surrounded by a thin sheath of cytoplasm, externally limited by a cell membrane. The cytoplasm of the sperm cell contains cell organelles like mitochondria, endoplasmic reticulum (ER), Golgi bodies, ribosomes, small vacuoles and plastids with reduced lamellar structure. In addition, microtubules and numerous microfilaments are arranged parallel to the longitudinal axis of the cell. A cell wall around the round cell is absent.

The two- and three-celled pollen grains exhibit physiological differences in their behaviour (Johri and Shivanna, 1977). The two-celled pollen grains remain viable for longer periods and show a high percentage of germination *in vitro*, as compared to the three-celled grains. The families showing bicellular and tricellular pollen grains are given in Table 4.2.

**TABLE 4.2**
**Families Showing Bicellular and Tricellular Pollen Grains**

| Bicellular | Tricellular |
|---|---|
| Betulaceae | Asteraceae |
| Rosaceae | Brassicaceae |
| Solanaceae | Caryophyllaceae |
| | Poaceae |

The formation of supernumerary sperm nuclei, through the division of sperm cells, is an abnormality and is known as polyspermy.

## ISOLATION OF SPERM CELLS

Sperm cells can be isolated from the mature pollen grains. Isolating sperm cells is generally more difficult in species with bicellular pollen because the generative cell of bicellular pollen needs to undergo mitosis to form two sperm cells within a pollen tube. Thus, to isolate the sperm cells of bicellular pollen, it is necessary to induce pollen tube growth. On the other hand, it is relatively easy to isolate sperm cells from species with tricellular pollen. When tricellular pollen grains are incubated in an osmotic solution, the grains hydrate and rupture, thereby releasing their contents, including the two sperm cells. Isolating the sperm and egg cells avoids any interference from the somatic cells of the stigma and style tissues. Thus, the sperm and egg cells can then be directly fused to produce a hybridized zygote that may not be obtainable using typical hybridization techniques.

Yang et al. (2015) isolated sperm cells of *Solanum verbescifolium* by using both *in vivo* and *in vitro* methods. Deng et al. (2020) isolated sperm cells from the pollen grains of *Oryza officinalis* (wild rice) using an osmotic shock method. The isolation of sperm cells is a prerequisite for the *in vitro* hybridization of distantly related cultivated wild rice lines.

### Isolation Method

In *Solanum verbescifolium* (Yang et al., 2015) hand-pollinated styles were first grown *in vivo* for several hours, then cut off from their base and cultured *in vitro* until pollen tubes grew out from the cut end. When the pollen tubes were transferred to broken solution, the sperm cells were released from broken tubes (Figure 4.5). The 10% mannitol solution was considered optimal for isolating generative cells.

In *Oryza officinalis* (Deng et al., 2020) at the time of anthesis, the mature rice pollen grains are tricellular and consist of a vegetative cell and two sperm cells. When pollen grains were treated with osmotically active solutions (e.g., incubated in 4–12% mannitol solution), the pollen grains took up water. The increase in turgor pressure caused the pollen grains to rupture and release the contents, including two sperm cells which were collected with a micromanipulator.

The osmotic solutions must be optimized for isolating sperm cells. If the osmotic pressure of the solution is low, more pollen grains will rupture, but the released sperm cells will also rapidly take up water and rupture. When distilled water without mannitol was used, over 80% of the pollen grains ruptured, but viable sperm cells were undetectable. In contrast, a treatment with a 6% mannitol solution resulted in 63.21% of the pollen grains rupturing. If the osmotic pressure of the solution is high, fewer pollen grains will rupture, but the released sperm cells will remain intact for a relatively longer period. Therefore, a 10% mannitol solution appears to be optimal for isolating medicinal wild rice sperm cells.

## MALE GERM UNIT

The concept of the male germ unit was first introduced by Dumas et al. (1985). In flowering plants, the vegetative nucleus and the two sperm cells form a functional assemblage termed the male germ unit (MGU). The sperm cell dimorphism observed in many species suggests that the MGU is a "polarized fertilization unit" in which the sperm cells are predetermined to fuse with the egg nucleus or the two polar nuclei.

In the MGU, one sperm cell is connected to the vegetative nucleus via a "tail" or cell extension, containing forked arrays of microtubules. This tail penetrates the highly convoluted vegetative nucleus (McConchie et al., 1985), whereas the two sperm cells are enclosed within a vegetative membrane bound compartment and are linked by a transverse cell wall and evaginations of their plasma membrane (Charzinska et al., 1989).

Rusell (1984) made a detailed study of the structure of sperm cells in *Plumbago zeylanica*. One sperm cell consistently is attached to the vegetative cell by a projection, while the other is linked to the first by plasmodesmata connections. Figure 4.6 shows two sperms, Svn and Sua, and a vegetative nucleus VN. Svn has the majority of mitochondria (m) with very few plastids, while Sua contains plastids (p) and only a few mitochondria. In *Zea mays* (Rusche and Mogensen, 1988) each sperm is a long, tapering cell with a heterochromatic nucleus and two cytoplasmic extensions. The length of the extensions in sperm 1, the larger sperm, are approximately equal; in sperm 2, the smaller sperm, one extension is twice the length of the other. Within the thin layer of cytoplasm surrounding the nuclei are the 14 mitochondria of sperm 2 and 3 of the 5 mitochondria of sperm 1. The endoplasmic reticulum associated with the mitochondria continues into the cytoplasmic extensions where parallel ER sheets are observed. Few dictyosomes and no plastids are found in the sperm cells. Cytoplasmic inclusions of unknown composition occur in the sperm cell cytoplasm. No physical association of sperm cells with each other or with the vegetative nucleus is observed. Minimum distances between the two sperm cells, between sperm 1 and the vegetative nucleus and between sperm 2 and the vegetative nucleus are 0.5 μm, 4.3 μm and 3.5 μm, respectively.

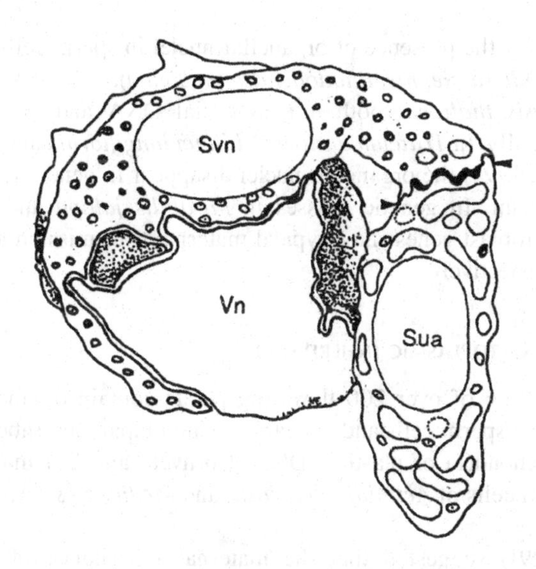


**FIGURE 4.6** *Plumbago zeylanica.* Vegetative nucleus (VN) and two sperm cells superimposed profile of mitochondria (m) and plastids (p), in pollen grain. One sperm cell ($S_{vn}$) contains the majority of mitochondria and two plastids near the sperm cross wall (arrowheads). The other sperm cell $S_{ua}$ contains most of the plastids and significantly fewer mitochondria. Note the gradation between the main body of $S^{vn}$ and its cellular projection (after Russel, 1984).

Lalanne and Twell (2002) studied the genetic content of MGU organisation in *Arabidopsis* (Brassicaceae) and reported two classes of mutations that affect the integrity and/or the positioning of the MGU in the mature pollen grain.

The formation of MGU is of wide occurrence in angiosperms. However, the appearance of MGU may vary from species to species. The MGU is organised in the pollen tube in *Gossypium hirsutum*, *Petunia hybrida*, *Hordeum vulgare* and *Spinacea oleracea*. After its formation, MGU travels through the pollen tube as a single entity. As soon as the pollen tube comes in contact with the ovule, the vegetative nucleus separates from the sperm followed by separation of the two sperms. According to Mathys-Rochon and Dumas (1988) the quality and spatial disposition of the sperms in the MGU allows their targeted fusion with the female gametophyte and the polar nuclei.

## INHERITANCE OF CYTOPLASMIC TRAITS

The process by which organelles are partitioned into gametes and transmitted to zygotes or fertilized eggs in called cytoplasmic inheritance (Kuroiwa and Uchida, 1996). Two basic types of organelle DNA behaviour in generative and sperm cells of pollen grains have been reported (Miyamura et al., 1987). The first type is characterised by the disappearance of organellar nuclei in the sperm cells of pollen grains (e.g., *Nicotiana tabacum*, *Mirabilis jalapa*). The second type of behaviour is characterised by the presence of organellar nuclei in sperm cells in the pollen tubes (e.g., *Oenothera biennis, Rhododendron indicum*).

In *Arabidopsis thaliana*, both mitochondrial DNA and plastid DNA are inherited maternally. In *Triticum aestivum, Lilium longiflorum* and *Zea mays* (Li and Sodmergen, 1995) the organellar nuclei disappear in generative cells during pollen development. In genetic crosses in *Mirabilis jalapa* and *Pelargonium zonale*, the chloroplast genes show typical maternal transmission and biparental transmission, respectively.

### Mechanism of Cytoplasmic Inheritance

Approximately 20% of over 200 flowering plants contain organellar nuclei in mature generative/sperm cells and appear to exhibit biparental inheritance of the organelle (mitochondria or plastids) DNA (Cornvelu and Coleman, 1988). The generative/sperm cells of *Actinidia deliciosa*, and *Medicago sativa*) contain only plastid nuclei.

Kuroiwa (1991) suggested that the maternal inheritance of plastid DNA is responsible for the selective disappearance of organellar nuclei in mature generative cells or sperm cells in higher plants. Nagata et al. (1999) found that the presence or absence of organellar nuclei in mature generative cells, which defines the mode of organellar inheritance, is determined by the behaviour of organellar nuclei in young generative cells just after pollen mitosis I (PMI).

Where the organellar nucleus was absent from mature generative cells (and exhibited maternal inheritance), the DNA content of the organelles in generative cells began to decrease immediately after PMI. In these cases, the organellar nucleus was degraded more rapidly in generative cells than in vegetative cells, suggesting active degradation of organellar DNA in the generative cells.

The inheritance of mitochondria and plastids in flowering plants has been categorized into three modes: maternal, biparental and paternal. Several mechanisms have been proposed for maternal inheritance, including (i) physical exclusion of the organelle itself during pollen mitosis I (PMI); (ii) elimination of the organelle by formation of enucleated cytoplasmic bodies (ECB); (iii) autophagic degradation of organelles during male gametophyte development; (iv) digestion of the organelle after fertilization; and (v) the most likely possibility – digestion of organellar DNA in generative cells just after PMI (Nagata, 2010).

## POLLEN WALL

The pollen wall is a complex and robust structure surrounding the microspore cytoplasm that helps to resist various severe environments and is also involved in pollination and pollen–stigma recognition. The mature pollen wall is composed of two principal layers: the inner one is called intine, and the outer is called exine (Figure 4.7). The terms exine and intine were first proposed by Fritzsche (1837). The intine is composed of pectin and cellulose. The cellulose component is microfibrillar, with the microfibrils oriented in a plane parallel to the surface.

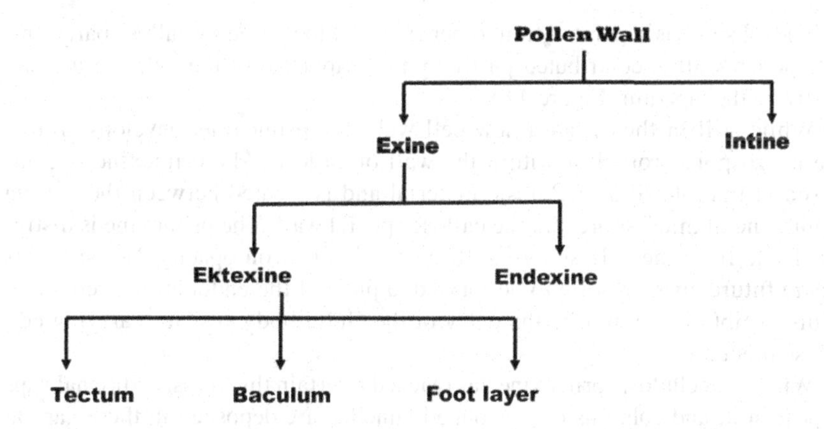


**FIGURE 4.7** Pollen wall stratification.

The intine is characterised by the presence of beads, ribbons, or plates of enzymatic proteins, particularly in the vicinity of the germ pores.

The principal component of the exine is sporopollenin which gives durability to the pollen grain wall. Sporopollenin is derived from oxidative polymers of the carotenoids and/or carotenoid esters (Shaw, 1971). Sporopollenin is resistant to biodegradation; for this reason, pollen grain walls are well preserved in fossil forms. The pollen wall also has a protective function during the hazardous journey of pollen from anther to stigma.

According to Heslop-Harrison (1975), at the time of fertilization, enzymes and enzyme precursors are released from the intine. These enzymes degrade the cuticle of the stigma papillae after appropriate activation and enable the pollen tube to enter the style.

The exine appears first as a thin layer but later becomes considerably thick and can be differentiated into two layers – the ektexine (also called sexine) and endexine (also called nexine). The ektexine is further differentiated into three layers: a basal foot layer, a middle bacula and an upper tectum.

The endexine, which completely covers the intine, usually forms a smooth layer. The sculpturing of the sexine results from radially directed rods, the bacula (singular baculum) with enlarged heads. The bacula differ in size and may be either isolated or clustered in groups. In many genera, the heads of the bacula are fused to form a roof or tectum which may again be perforated or sculptured in characteristic ways.

In *Heliconia*, many tropical Lauraceae and members of Scitamineae, the exine of the pollen grains is greatly reduced while the intine is quite thick.

## DEVELOPMENT OF THE POLLEN WALL

During microsporogenesis, after meiosis four haploid microspores are enclosed in a common callose wall, and the microspores lack a wall of their own. The individual spore is separated from other spores of the tetrad by callose partitions. The pollen wall is contributed partly by the cytoplasm of the pollen grain and partly by the tapetum (Figure 4.8).

While still in the tetrad a new cell wall, the primexine, develops around the microspore protoplast within the wall of callose. The primexine is composed of microfibrillar cellulosic material and is located between the plasma membrane of microspore and the callose special wall. The primexine is distinguishable from the callose by its difference in electron opacity. In the places where future apertures are to be formed, a plate of the endoplasmic reticulum is discernible which blocks the entry of the Golgi body vesicles carrying cellulose precursors.

When the cellulosic primexine has reached a certain thickness, additional gaps appear in it, and columns of convoluted lamellae are deposited in these gaps at the surface of the plasmalemma. These columns are called probacula. Now the cytoplasm of the spore starts synthesizing the precursors of sporopollenin, which are polymerized and deposited on the surface of the lamellae. The columns are now called bacula.

The bacula enlarge and increase due to the deposition of sporopollenin and their heads extend laterally to form tectum. In some species, the bases of bacula expand laterally to form a foot layer. The foot layer forms a sort of floor on which

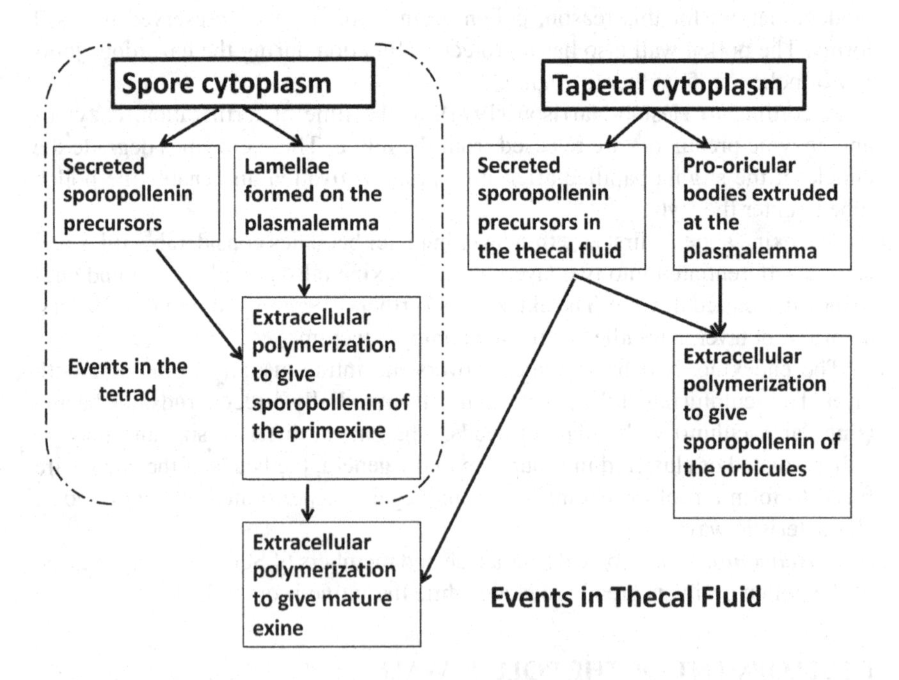


**FIGURE 4.8** Scheme for deposition of sporopollenin in the orbicules and exine, based upon the evidence from *Lilium* and other genera with a tapetum of the parietal, secretory type (modified from Heslop-Harrison and Dickinson, 1969).

are raised the columns (or bacula). Further deposition of the sporopollenin is continued. The callose wall disappears, and the pollen grains lie free in the pollen sac.

The endexine is deposited below the ektexine. During the formation of endexine, a number of very thin electron transparent lamellae, which appear to arise from the cytoplasm and provide a locus around which sporopollenin are deposited. As the deposition proceeds, the lamellae thicken and merge with each other to form endexine. Below endexine, intine is formed (Figure 4.9, Figure 4.10).

Intine is an essential wall layer for the pollen grains. The intine is formed by the dictyosomes. During the formation of intine, two processes are involved: (i) during early growth of the intine, the thickening of the innermost layer of the exine (endexine) continues, and the lamellar material and sporopollenin precursors, which are contributed by the spore cytoplasm, must pass through the developing intine; (ii) certain proteinaceous plates or ribbons are incorporated in the intine in the vicinity of the germ pores, and these proteins show enzymatic activity.

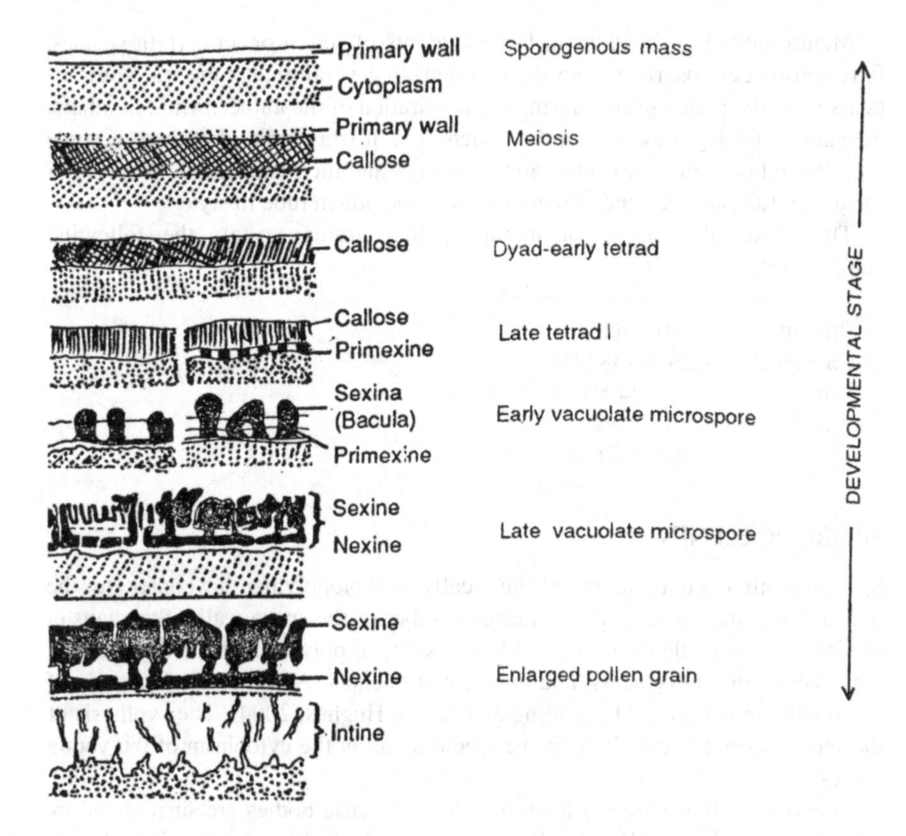


**FIGURE 4.9** Summary of the pollen wall (sporoderm) ontogeny in *Sorghyum bicolour.* Corresponding developing stages in the anther loculus, are also mentioned opposite to each figure (adapted from Christensen et al., 1972).

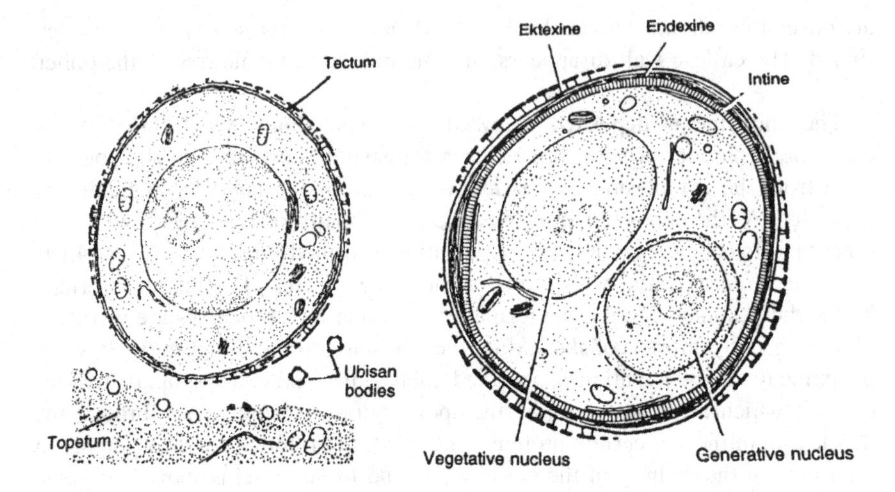


**FIGURE 4.10** Development of pollen wall (after Echlin, 1968).

Mature pollen grains contain large amounts of starch or, in certain species, fatty substances absorbed from the tapetum. In several plants, the starch disappears from the pollen grains during the maturation of the anther; while in others, the starch disintegrates only in the pollen tube. It is assumed that there is some correlation between the disintegration of starch and the high osmotic pressure of the pollen tubes at the time of penetration of the pollen tube in style.

The chemical analysis of mature pollen grains reveals the following composition:

| | |
|---|---|
| Proteins | 7.0–26.0% |
| Carbohydrates | 24.0–48.0% |
| Fats | 0.9–14.5% |
| Ash | 0.9–5.4% |
| Water | 7.0–16.0% |

## SPOROPOLLENIN

Sporopollenin is a ubiquitous and chemically inert biopolymer that constitutes the pollen exine and exosporium of a large number of the spore walls (Steemans et al., 2010). Sporopollenin is derived from oxidative polymers of the carotenoids, polyunsatutrated fatty acids, and conjugated phenols. A generalised formula of sporopollenin is $C_{90}H_{142}O_{36}$ (Riding and Kyffin-Hughes, 2004). The synthesis of the sporopollenin occurs both in the tapetum and in the cytoplasm of the young spores.

In anther with a secretory tapetum, electron dense bodies are surrounded by a limiting membrane, the pro-Ubisch bodies, which are primary sites of sporopollenin polymerization. In early stages of the pollen wall development, the sporopollenin is formed from precursors within the pollen grain cytoplasm.

The medium electron dense spherosome-like structures reported by Echlin and Godwin (1968a,b) represent deposits or synthetic sites of such precursors. The cellulosic primexine forms a template on which the sporopollenin is deposited. According to Jonkar (1971), synthesis of sporopollenin in mitochondria of the tapetal protoplasts is later transported to the exterior of the tapetal cells.

## POLLENKITT AND TRYPHINE

The term pollenkitt is applied to various substances responsible for imparting stickiness to the pollen. Pollenkitt is the final product of the degeneration of the secretory and amoeboid tapetum. It covers the pollen surface just before the opening of the anther, consisting only of hydrophobic material derived from blends of elaioplast and cytoplasmic lipids. In *Lilium longifolium*, Heslop-Harrison (1968a,b) observed that pollenkitt material is of tapetal origin and comprised mainly of lipids and carotenoids. In *Artemisia mutellina* (Hesse, 1979), soon after degeneration of the tapetum, lumps of pollenkitt material composed of lipid droplets of various shapes and sizes combine together with other cellular organelles floating in the anther locule.

Several functions are ascribed to pollenkitt according to the pollination strategy and pollen grain size (Table 4.3). Some functions are due to the viscous nature of pollenkitt, namely to stick pollen grains together in the anther until the arrival of pollinators and during dispersal. Other properties depend on the chemical nature of pollenkitt, namely protection against solar radiation and water loss during exposure. Others are due to colour and smell which attract pollinators.

**TABLE 4.3**
**Functions of Pollenkitt (after Pacini and Franchi, 1993)**

| Primary Function | Secondary Function |
|---|---|
| 1. To hold pollen grains together | - in anther until collection by insects<br>- in anther and in air. Physical-chemical properties of<br>- pollenkitt, and pollen grain size determine number of grains per group. Deposition of whole group on stigma surface; this occurs in both ento-and anemophilous species<br>- in air, sticking them to the insect during flight |
| 2. To protect against effects of solar radiation, i.e. | - damage to vegetative cell cytoplasm<br>- mutations in generative and sperm cells |
| 3. To protect pollen from further water loss | |
| 4. To determine pollen colour | |
| 5. To maintain sporophytic proteins inside exine cavities | |
| 6. To attract potential insect pollinators | |

Distinction is made between the pollenkitt and tryphine, though these are similar in many respects (Dickinson, 1973). The former is a synthetic activity while the latter is formed from the remains of degenerated tapetum. The pollenkitt contains hydrophobic lipids whereas tryphine includes hydrophilic lipids.

## UBISCH BODIES (ORBICULES)

The secretory tapetum is characterized by the development of the Ubisch bodies or orbicules (the term was coined by Heslop-Harrison, 1968 in preference to the Ubisch bodies) along the tangential surface facing the anther locule.

Ubisch bodies may remain individually separate or, during the course of the development, fuse to form larger aggregations or compound bodies. With the help of an electron microscope, Echlin and Godwin (1968) traced the origin and the development of Ubisch bodies in *Helleborus foetidus*. At the sporogenous stage, numerous pro-Ubisch bodies appear in the tapetal cytoplasm, prominently aggregated towards the anther locule. Endoplasmic reticulum cisternae and ribosomes are in close association with the developing pro-Ubisch bodies at the tetrad stage.

During the extrusion of pro-Ubisch bodies, the surrounding membrane fuses with the plasmalemma releasing them into the space between plasma membrane and the tapetal cell wall. Immediately after that, the pro-Ubisch bodies become covered with sporopollenin, which is deposited on either side of a unit membrane.

The role played by Ubisch bodies in deposition of pollen exine is not clearly understood. Maheshwari (1950) speculated that these granules contribute in the formation of exine which was confirmed later on by a number of researchers.

## POLLEN DEVELOPMENT IN CYPERACEAE

In the majority of angiosperms, the pollen mother cells after meiotic division give rise to four haploid spores, but in the members of the family Cyperaceae, a special mode of pollen development is found. Of the four microspore nuclei produced after meiosis, only one becomes functional. The other three non-functional nuclei are cut off on one side of the cell (Figures 4.11A, B). In *Cyperus*, *Kyllinga* and *Scirpus*, the non-functional nuclei are separated from the functional nucleus by a wall. The functional nucleus, which lies in the centre, divides with its spindle oriented in the direction of the long axis of the cell (Figures 4.11C, D). The cell plate, which is laid down between the vegetative nucleus and generative nucleus, extends around the latter so as to give rise to a continuous plasma membrane. The generative cell (Figure 4.10E) soon becomes spindle-shaped and divides to form two sperm cells (Figure 4.11F). The abortive nuclei get further separated from each other by the delicate membranes and seem to have a feeble capacity for further divisions, but eventually all the daughters coalesce together forming an irregular darkly stained black mass.

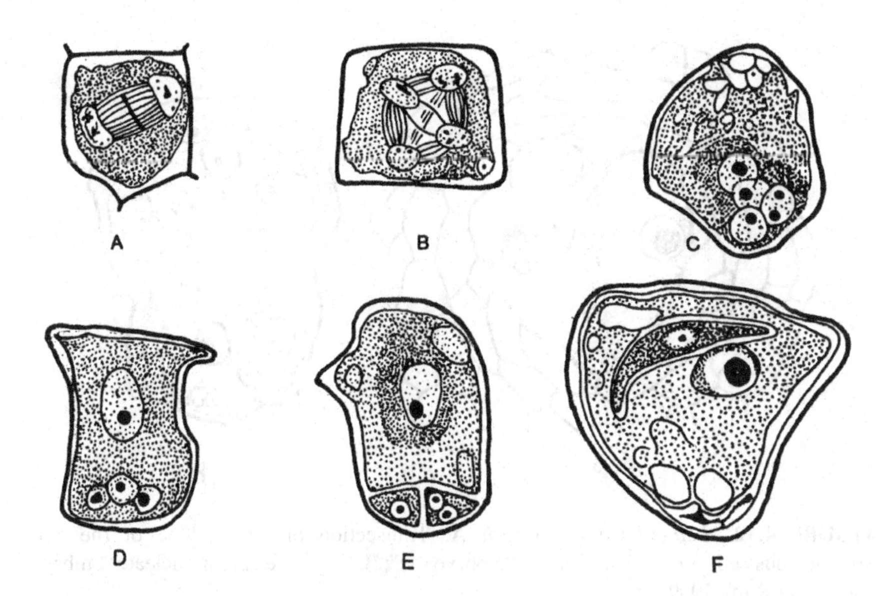


**FIGURE 4.11** Development of pollen grain in *Carex wallichiana* (Cyperacese). A, B. Meiosis I and II; C. A tetrad, all the four microspore nuclei have migrated to one side; D. Same, three of the microspore nuclei remain at one end of pollen nucleus from the three micronuclei by a septum; E. Delimitation of pollen nucleus from the three micronuclei by a septum; F. Two-celled pollen grains (after Shah, 1962).

## POLLEN EMBRYO SACS (NEMEC PHENOMENON)

Nemec (1898) reported, for the first time, pollen embryo sac-like structures in the petaloid anthers of *Hyacinthus orientalis*, generally referred to as the Nemec phenomenon.

In *H. orientalis*, the pollen grains are of two types: small and large. The small uninucleate pollen is devoid of reserve food and rarely develops into an embryo sac. The large pollen grain contains abundant food reserves. It may become binucleate, or the vegetative nucleus enlarges and forms an uninucleate, binucleate, four nucleate or six nucleate pollen embryo sacs.

Stow (1930) induced pollen embryo sacs in *H. orientalis* by subjecting the bulbs to a temperature of 17–20°C during meiosis of the pollen mother cells. In *Leptomeria billardierii* (Ram, 1959), some of the sporogenous cells enlarge and develop into an uni- or binucleate embryo sac-like structure. Two subsequent mitotic divisions result in the formation of an eight-nucleate embryo sac. These nuclei organize like a typical embryo sac with a three-celled egg apparatus, two polar nuclei and three antipodal cells (Figure 4.12).

In *Heuchera micrantha*, Vijayaraghavan and Ratnaparkhi (1977) observed exine all around the pollen embryo sac. Stow (1930), Naithani (1937) and Geitler (1941), however, observed exine only at one end. Stow considers the cells at the exine end to be equivalent to the egg and synergids and the three nuclei at the other end to represent the antipodals. A pollen embryo sac may arise from the divisions of vegetative nucleus or directly from microspore mother cells.

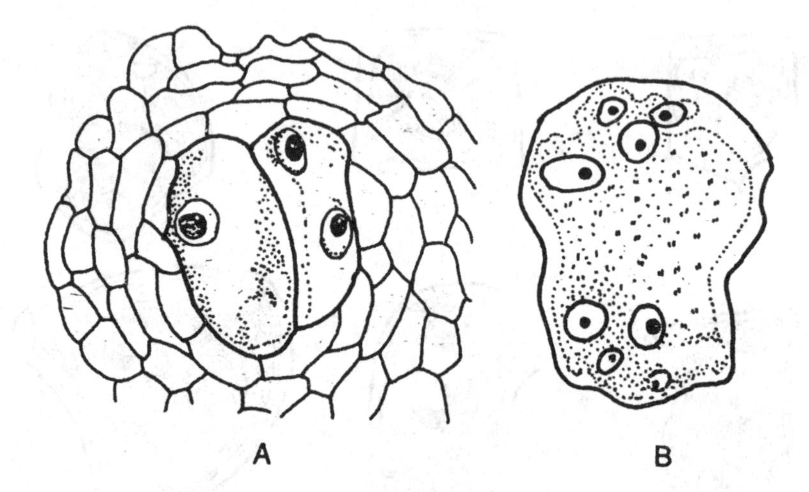


**FIGURE 4.12** *Leptomeria billardierii* A. Transection of anther lobe of the two sporogenous cells. One is a binucleate "embryo sac"; B. Organized eight-nucleate "embryo sac" (after Ram, 1959).

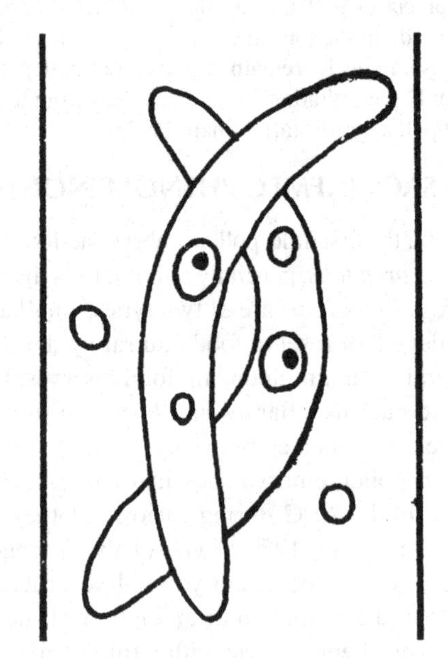

**FIGURE 4.13** Filiform pollen of sea-grass *Halodule.*

## POLLEN GRAINS IN AQUATIC PLANTS

Marine plants, like sea grasses, show variations in the shape of the pollen grains. *Halophila*, *Thalassia*, *Halodule* and *Zostera* show filiform (thread-like) pollen grains (Figure 4.13). The microspores after their release from the tetrads elongate and become filamentous. The pollen grains are 1–5mm long and lack a well-defined exine.

## SUMMARY

- The microspore is the first cell of the gametophytic generation. The first division of the microspore gives rise to a small generative cell and a large vegetative cell.
- Two sperm cells are formed by the mitotic division of the generative cell.
- The two sperm cells frequently remain connected to each other and become located near the vegetative nucleus, forming the male germ unit (MGU).
- The process by which organelles are partitioned into gametes and transmitted to zygotes or fertilized eggs is called cytoplasmic inheritance.
- The mature pollen wall is composed of two principal layers: the inner one is called intine and the outer is called exine. The exine is composed of sporopollenin, and the intine is pectocellulosic in nature.
- Pollenkitt is responsible for imparting stickiness to the pollen and is formed after the degeneration of tapetum.
- The embryo sac-like structures in the petaloid anthers are referred to as the Nemec phenomenon.

## SUGGESTED READING

Cresti, M. and A. Tiezzi. 1990. Germination and pollen tube formation. In *Microspores, Evolution and Ontogeny*, eds. S. Blackmore and R.B. Knox, 239–263. London: Academic Press.

Dickinson, H.G. 1973. The role of plastids in the formation of pollen grain coatings. *Cytobios* 8: 25–40.

Dickinson, H.G. 1981. The structure and chemistry of plastid differentiation during male meiosis in *Lilium heryi*. *J. Cell Sci.* 52: 223–241.

Dumas, C. and R.B. Knox. 1983. Callose and determination of pistil viability and incompatibility. *Theor. Appl. Genet.* 67: 1–10.

Dumas, C., R.B. Knox and T. Gaude. 1985. The spatial association of the sperm cells and vegetative nucleus in the pollen grain of *Brassica*. *Protoplasma* 124: 168–174.

Echlin, P. and H. Godwin. 1968a. The ultrastructure and ontogeny of pollen in *Helleborus foetidus* L. I. The development of the tapetum and Ubisch bodies. *J. Cell Sci.* 3: 161–174.

Echlin, P. and H. Godwin. 1968b. The ultrastructure and ontogeny of pollen of *Helleborus foetidus* L. II. Pollen grain development through the callose special wall stage. *J. Cell Sci.* 3: 175–186.

Fritzsche, J. 1837. *Über den Pollen*. St. Petersburg: Academie der Wissenschaften.

Geitler, L. 1941. Das Wachstum des Zellkerns in tierischen und pflanzlichen Geweben. *Ergebn. Biol.* 18: 1–54.

Heslop-Harrison, J. 1968a. Wall development within the microspore tetrad of *Lilium longiflorum*. *Can. J. Bot.* 46: 1185–1191.

Heslop-Harrison, J. 1968b. Tapetal origin of pollen coat substances in *Lilium*. *New Phytol.* 67: 779–786.

Heslop-Harrison, J. 1969. An acetolysis-resistant membrane investing tapetum and sporogenous tissue in the anthers of certain Compositae. *Can. J. Bot.* 47: 541–542.

Heslop-Harrison, J. 1975. Male gametophyte selection and the pollen-stigma interaction. In *Gamete Competition in Plants and Animals*, ed. D.L. Mulcahy, 177–190. Amsterdam: North Holland.

Heslop-Harrison, J. 1975. The physiology of pollen grain surface. *Proc. R. Soc. Lond. Ser B* 190: 275–299.

Hesse, H.L. 1979. Zur Artefaktbildung bei der Pripantion hochviskoser lipider Substanzen (Pollenkitt). - Beitr. Elektronenmikr. *Direktabb. Obertl.* 12: 17–20.

Huynh, K.L. 1976. Arrangement of some monosulcate, disulcate, trisulcate, dicolpate, and tricolpate pollen types in the tetrads, and some aspects of evolution in the angiosperms. In *The Evolutionary Significance of the Exine*, eds. I.K. Ferguson and J. Muller, 101–124. London: Academic Press.

Knox, R.B. 1984. The pollen grain. In *Embryology of Angiosperms*, ed. B.M. Johri, 197–271. Berlin: Springer-Verlag.

Kuroiwa, T., Kawano, S., Watanabe, M. and Hori, T. 1991. Preferential digestion of chloroplast DNA in male gametangia during late stage of the gametogenesis in the anisogamous alga *Bryopsis maxima*. *Protoplasma* 163: 102–113.

Kuroiwa, T. and U. Hidenobu. 1996. Organelle division and cytoplasmic inheritance: The origin and basis of the transmission of organelle genomes. *BioScience* 46 (11): 827–835.

Lalanne, E. and D. Twell. 2002. Genetic control of male germ unit organisation in *Arabidopsis*. *Plant Physiol.* 129 (2): 865–875.

Maheshwari, P. 1950. *An Introduction to the Embryology of Angiosperms*. New York: McGraw-Hill Book Company, Inc.

McConchie, C.A., S. Jobson and R.B. Knox. 1985. Computer-assisted reconstruction of the male germ unit in pollen of *Brassica campestris*. *Protoplasma* 127: 57–63.

McConchie, C.A. and R.B. Knox. 1986. The male germ unit and prospects for biotechnology. In *Biotechnology and Ecology of Pollen*, eds. Mulcahy, G.M. Bergamini Mulcahy and E. Ottaviano, 289–296. Berlin: Springer.

McCue, A.D., M. Cresti, J.A. Feigo and R.K. Slotkin. 2011. Cytoplasmic connection of sperm cells to the pollen vegetative cell nucleus: Potential roles of the male germ unit revisited. *J. Exp. Bot.* 62 (5): 1621–1631.

Miyamura, S., T. Kuroiwa and T. Nagata. 1987. Disappearance of plastid and mitochondrial nucleoids during the formation of generative cells of higher plants revealed by fluorescence microscopy. *Protoplasma* 141: 149–159.

Mogensen, H.L. 1992. The male germ unit: Concept, composition and significance. *Int. Rev. Cyto.* 140: 129–147.

Nagata, N., C. Saito, A. Sakai, H. Kuroiwa and T. Kuroiwa. 1999. A control of selective increase and decrease of organellar DNA in generative cells just after pollen mitosis one concerning cytoplasmic inheritance. *Planta* 209: 53–65.

Naithani, S.P. 1937. Chromosome studies in *Hyacinthus orientalis* L. III. Reversal of sexual state in the anthers of *H. orientalis* L. var. "Yellow hammer". *Ann. Bot.* 1: 369–377.

Němec, B. 1898. Über den Pollen der petaloiden Antheren von *Hyacinthus orientalis* L. *Bull. Intl. Acad. Sci. Bohème* 5: 17–23.

Pandolfi, T., E. Pacini, and D.M. Calder. 1993. Ontogenesis of monad pollen *Pterostylis plumosa* (Orchidaceae Neottioideae). *Plant Syst. Evol.* 186: 175–185.

Ram, M. 1959. Occurrence of embryo sac-like structures in the microsporangia of *Leptomeria biliardieri*. *Nature* 184: 912–915.

Riding, J.B. and Kyffin-Hughes, J.E. 2004. A review of the laboratory preparation of palynomorphs with a description of an effective non-acid technique. *Revista Brasileira de Paleontologia* 7 (1): 13–44.

Rusche, M.L. and Mogensen, H.L. 1988. The male germ unit of *Zea mays*: Quantitative ultrastructure and three dimensional analysis. In *Sexual Reproduction in Higher Plants*, eds. M. Cresti, P. Gori and E. Pacini, 221–226. Berlin: Springer-Verlag.

Russel, S.D. and D.D. Cass. 1981. Ultrastructure of the sperms of *Plumbago zeylanica*. 1. Cytology and association with the vegetative nucleus. *Protoplasma* 107: 85–107.

Russell, S.D. 1984. Ultrastructure of the sperm of *Plumbago zeylanica*. *Planta* 162: 385–391.

Sanger, J.M. and W.T. Jackson. 1971. Fine structure study of pollen development in *Haemanthus katherinae* Baker. I. Formation of vegetative and generative cells. *J. Cell Sci.* 8 (2): 289–301.

Shah, C.K. 1963. The life history of *Juncus bufonius* Linn. *J. Indian Bot. Soc.* 42: 238–251.

Shaw, G. 1971. The chemistry of sporopollenin. In *Sporopollenin*, eds. Brooks, J., Grant, P.R., Muir, M., Gijzel, P. van and Shaw, G., 305–350. London: Academic Press.

Stow, I. 1930. Experimental studies on the formation of embryo sac-like giant pollen grain in the anthers of *Hyacinthus orientalis*. *Cytologia* 1: 417–439.

Vijayaraghavan, M.R. and S. Ratnaparkhi. 1977. Pollen embryo sacs in *Heuchera micrantha* Dougl. *Caryologia* 30: 105–119.

Yang, S.J., D.M. Wei and H.Q. Tian. 2015. Isolation of sperm cells, egg cells, synergids and central cells from *Solanum verbascifolium* L. *J. Plant Biochem. Biotechnol.* 24: 400–407.

Yu, H.-S., S.-Y. Hu and C. Shu. 1989. Ultrastructure of sperm cells and the male germ unit of *Nicotiana tabacum*. *Protoplasma* 152: 29–36.

# 5 Palynology
## *Pollen Morphology*

## PALYNOLOGY

Palynology (*palynos*, dust) is the science of pollen and spores. The study of the external morphological features of mature pollen grains is referred to as palynology. Hyde and Williams (1955) used the term palynology for the first time. Pollen grains are the male reproductive structures produced by flowering plants (angiosperms) and gymnosperms (naked seeded plants). A pollen grain contains the male gamete of the angiosperm plant. Pollen has two functions – reproduction and reward of visitors. The outer layer of pollen and spores often contains a special compound, sporopollenin, which resists degradation by various chemicals, bacteria and fungi.

## POLLEN MORPHOLOGY

Pollen grains come in a wide variety of shapes, sizes and surface markings characteristic of the species.

**Shape:** The shape of the pollen grains is variable. Pollen grains may be spheroidal (globose or ball-shaped, as in Rubiaceae and Cucurbitaceae), oblate (compressed along the polar axis, as in Apocynaceae and Euphorbiaceae) and prolate (elongated along the polar axis, as in Euphorbiaceae and Rhizophoraceae).

**Size:** Pollen size varies in different taxa and is measured in terms of both the polar diameter and the equatorial diameter. The diameter of the pollen varies from 18 µm (*Myosotis*), 30–50 µm (*Acacia*, polyad), 100–200 µm (Cucurbitaceae) and 150–200 µm (Malvaceae).

## POLLEN POLARITY

Pollen polarity refers to the arrangement of microspores in the tetrad (Figure 5.1). Each pollen grain has two poles at opposite ends (Figure 5.2). It is known as polar axis. The pole closest to the centre of the tetrad is the proximal pole and farthest from the centre is the distal pole. The axis at a right angle to the polar axis is the equatorial axis. The two halves of the pollen are demarcated by the equatorial axis. When proximal and distal poles are similar, the pollen grain is called isopolar (e.g., Cucurbitaceae); and if different, it is described as heteropolar. In heteropolar grains, one face has an aperture while the other has none.

DOI: 10.1201/9781003260097-5

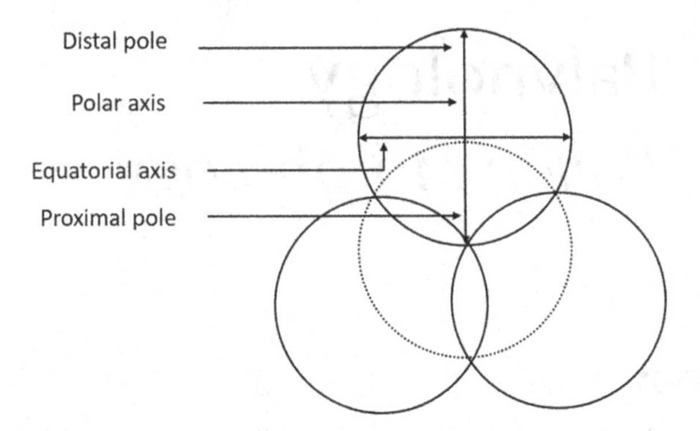


**FIGURE 5.1** Diagram of a tetrahedral tetrad of microspores to show the two poles and two axes.

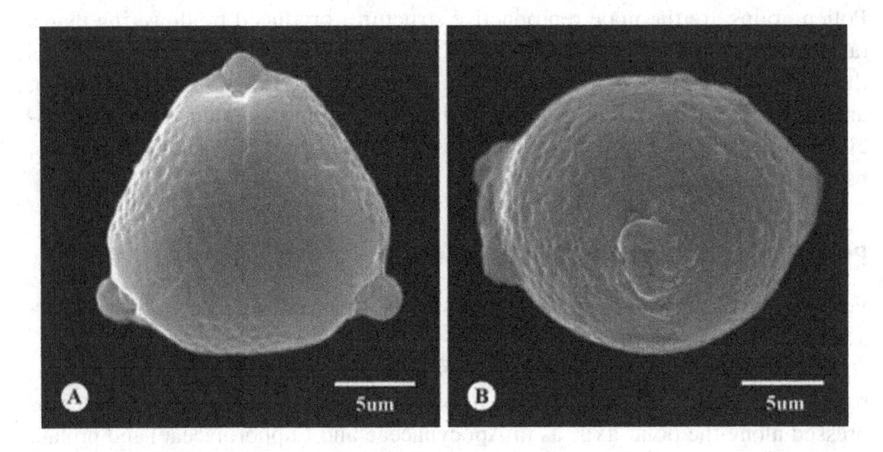


**FIGURE 5.2** *Fenghwaia gardeniicarpa.* A. Polar view; B. Equatorial view (after Wang et al., 2021).

**P/E Ratio:** This refers to the length of the polar axis (P) between the two poles compared to the equatorial diameter (E).
**Palynogram:** A schematic illustration of a pollen grain.

## POLLEN APERTURE

An aperture is a specially delimited region or an opening in a pollen grain wall. The aperture is directly or indirectly associated with pollen germination. Apertures are the points from where the pollen tube grows out. An aperture may be elongated (colpus, colpi) or rounded (pores). The apertures may be simple (colpate, porate) or compound (colporate). A compound aperture consists of two regions: (i) the central region called poral and (ii) an outer region called colpal.

**TABLE 5.1**
**Types of Apertures**

| Aperture Pattern | Family |
|---|---|
| Monoporate | Poaceae |
| Monocolpate | Magnoliaceae |
| Tricolpate | Acanthaceae |
| Triporate | Cucurbitaceae |
| Tricolporate | Goodeniaceae |
| Polyporate | Amaranthaceae, Convolvulaceae |
| Pantoporate | Papveraceae, Costaceae |

Different types of apertures are given in Table 5.1. Pollen grains lacking apertures are called inaperturate.

**Porus:** A circular aperture, if situated equatorially or globally, is called porus.
**Ulcus:** A circular aperture, if situated distally, is called ulcus.
**Colpus:** An elongated aperture, if situated equatorially or globally, is called colpus.
**Sulcus:** An elongated aperture, if situated distally, is termed a sulcus.
**Colporus:** A combination of porous and colpus.

## NPC SYSTEM

NPC is an artificial system of classification of pollen and spores based on three features of apertures: Number (N), Position (P) and Character (C) (Figure 5.3). In this system, aperture has been replaced by treme. The NPC classification was introduced by Erdtman and Straka (1961).

### Number

Pollen grains without apertures are called atreme. Such pollen grains are represented as $N_0$. Depending upon the number of apertures, the pollen grains are classified as monotreme ($N_1$), ditreme ($N_2$), tritreme ($N_3$), tetratreme ($N_4$), pentatreme ($N_5$) or hexatreme ($N_6$). Pollen grains with more than 6 apertures (7 or more) are called polytreme and is represented as $N_7$. When the apertures, irrespective of their number, are irregular or irregularly distributed, the pollen grains are described as anomotreme ($N_8$).

### Position

Depending upon the position of the apertures, the pollen grains are classified into seven groups ($P_0$–$P_6$). When the aperture is on the proximal face, it is

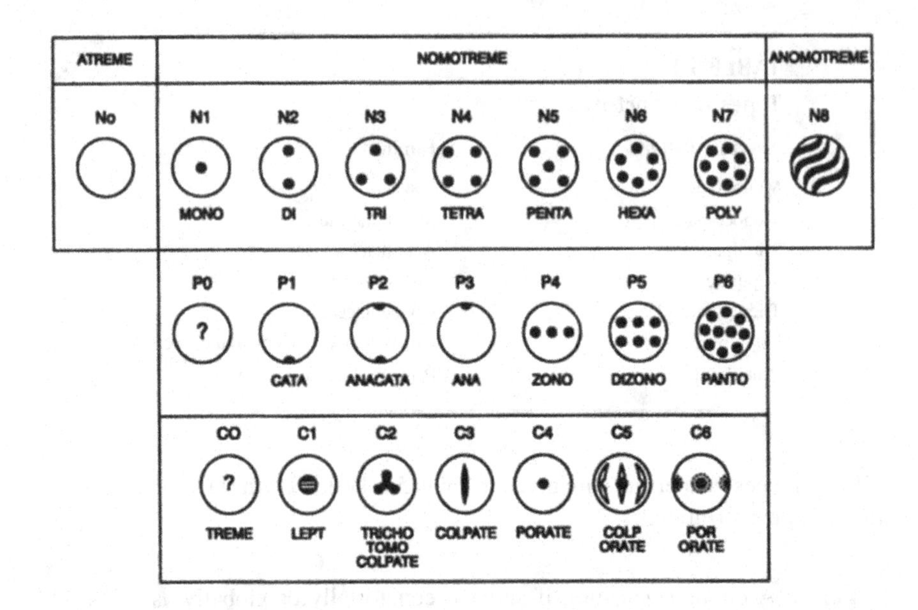


**FIGURE 5.3** The NPC system. Diagram showing number (N), position (P), and character (C) of apertures (after Erdtman, 1969).

designated as catatreme ($P_1$). If the aperture is on the distal face, it is referred to as anatreme ($P_3$). When the centres of apertures are located on the equator, the pollen grains are designated as zonotreme ($P_4$). When the apertures are more or less uniformly distributed all over the pollen surface, the condition is known as pantotreme ($P_6$).

### Character

Based on the character of the apertures, pollen grains are classified into seven groups ($C_0$–$C_6$). When the character is not known, it is referred to as $C_0$. When the pollen grains have an aperture-like thin area, or leptoma, it is designated as $C_1$. Pollen grains with one leptoma are called monolept. If the leptoma is on the proximal face, it is called catalept, and when on distal face, it is called analept. The grains having a 3-slit corpus are grouped in the $C_2$ category and are called trichotomocolpate. Other characters include $C_3$ (colpate), $C_4$ (porate), $C_5$ (colporate) and $C_6$ (pororate).

**Stenopalynous:** When pollen grains of a taxon are characteristic and similar among species they are termed stenopalynous (e.g., Poaceae, Lamiaceae, Brassicaceae).
**Eurypalynous:** Taxa are heterogeneous, and pollen grains can vary among others in size, aperture and in exine stratification (e.g., Acanthaceae, Araceae).

## Pollen Wall Features

Pollen grains consist of a hard outer wall (exine) and a soft inner wall (intine) which encloses the cytoplasm with its cells (nuclei) and organelles. The intine is pectocellulosic in nature and usually destroyed during acetolysis. The outer layer exine is resistant to acetolysis and also to physical and biological degradation. The pollen wall is designated to protect the sperm nucleus from desiccation and irradiation during transport from the anther to the stigma.

### Exine Stratification

The exine has two distinct layers: (i) sexine and (ii) nexine. Sexine is an outer sculptured layer, whereas nexine is non-sculptured layer. The sexine is further divisible into two layers, columellae (bacula) and tectum. The columellae are an internal layer of upright rod-like elements covered over by a roof-like layer, the tectum. The exine is composed of three layers: tectum, columellae and nexine (Figure 5.4). The tectum may be smooth or may have various types of processes. The exine is differentiated into two layers: ektexine (outer layer) and endexine (inner layer). Outer layer of nexine is also termed as the foot layer.

### Exine Structure

The exine of mature pollen grains is of three basic types: tectate, semitectate and intectate (atectate). In the tectate and semitectate types, the exine comprises of nexine, columellae and tectum. In tectate exine, the tectum is continuous and roof-like (Figure 5.4). In the semitectate type, the exine is perforated; the diameter of the perforation is greater than the breadth of the pollen wall between them. In semitectate exine, the reticulum is open. If tectum is absent and the columellae are free and exposed, such pollen are called intectate.

### Pollen Sculpturing

Pollen sculpturing is the exposed surface of the pollen grain. Different types of pollen sculpturing are given in Table 5.2.

### Harmomegathy

Pollen grains exist in two morphologically distinct conditions, dry and hydrated. Pollen grains are able to absorb and release water (and various other liquids). The infoldings of the pollen wall accommodate the change of the osmotic pressure in

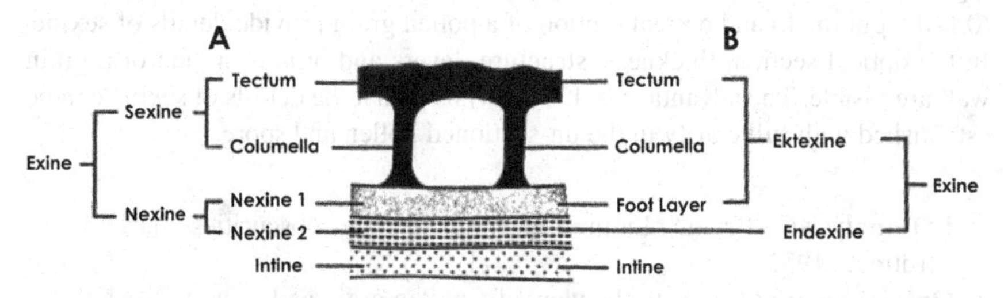


**FIGURE 5.4** Pollen wall architecture. Two sets of terminologies are used to describe exine layers: A. Sexine and Nexine; B. Ektexine and Enexine

**TABLE 5.2**
**Exine Sculpturing**

| Sculpturing Type | Pattern |
|---|---|
| Baculate | Rod-like projections whose greatest diameter is less than the height |
| Chagrenate | Shiny, smooth, translucent |
| Echinate | Spine-like projections tapering from a base to a sharp point |
| Fossulate | Grooved; possessing a negative reticulum. Surface with irregular grooves |
| Foveolate | Pitted surface; circular pits whose diameter is less than the diameter of ridges separating them |
| Gemmate | A spherical projection with constricted base |
| Pilate | Rod-like elements with swollen tips |
| Psilate | Smooth, a mat-like surface, may possess small depressions less than 1 μm in diameter |
| Punctate | Minute perforations |
| Reticulate | Net-like, ridges forming a net |
| Rugulate | Elongate elements irregularly distributed tangentially over the surface |
| Scabrate | Very fine projections, less than 1 μm in diameter |
| Striate | Elongate, more or less parallel elements distributed tangentially over the surface |
| Verrucate | Wart-like, a rounded projection of broader than high which is not constricted at base |

the cytoplasm during hydration or dehydration. This process is called the hermomegathic mechanism. The main purpose of the hermomethic effect (also called the "Wodehouse effect") is to protect the male gametophyte against desiccation during pollen presentation and dispersal (Halbritter et al., 2018).

## LO-ANALYSIS

LO-analysis is a palynological method for analyzing the sculpture and structure of the sexine by comparing the elements in high and low focus. The sporoderm pattern appears as bright islands separated by dark channels when viewed at a higher plane of focus. At low focus, this pattern is reversed. The ultra-thin section (0.1–0.5 μm thick) and optical section of a pollen grain provide details of sexine. In the optical section, thickness, structure, layers and ornamentation of a grain wall are visible. The advantage of LO-analysis is that the details of sexine can be established with full clarity in the un-sectioned pollen and spore.

**LO-analysis:** (from Latin = Lux = light and obscurities = darkness (Erdtman, 1952)
**Optical section:** Refers to the view of a pollen grain and spore under light microscope when the focal plane is halfway through the grain.

## Applied Palynology

Palynology finds applications in various fields including palaeobotany, taxonomy, plant evolution, aeropalynology, pollen allergy, fossil fuel exploration industry, food industry, forensic science, melittopalynology, etc. Pollen data provide information of changes in vegetation, climate, biogeography, migration and human disturbance of terrestrial ecosystems. Pollen morphology is also important in understanding the functional aspects of pollen such as pollination biology and pollen–pistil interaction.

1. **Palynology and systematics:** Palynological data may be helpful at all levels of systematics, especially in angiosperms (Stuessy, 2009). Palynological features are very valuable in delimiting taxa and phylogenetic analysis. The palynological features of a spore or pollen grain can often be used to identify a particular taxon. Pollen identification is largely based on the structure and surface sculpturing of the exine.
2. **Petroleum Exploration:** One of the most important applications of palynology is in the field of petroleum exploration. When analyzed, fossilized pollen collected from various depths of oil wells provide useful information about the availability of oil and gas.
3. **Archaeology:** Pollen analysis is important in archaeological studies and helps in understanding past climatic changes, prehistoric cultures, and the spread of agriculture (Bryant, 1990). Analysis of pollen collected from archaeological sites has helped in dating specific archaeological events.
4. **Criminology:** Analysis of pollen grains collected from crime sites has been effectively used in solving crimes (Bryant, 1990).
5. **Testing purity of honey:** The analysis of pollen present in honey provides data on the geographical location and plant species that the honey bees visited to collect nectar.

**Aeropalynology:** Study of biological particles such as pollen, fungal spores, dust mites, insect debris and organic dusts present in the air.
**Melittopalynology:** Study of pollen grains in honey samples and its application in apiculture.

## Pollen Wall Proteins

Both exine and intine contain proteins. Exine proteins are sporophytic in origin, as they originate from the surrounding tapetum. In tectate pollen, exine proteins are located in the interporal regions in exine cavities; but in pilate pollen, proteins are present in surface depressions. The intine proteins originate in the pollen cytoplasm and are generally present in the form of radially oriented tubules, the concentration being more in the germ pore region. The intine proteins are gametophytic in origin.

During intine development, the plasma membrane of the pollen produces radially oriented tubules into the developing intine. Later these tubules, with their protein inclusions, are cut off from the plasma membrane and become incorporated in the intine.

**Tectum:** Outer layer of the exine; can be tectate, semitectate, or intectate.
**Tectate Pollen:** Pollen grain with a continuous tectum; also known as eutectate.
**Pilate Pollen:** Pilum is a blunt, drumstick-shaped element. It has a swollen, more or less round or elongated top called a caput and a rod-like part that holds the caput-termed baculum (Erdtman, 1952). A pilum is taller than one micron.

Pollen wall proteins contain enzymes such as esterases, amylases and ribonucleases (Knox and Heslop-Harrison, 1970). These enzymes can be used as a marker enzyme for both exine and intine proteins. Esterases mainly occur in the exine and acid phosphatases in the intine. Proteins which cause pollen allergy are also present in the pollen wall.

The tapetally derived proteins are absent in many species belonging to Cannaceae, Zingiberaceae and marine angiosperms, in which exine is either absent or highly reduced. In such cases, the intine is generally thick and often differentiated into many layers. The wall proteins are thus exclusively confined to the intine (Knox, 1984). In *Olea* (Pacini et al., 1981) and *Pterostylis* (Pandolfi et al., 1993), the tapetal proteins are also deposited in the poral region. The interporal exine proteins are deposited after tapetal breakdown but the poral proteins are deposited when the tapetal cells are still intact.

## Pollen Preparation: Acetolysis Method

In order to study pollen morphology, the mature pollen grains are acetolyzed. Acetolysis is a standard method of pollen preparation introduced by G. Erdtman (1952, 1969). The pollen grains are acetolyzed using the following method (see Nair, 1960).

### Pre-treatment and Staining

1. Fix the pollen material (fresh or dry anthers) in 70% alcohol in glass vials. if anthers are dry, fix for at least 24 hours and if fresh, at least for one hour.
2. Transfer the material to a centrifuge tube and crush the polliniferous material with a glass rod.
3. Sieve through a brass mesh with 48 divisions/sq. cm. and collect it in two separate tubes, marked “A” and “B”.

4. Centrifuge the contents of tube A, decant the alcohol and add about two drops of safranin (5% in water). Wait for 5–10 minutes.
5. Wash the stained sediment with 70% alcohol, centrifuge it, decant off the water, wash with water at least twice or thrice by centrifuging until the supernatant becomes colourless.
6. Add 2 ml of dilute glycerine (50% in water) and keep this centrifuge tube A aside.

**Acetolysis**

7. Centrifuge the second tube, B, and pour out alcohol.
8. Add 5 ml glacial acetic acid. Centrifuge and decant off the acid.
9. Add about 6 ml of acetolysis mixture (acetic anhydride and conc. sulphuric acid in a ratio of 9:1). Keep tube B in a water bath and heat the water from 70°C to boiling point. Centrifuge tube B and decant off the waste acetolysis mixture.
10. Add 10 ml of glacial acetic acid in tube B, centrifuge it and decant off the acid.
11. Add water once or twice, centrifuge it every time and decant off the water. Disperse the grains in water.
12. Transfer one-half of the sediment in tube B to another centrifuge tube marked “C”. Leave the centrifuge tubes B and A in the rack.

**Chlorination or Oxidation**

13. Take the contents of tube C, centrifuge and decant off the water.
14. Add about 5 ml glacial acetic acid followed by 2–4 drops of saturated solution of sodium chlorate or potassium chlorate in water and then one or two drops of concentrated hydrochloric acid. Wait for about five minutes, centrifuge and decant.
15. Wash with water once or twice and decant off the water.
16. Add a few drops of methyl green on the sediment and wait for about five minutes.
17. Wash the sediment with water twice or thrice until the water becomes colourless and centrifuge every time.

**Mounting**

18. Transfer the pollen grains from centrifuge tube A to B and then from tube B to tube C. Centrifuge the mixture of tubes A, B and C.
19. Place the pollen grains in glycerine jelly. Place a cover slip gradually and seal with wax.
20. The acetolyzed pollen grains look brown in colour and unacetolyzed grains take the red stain of safranin. The chlorinated or oxidized grains are green in colour.

## SUMMARY

- The study of external morphological features of mature pollen grains and spores is referred to as palynology.
- Each pollen grain has two poles at opposite ends. It is known as polar axis. Pollen polarity refers to the arrangement of microspores in tetrad.
- An aperture is a specially delimited region or an opening in a pollen grain wall.
- NPC is an artificial system of classification of pollen and spores based on number, position and character.
- The exine is distinguishable in to sexine and nexine. The exine of mature pollen grains is of three basic types: tectate, semitectate and intectate (atectate).
- LO-analysis is a palynological method for analyzing the sculpture and structure of the sexine.
- Palynology finds applications in various fields including palaeobotany, taxonomy, plant evolution, aeropalynology, pollen allergy, fossil fuel exploration industry, food industry, forensic science, melittopalynology.

## SUGGESTED READING

Agashe, S.N. 2006. *Palynology and Its Applications.* New Delhi: Oxford & IBH Publication Co. Pvt Ltd.

Agashe, S.N. and E. Coulton. 2009. *Pollen and Spores. Applications with Special Emphasis on Aerobiology and Allergy.* Boca Raton, FL: CRC Press.

Bryant, V.M.Jr. 1990. *Pollen: Nature's Fingerprints of Plants. Encyclopedia Britannica, Inc.*, 93–111.

Erdtman, G. 1952. *Pollen Morphology and Plant Taxonomy: Angiosperms.* Stockholm: Almqvist and Wiksell.

Erdtman, G. 1957. *Pollen and Spore Morphology. Plant Taxonomy. Gymnospermae, Pteridophyta, Bryophyta.* Stockholm: Almqvist & Wiksell.

Erdtman, G. 1969. *Handbook of Palynology.* New York: Hafner Publication Co.

Erdtman, G. and H. Straka. 1961. Cormophyte spore classification. *Geol. Fören. Förenhandl.* 83: 65–78.

Halbritter, H., S. Ulrich, F. Grimsson, M. Weber, R. Zetter, M. Hesse, R. Buchner, M. Svojtka and A. Frosch-Radivo. 2018. *Illustrated Pollen Terminology.* 2nd ed. Switzerland: Springer International Publishing AG.

Hyde H.A. and D.A. Williams. 1944. The right word. *Pollen Analysis Circulars* 8: 6.

Knox, R.B. and J. Heslop-Harrison. 1970. Pollen wall proteins: Localization and enzymic activity. *J. Cell Sci.* 6: 1–27.

Knox, R.B. 1984. The pollen grain. In *Embryology of Angiosperms*, ed. B.M. Johri, 197–272. Berlin: Springer-Verlag.

Nair, P.K.K. 1966. *Essentials of Palynology.* Bombay: Asia Publishing House.

Pacini, E., G. Franchiand and G. Sarfatti. 1981. On the widespread occurrence of poral sporophytic proteins in pollen of dicotyledons. *Ann. Bot.* 47: 405–408.

Pandolfi, T., E. Pacini and D.M. Calder. 1993. Ontogenesis of monod pollen in Pterostylis plumose (Orchidaceae, Neottioideae). *Plant Syst. Evol.* 186: 175–184.

Punt, W., P.P. Hoen, S. Blackmore, S. Nilsson and Thomas, A.L. 2007. Glossary of pollen and spore terminology. *Rev. Palaeobot. Palyno.* 143: 1–81.

Sarkar, B., S. Basak, A.P. Das, S. Siddhanta, D.B. Maity and S. Bera. 2021. Pollen morphology of some Eastern Himalayan species of Maesa (Primulaceae) and its taxonomic significance. *Proc. Natl. Acad. Sci., India.* https://doi.org/10.1007/s40011-020-01217-8.

Stuessy, T.F. 2009. *Plant Taxonomy: The Systematic Evaluation of Comparative Data.* New York: Columbia University Press.

Walker, J.W. 1976. Evolutionary significance of the exine in the pollen of primitive angiosperms. In *The Evolutionary Significance of Exine*, eds. Ferguson, I.K. and Muller, J., 251–308. Linn. Soc. Symp, London.

Wang, G.T., J.P. Shu, G.B. Jiang, Y.Q. Chen and R.J. Wang. 2021. Morphology and molecules support the new monotypic genus Fenghwaia (Rhamnaceae) from south China. *PhytoKeys* 171: 25–35.

Wodehouse, R.P. 1935. *Pollen Grains: Their Structure, Identification and Significance in Science and Medicine.* New York: McGraw-Hill.

# 6 Megasporangium

The megasporangium (or ovule) consists of the nucellus, which is surrounded by one or two integuments (Figure 6.1). The ovule is attached to the placenta by means of a small or elongated stalk known as the funiculus. The place where the funiculus joins the ovule is known as the hilum. The base of the nucellus, which is not easy to delimit because it merges with the base of the integument, is called the chalaza. A narrow tubular opening formed by the integument(s) is known as the micropyle. In certain cases, the hilum and chalaza lie close to each other, while in others, they are separated by an intervening tissue which is funiculus-like. This is known as the raphe. Sometimes, in bent ovules, an ingrowth develops from the chalazal end and projects into the bent portion of the ovule. This projection is termed as basal body.

## TYPES OF OVULES

Mature ovules may be of different forms and are characteristic of certain groups of flowering plants. The classification of ovules is based primarily on the position of the micropyle with respect to the funiculus (Figure 6.2).

### Orthotropous or Atropous

When the micropyle and hilum lie in a straight line, such ovules are termed as orthotropous (e.g., Juglandaceae, Polygonaceae, Urticaceae, Piperaceae and Najadaceae).

### Anatropous

In this type, the body of the ovule becomes completely inverted towards the base of the funiculus so that the micropyle and hilum come to lie very closely (e.g., Asteraceae, families included in Gamopetalae). According to Davis (1966) 82% of the angiosperm families show anatropous ovules.

### Campylotropous

In this type, the ovule is curved in so that the chalaza and micropyle lie in one line parallel to the long axis of the body of ovule (e.g., Fabaceae, Resedaceae).

### Hemianatropous or Hemitropous

In this type of ovule, the nucellus and integuments lie nearly at right angles to the funiculus, which is separated from the chalaza by the presence of raphe (e.g., Malpighiaceae, Primulaceae).

DOI: 10.1201/9781003260097-6

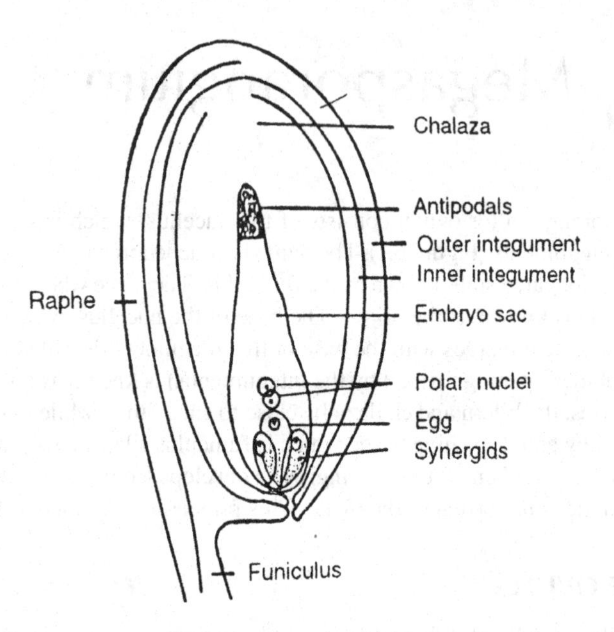


**FIGURE 6.1** Longitudinal section of an anatropous ovule.

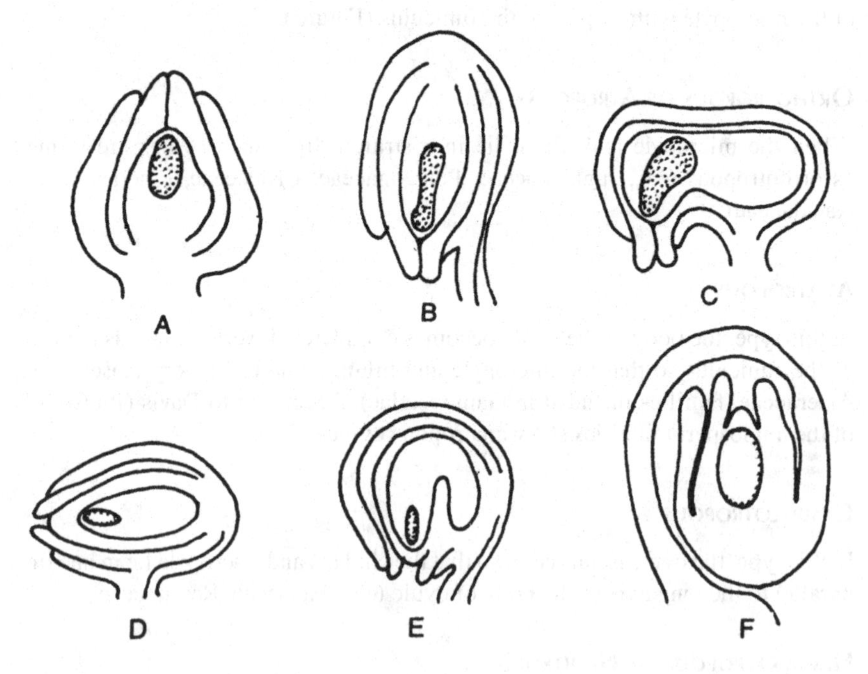


**FIGURE 6.2** Types of ovules. A. Orthotropous; B. Anatropous; C. Campylotropous; D. Hemianatropous; E. Amphitropous; F. Circinotropous (adapted from Maheshwari, 1950).

**TABLE 6.1**
**Data on the Occurrence of Various Features of Ovules in Angiosperms (after Davis, 1966)**

| Features | No. of Families Showing** | Examples |
|---|---|---|
| **OVULE** | | |
| Anatropous | 204 | Gamopetalae |
| Orthotropous | 20 | Polygonaceae, Piperaceae |
| Hemianatropous | 13 | Malpighiaceae, Primulaceae |
| Campylotropous | 5 | Caprifoliaceae, Chenopodiaceae |
| Amphitropous | 4 | Crossosomataceae, Leitneriaceae |
| Circinotropous | 1 | Cactaceae |
| **OVULE** | | |
| Bitegmic | 208 | Euphorbiaceae, Amaryllidaceae |
| Unitegmic | 90 | Acanthaceae, Solanaceae |
| **MICROPYLE FORMED BY** | | |
| Inner integument | 88 | Centrospermales, Plumbaginales |
| Outer integument | 4 | Podostemaceae, Euphorbiaceae |
| Both integuments | 74 | Pontederiaceae |

**Only those families are taken into account for which the feature is known to form a family character.

### Amphitropous

An ovule with the nucellus and the integuments situated more or less at right angles to the funiculus is called amphitropous. In this type, curvature of the ovule is so pronounced and affects the nucellus so much that the latter becomes horse-shoe-shaped, (e.g., Alismaceae, Butomaceae, Crossosomataceae).

### Circinotropous

In this type, the funiculus is very long, and it surrounds the ovule in such a way that its micropylar end takes another turn and faces upwards (e.g., Cactaceae, Plumbaginaceae) (Table 6.1).

## DEVELOPMENT OF OVULES

The ovules develop from the placentae of the ovary. Localized periclinal divisions in the hypothermal layers of placenta give rise to small dome-shaped structures. Further development of each dome proceeds more quickly on one side than the other so that the protuberance (young nucellus) shows varying degrees of curvature. Near the basal region of the nucellus, one or two to rim-like outgrowths develop and, as they grow, envelope the nucellus closely. These envelopes are known as integuments (Figure 6.3).

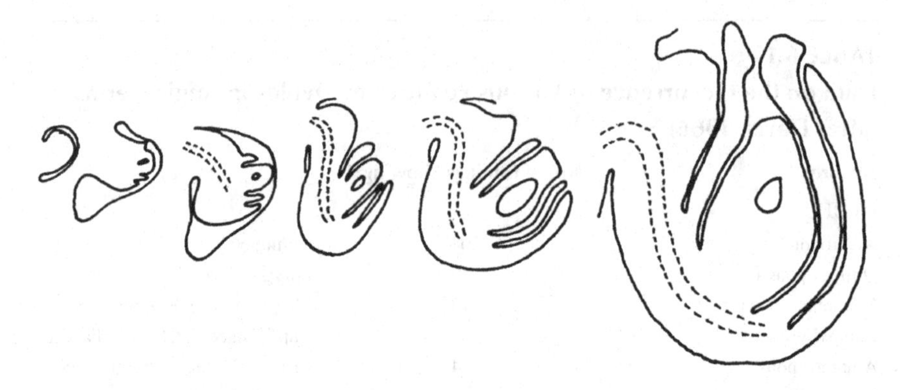

**FIGURE 6.3** *Skimmia japonica*. Stages in the ontogeny of an ovule (after Boesewinkel, 1972).

Except for the family Euphorbiaceae, where the inner integument is initiated sub-dermally, in all others it is dermal in origin (Bouman, 1984). The outer integument is initiated either dermally or sub-dermally. With the differentiation of the integuments, the ovule begins to curve. By the time megaspore tetrad stage is attained, the ovule assumes its final shape.

## INTEGUMENTS

Ordinarily the ovule has either one or two integuments. The ovules having only one integument are known as unitegmic and those with two integuments as bitegmic. The former type of ovule is found in Gamopetalae, while the latter are common in Polypetalae and monocots.

In bitegmic ovules, the inner integument differentiates earlier than the outer integument; but in later stages, the outer integument overgrows the inner integument. In a mature ovule, the outer integument is thicker compared to the inner integument. Occasionally in the members of the families Annonaceae, Proteaceae and Trapaceae, the inner integument projects well beyond the outer integument.

In some taxa of the family Olacaceae (*Olax*, *Liriosoma*), the ovules do not possess an integument. Such ovules are termed ategmic (Davis, 1966). The family Ranunculaceae is of interest, as it shows unitegmic (*Anemone*, *Clematis*, *Ranunculus*), bitegmic (*Thalictrum*, *Nigella*, *Adonis*) and ategmic (sterile ovules of *Adonis*, *Anemone*, *Clematis*) ovules (Vijayraghavan, 1970). In *Optunia* (Cactaceae), the funiculus grows and completely surrounds the ovule (Figure 6.4), appearing as a third integument (Circinotropous). In *Dendrocalamus hamiltonii*, the integuments remain short and do not cover the micropyle.

In Cactaceae, some members show air space present between the two integuments in the chalazal region, a feature also observed in the species of *Bassia*, *Tetragonia* and *Trianthema*. The presence of stomata on the outer integument has been reported in *Cleome* and *Magnolia*. Abundance of chlorophyll has been observed in *Gladiolus*, *Lilium* and *Amaryllis*.

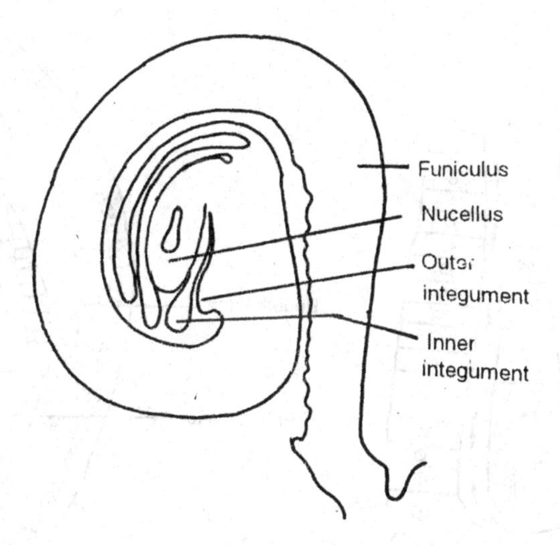


**FIGURE 6.4** *Optunia aurantiaca.* Longitudinal section of circinotropus ovule showing excessive funicular growth (after Archibald, 1939).

## ENDOTHELIUM

In most plants which possess unitegmic and tenuinucellate ovules, the inner epidermis of the integument during early stages of ovule development elongate radially, acquire dense cytoplasm, show prominent nuclei and differentiate into the endothelium (Figure 6.5). As the ovule grows, the nucellus degenerates and the fully organized female gametophyte comes in direct contact with the endothelium. Normally the endothelium covers the entire length of the embryo sac but it may remain confined to the lower two-thirds or the chalazal half, or to the micropylar end of the embryo sac. Occurrence of endothelium is a common feature in the Gamopetalae (e.g., Scrophulariaceae, Asteraceae). However, in Acanthaceae and Loganiaceae, the endothelium develops in some members. Davis (1966) reported presence of endothelium in 57 families of angiosperms.

The endothelium usually comprises a single layer of uninucleate cells. Multiseriate and multinucleate endothelium have been observed in many Asteraceae. Pandey (1980) reported four categories of endothelium in Asteraceae: (i) the endothelium remains single layered throughout the course of seed development (e.g., *Vernonia altissima*); (ii) the endothelium divides only on the chalazal side (e.g., *Cosmos bipinnatus*); (iii) the endothelium divides both on the micropylar and chalazal side (e.g., *Coreopsis auriculata*); and (iv) the endothelium divides throughout the length of the embryo sac (e.g., *Dahalia pinnata*).

The endothelial cells are connected with each other and with those of the integumentary cells through plasmodesmatal connections. They contain proteins, RNA, carbohydrates, ascorbic acid and enzymes. They often become polyploid. In *Pedicularis palustris*, the endothelial cells show ploidy up to 32 n. In *Balanites*, the endothelial cells become multinucleate.

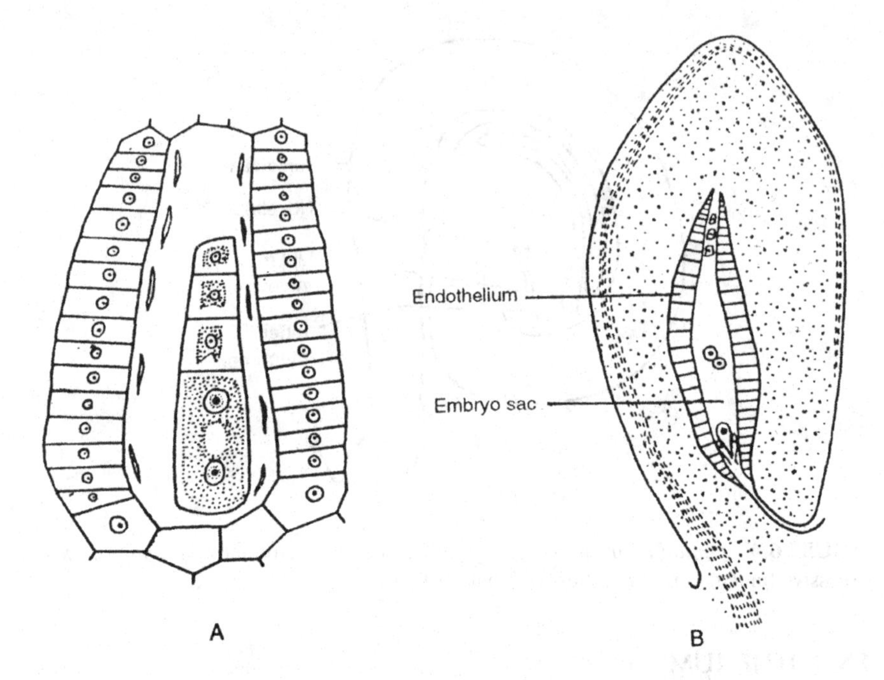


**FIGURE 6.5** Endothelium. A. *Youngia japonica* Longitudinal section of ovule at organized female gametophyte stage (after Pandey, 1980); B. *Calceolaria mexicana*. Endothelium around a two-nucleate embryo sac (after Taneja, 1970).

Studies have revealed that endothelium is nutritive in function. The flow of nutrients from the integumentary cells to the embryo sac depends on the chemical nature of the intervening cell walls of endothelial cells and the embryo sac. The flow may be restricted to the micropylar, submicropylar or chalazal region (Kapil and Tiwari, 1978).

## OBTURATORS

Obturators are specialized glandular structures involved in directing the pollen tube towards the micropyle. They may originate from placenta or funiculus or both (Table 6.2, Figure 6.6). An obturator varies greatly in form and is of common occurrence in the family Euphorbiaceae. Outgrowths arising from the base of the stylar canal and reaching the micropyle have also been termed as obturator (e.g., Thymelaeaceae). In *Gerbera jamisonii* (Pandey and Chopra, 1979), some cells of the integument, lining the micropylar canal on the raphe side, elongate radially and function as an obturator (Figure 6.7).

In *Ottelia alismoides* (Hydrocharitaceae), the obturator is composed of a group of gland cells situated above the placentum. The cells are positioned in such a way that the micropyle of the ovule comes closer to the obturator. The cells of the obturator are invariably uninucleate. The number of cells constituting a single obturator varies from 10 to 70 (Indra and Krishnamurthy, 1984). Normally,

**TABLE 6.2**
**Families Showing Presence of Obturator**

| | |
|---|---|
| Funicular obturator | Acanthaceae, Boraginaceous, Cyperaceae, Lamiaceae, Linaceae, Magnoliaceae, Rubiaceae |
| Placental obturator | Cuscutaceae, Euphorbiaceae, Meliaceae |
| Both funicular and placental obturator | Garryaceae, Molluginaceae |
| Integumentary obturator | Urticaceae |
| Ovary wall obturator | Hydrocharitaceae, Phytolacaceae |

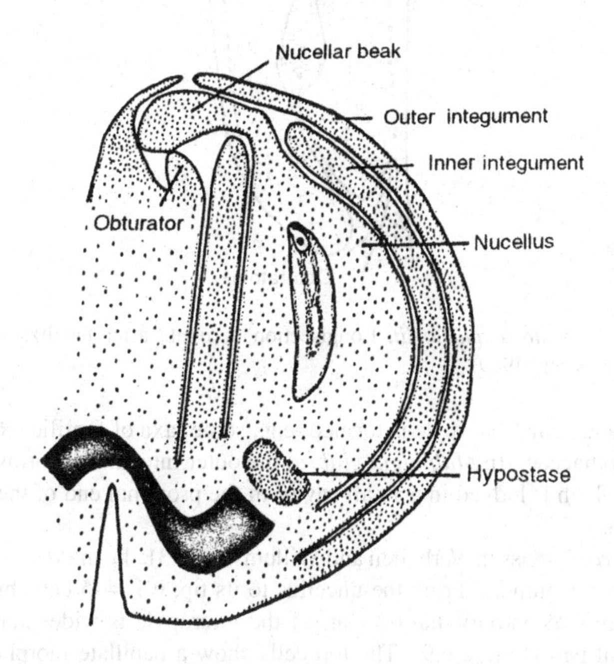


**FIGURE 6.6** *Phyllanthus niruii.* Longitudinal section of ovule at zygote stage showing nucellar beak and obturator (after R. P. Singh, 1959).

obturators lose their function after directing the pollen tubes, but in *Ottelia alismoides*, they persist until fruit maturity.

## MICROPYLE

In bitegmic ovules, the micropyle may either be formed by the inner integument or by both outer and inner integuments. Occasionally the micropyle is formed by the outer integument alone (e.g., Euphorbiaceae, Podostemaceae).

In cases where micropyle is formed by both the integuments, the passage thus formed by the outer integument is known as exostome and that by the inner integument is called endostome. When exostome and endostome are not in the

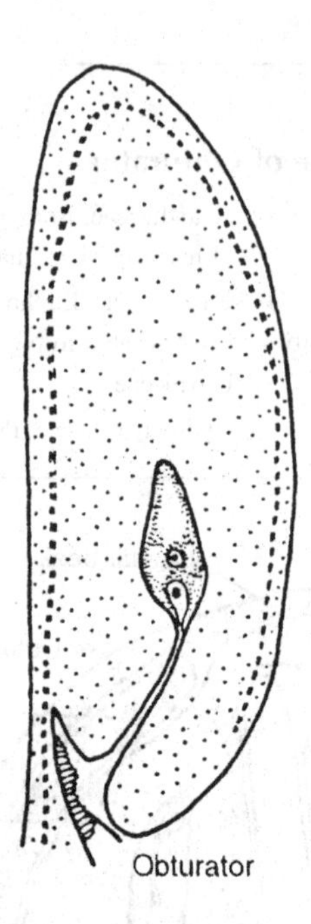


**FIGURE 6.7** *Ipomoea quamoclit.* Longisection of ovule after fertilization showing obturator (after Kaur, 1969).

same line, a zig-zag micropyle is formed (e.g., some taxa of families Resedaceae and Melastomaceae). In *Malphigia glabra*, the outer integument forms a sinuous micropyle which is lodged in a small cavity in the proximal end of the funiculus (Figure 6.8).

In *Gasteria* (Franssen-Verheijen and Willemse, 1993), the micropyle is formed by the inner integument. From the nucellus to its tip, it is 4–5 cells high. At the time of stigma receptivity, the top part of the micropyle is wider than the very narrow basal part (Figure 6.9). The top cells show a papillate morphology, and the widening of this part of the micropyle seems to result from the swelling of these cells.

Micropylar substances found in the micropyle of *Paspalum*, *Ornithogalum*, *Beta* and *Gasteria* are involved in pollen tube guidance. The origin, composition, and amount of these substances differ in different species. The micropyle in *Ornithogalum* becomes filled with micropylar exudate produced by the nucellar cap and the inner integument. An interesting feature is the formation of a thin sheet of material across the exostome, thus sealing the latter. Tilton (1979) proposed that it keeps chemotrophic agents within the micropyle. According to Franssen-Verheijen and Willemse (1993), it might also serve as a barrier that prevents the placental fluid from penetrating deep inside the micropyle. Malik and Vermani (1975) suggest that micropylar exudate serves as a second nutritional source for the pollen tube. Micropylar exudate can also function as a signal for pollen tubes to enter the micropyle.

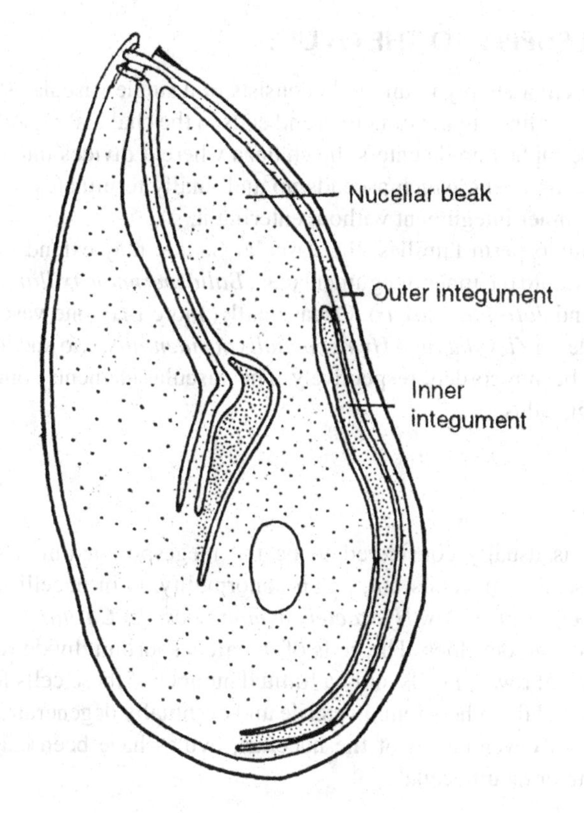


**FIGURE 6.8** *Malphighia glabra*. Longitudinal section of mature ovule ready for fertilization. Note prominent nucellar beak and micropyle formed by the outer integument alone (after B. Singh, 1961).

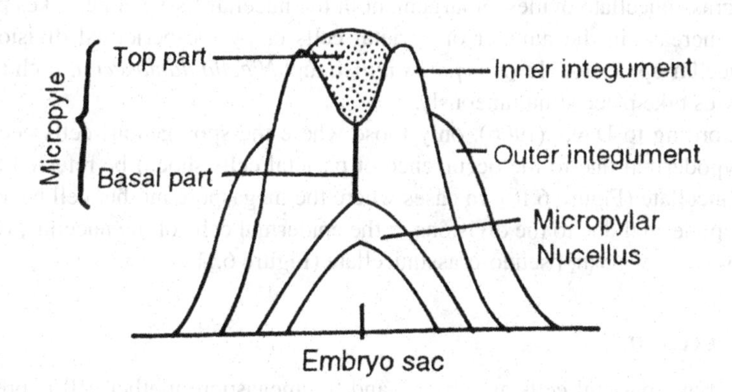


**FIGURE 6.9** Schematic drawing of median longitudinal section of the micropylar region of an ovule at SR of *Gasteria*. The shaded area represents the large amount of micropylar exudate (MEX) (after Franssen-Verheijen and Willemse, 1993).

## VASCULAR SUPPLY TO THE OVULE

Generally, vascular supply to an ovule consists of a single vascular strand which often remains confined to the funiculus and ends in the chalazal region. In *Ricinus communis*, the raphe bundle enters the chalaza where it divides into a number of branches forming a ring which ascends up and finally terminates near the inner margin of the inner integument without entering it.

In many angiosperm families, the vascular bundle may extend even into the outer or, rarely, into the inner integument (e.g., *Baliospermum axillaris*, *Codiaeum variegatum* and *Jatropha curcus*). Occasionally, more than one vascular bundle enters an ovule. In *Zizyphus* and *Hymenocaulis occidentalis*, two and four vascular bundles have been recorded, respectively. The vascular elements sometimes also occur in the nucellus.

## NUCELLUS

The nucellus is usually considered to be the megasporangium. Usually, each ovule possesses one nucellus but, as an abnormality, twin nucelli may also be found (e.g., *Aegle marmelos*, *Hydrocleis nymphoides*). In *Casuarina*, a multicellular archesporium develops. The cells of the archesporium divide repeatedly to form longitudinal rows of cells; only a limited number of these cells forms megaspore tetrads, and the others remain sterile and eventually degenerate. Depending on the extent of development of the nucellus, ovules have been categorized as crassinucellate or tenuinucellate.

### CRASSINUCELLATE

In this type, perietal tissue is well-developed, and the megaspore mother cell is separated from the nucellar epidermis by one or more layers of cells.

In crassinucellate ovules, enlargement in the nucellar tissue either takes place by an increase in the number of perietal cells or by the periclinal division in the nucellar epidermis. In *Quisqualis indica* and *Nigella damascena*, both these processes take place simultaneously.

According to Davis (1966), only those where the sporogenous cells become sub-hypodermal due to the occurrence of parietal cells should be referred to as crassinucellate (Figure 6.10). In cases where the megaspore mother cell becomes sub-hypodermal due to the divisions in the epidermal cells of the nucellus, Davis proposed a new term, pseudo-crassinucellate (Figure 6.11).

### TENUINUCELLATE

In this type, parietal cells are absent, and the megaspore mother cell is present just below the nucellar epidermis (Figure 6.12). Tenuinucellate condition of the ovule mostly occurs in advanced families viz., Scrophulariaceae, Asteraceae.

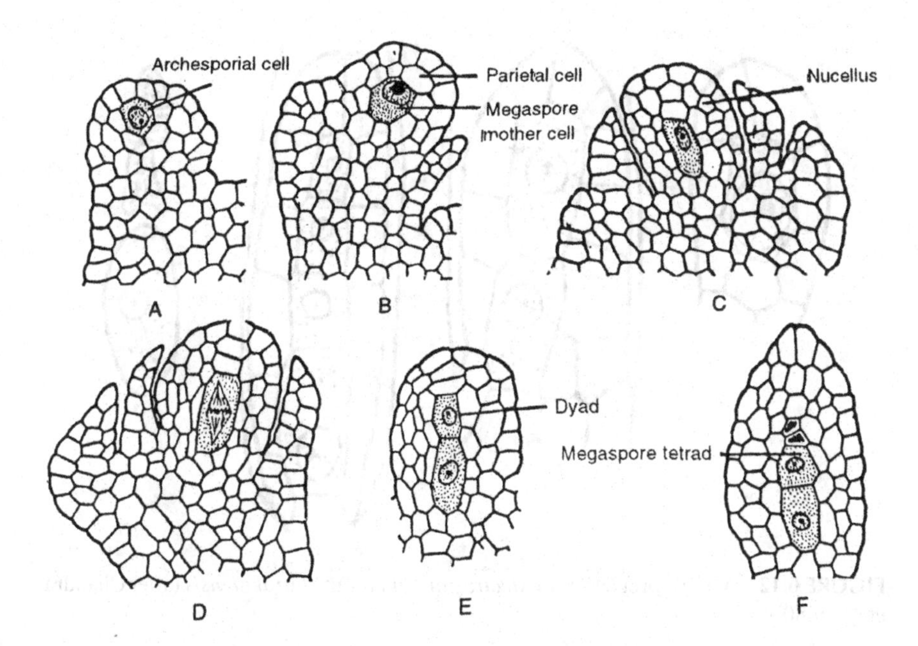


**FIGURE 6.10** Megasporogenesis in *Polygonatum cirrhifolium* (after Mohana Rao and Kaur, 1979).

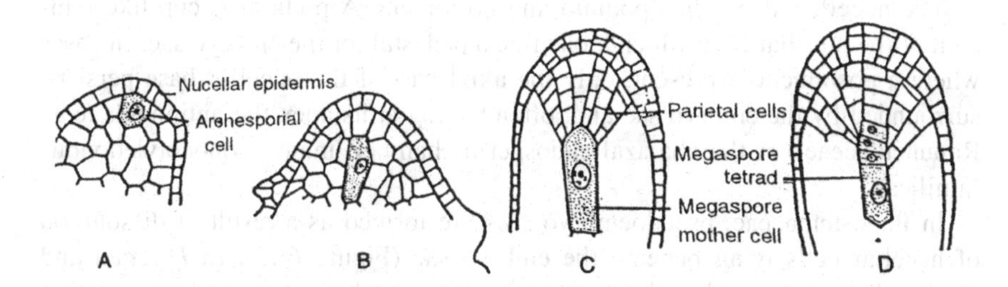


**FIGURE 6.11** *Glottidium vesicarium.* Stages in megasporogenesis. Note multi-layered nucellar epidermis and parietal cells of the nucellus (after Prakash and Herr, 1979).

According to Davis (1966), out of 314 families for which information is available, 179 families show crassinucellate ovules, 105 families show tenuinucellate ovules and 11 families bear pseudocrassinucellate ovules. The nucellus in the remaining 19 families forms only generic or specific characters.

In many taxa of the families Malpighiaceae, Nyctaginaceae, Euphorbiaceae and Cucurbitaceae, the nucellus forms a nucellar beak which reaches out into the micropyle (Figure 6.8).

As the development of the embryo sac proceeds, the nucellar cells are consumed. In tenuinucellate ovules, the nucellus degenerates during early stages of ovule development. In some plants, nucellus persists in the mature seeds

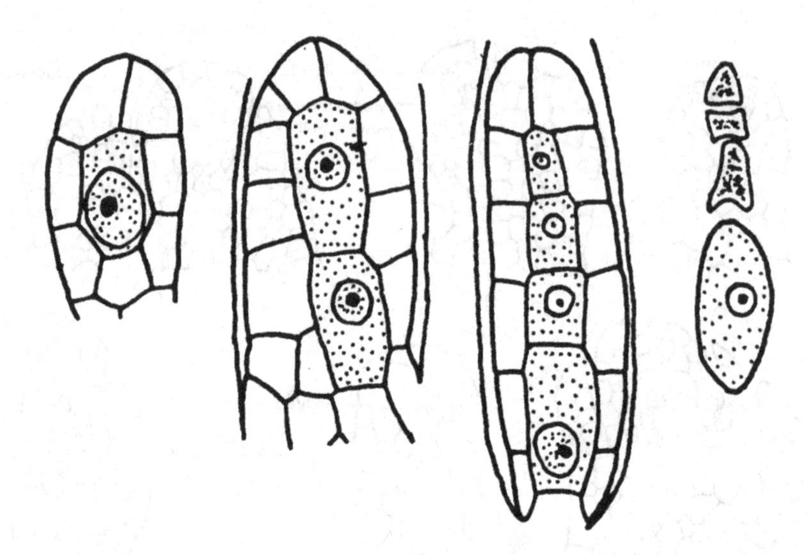

**FIGURE 6.12** Megasporogenesis in *Alectra parasitica* var. *chitrakutensis* (after Chandra et al., 1980).

and serves as nutritive tissue. Termed the perisperm, it is commonly found in Piperaceae, Amaranthaceae, Zingiberaceae and Cannaceae.

The nucellus also forms podium and postament. A podium is cup-like remnant of the nucellar base which looks like a pedestal for the embryo sac. In cases where a postament is present, only the axial part of the nucellar base persists, surrounded by the embryo sac and, often having at its apex the antipodals (e.g., Ranunculaceae), or the chalazal endosperm chamber (many monocotyledonous families).

In Podostemaceae, pseudoembryo sacs are formed as a result of dissolution of nucellar cells lying beneath the embryo sac (Figure 6.13). In *Dicraea* and *Zeylanidium*, the pseudoembryo sac is formed as early as the megaspore mother cell stage, while in *Indotristicha* and *Terniola*, it is formed only after fertilization. In *Hydrobryopsis sessilis*, pseudoembryo sac is formed when embryo sac attains the two-nucleate stage. Pseudoembryo sac contains dense cytoplasm and numerous nuclei. It provides nourishment to the growing embryo sac and embryo.

## HYPOSTASE

Hypostase is a group of cells present just at the level of origin of the two integuments and directly below the embryo sac (Figure 6.6). Hypostase consists of thick-walled lignified cells poor in cytoplasmic contents. Occasionally, the cells of the hypostase may surround a portion of the embryo sac and may even extend into the micropylar half of the ovule (Tilton, 1980). Occurrence of hypostase is reported in several families (e.g., Euphorbiaceae, Apiaceae, Elaegnaceae, Loranthaceae, Onagraceae).

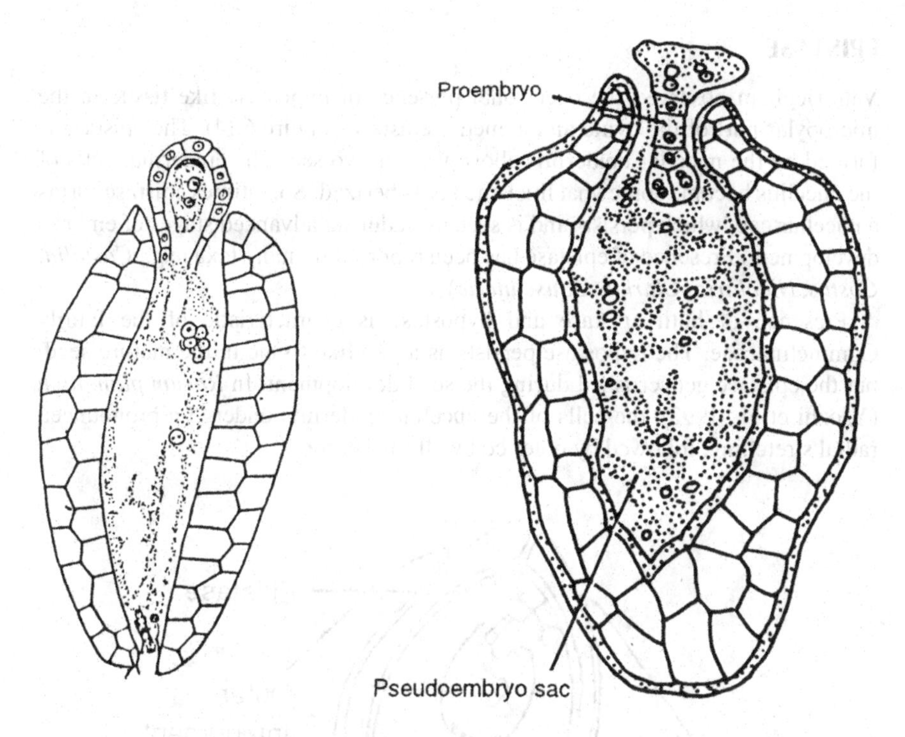


**FIGURE 6.13** Formation of pseudoembryosac. A. *Farmeria indica.* L.s. ovule showing organized nucellar primordia (after Govindappa and Nagendran, 1975); B. *Indostristicha ramosissima.* Proembryo at the quadrant stage; the nucellar cells have broken down to form pseudoembryosac (after Chopra and Mukkada, 1966).

## FUNCTIONS

Van Tieghem (1901), who coined the term hypostase, believed that it forms a sort of barrier or boundary for the growing embryo sac and prevents it from pushing into the base of the ovules. Johansen (1928) suggested that hypostase stabilizes the water balance of resting seeds over the long period of dormancy during hot, dry seasons. Goebel (1933) suggested that presence of hypostase directly above the termination of the vascular supply of the ovule indicates its relationship to the water economy of the embryo sac. According to Venkata Rao (1953), the hypostase serves to connect the vascular bundle in the funiculus with the embryo sac and helps in the rapid transport of food materials. Venkata Rao (1953) and Tilton (1980) opined that the chief function of hypostase is to translocate the nutrients. In some cases, however, hypostase may also act as a storage tissue. According to Savchenko (1973), hypostase supplies different enzymes, physiologically active and nutritive substances, to embryo sac. Bouman (1984) believes that hypostase may produce certain enzymes or hormones or play a protective role in mature seeds. Shamrov (1991) believes that the hypostase is a polyfunctional structural element of the ovule.

## EPISTASE

Van Tieghem observed the occasional presence of hypostase like tissue in the micropylar part of the ovule and named it epistase (Figure 6.14). The epistase is formed by the nucellar epidermis above the embryo sac. The epidermal cells of the nucellus become somewhat thickened or suberized. Sometimes epistase forms a nucellar cap, which persists and is seen even during advanced stages of embryo development. Presence of epistase has been reported in many taxa (e.g., *Castalia*, *Costus*, *Nicolaria*, *Clarkia*, *Zauschneria*).

Presence of both epistase and hypostase is characteristic of the family Commelinaceae. The hypostase persists as a pad-like tissue in the mature seed, but the epistase gets crushed during the seed development. In *Rheum plamatum* (Yaozhi et al., 1992), the cells of the nucellar epidermis undergo a pronounced radial stretching, followed by outer cell wall thickening.

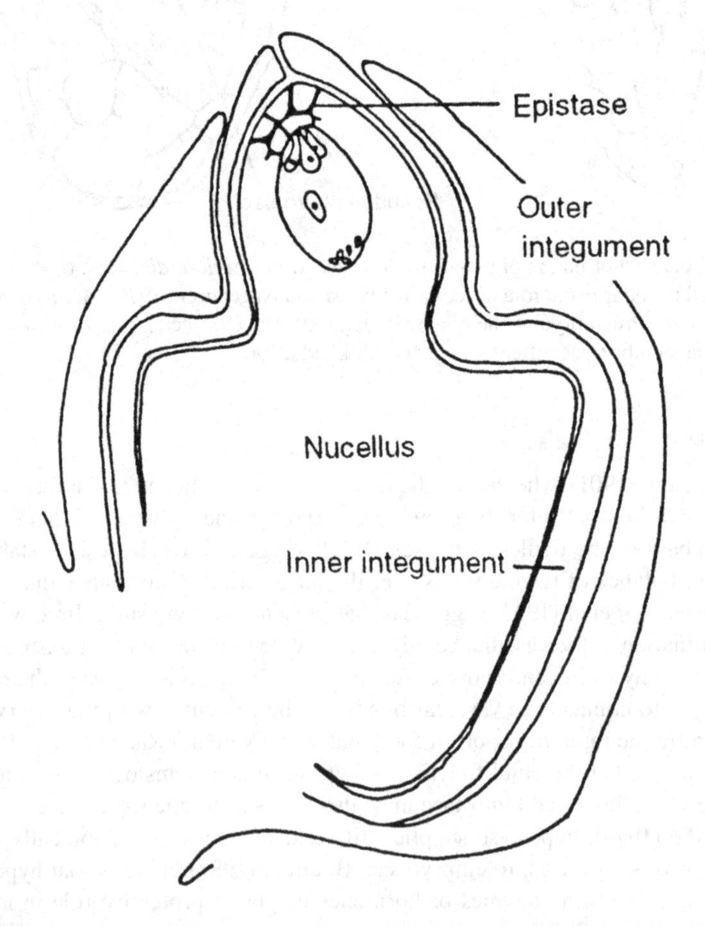


**FIGURE 6.14** *Tinantia figax*. Longitudinal section of ovule at organized female gametophyte stage showing epistase (after Chikkannaiah, 1962).

## MEGASPOROGENESIS

A single hypodermal cell of the nucellus present just below the epidermis enlarges considerably, acquires dense cytoplasmic contents, shows prominent nuclei and differentiates as the archesporium.

In tenuinucellate and pseudo-crassinucellate ovules, the archesporial cell directly functions as megaspore mother cell, but in crassinucellate ovules, the archesporial cell divides periclinally, forming an outer primary parietal cell and inner primary sporogenous cell (Figure 6.15). The former may remain undivided, or it may undergo periclinal and anticlinal divisions to form a variable number of wall layers. The primary sporogenous cell functions as a megaspore mother cell.

Usually, the archesporium is single-celled, but in Onagraceae, Paeoniaceae and Crossosomataceae, it consists of a group of cells. Of these, usually the central cell is functional. Sometimes multiple megaspore mother cells are formed by the multicellular archesporium which ultimately develops into bunches of megaspore tetrads (e.g., *Sedum*).

Megaspore mother cell (also called megasporocyte or meiocyte) divides meiotically to form four haploid megaspores. The first division of the megaspore mother cell is always transverse, resulting in the formation of two cells (dyad). The second division is also transverse, forming a tetrad of four megaspores (Figure 6.16). Megaspores may be arranged in linear fashion or T-shaped due to vertical division in the micropylar dyad cell and transverse division in the chalazal dyad cell. Linear as well as T-shaped megaspore tetrads may be formed in the same species (e.g., *Alectra parasitica, Drimys winteri, Orchis maculata*).

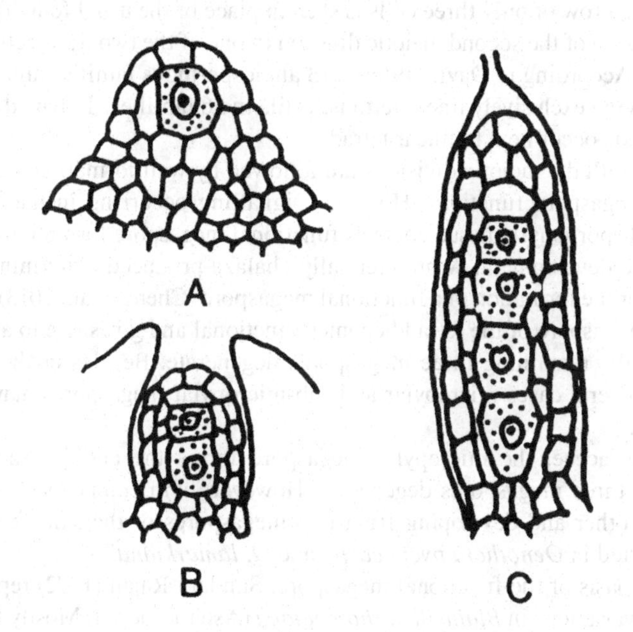


**FIGURE 6.15** Megasporogenesis in *Ipomoea*.

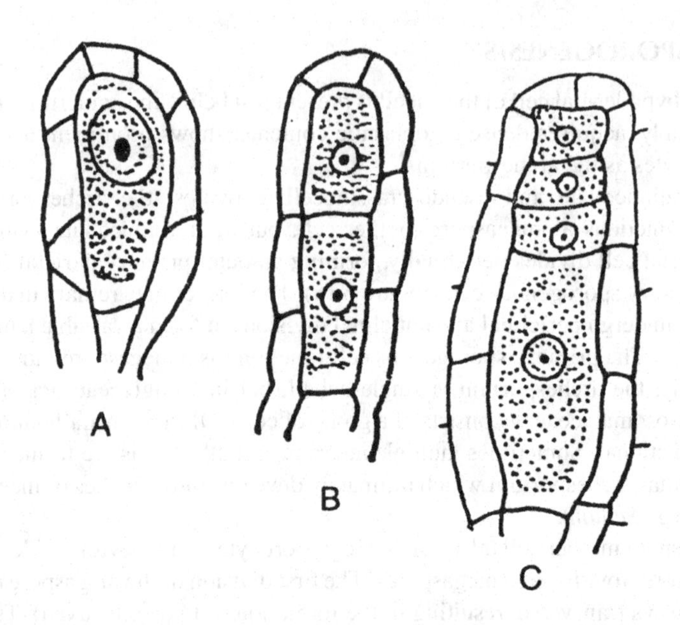


**FIGURE 6.16** Megasporogenesis in *Lobelia* (after Pandey et al., 1992).

The genus *Musa* is of special interest as tetrads of four different kinds (linear, T-shaped, I-shaped and isobilateral) may occur in the same species. In *Urginea indica*, the arrangement of a megaspore tetrad is linear, T-shaped and deccusate. Frequently, a row of only three cells is seen in place of the usual four. This is due to an omission of the second meiotic division in one of the two dyad cells, usually the upper. According to Davis (1966), 213 angiospermous families are characterized by having exclusively linear tetrads, while in 56 families, T-shaped tetrads to linear tetrads occur next to linear tetrad.

Where both the meiotic divisions are followed by wall formation, only one of the four megaspore functions. Hormone signalling occurring in the surrounding diploid sporophytic tissue controls functional megaspore formation and early embryo sac development. A preferentially chalaza-produced cytokinin signal is required for the formation of a functional megaspore (Cheng et al., 2013). Usually, chalazal megaspore of the tetrad becomes functional and gives rise to an embryo sac while the remaining three megaspores degenerate. Besides chalazal megaspore, in several cases micropylar and sub-micropylar megaspores may become functional.

In Onagraceae, the micropylar megaspore forms the embryo sac and the remaining three megaspores degenerate. However, twin embryo sacs lying one above the other and developing from two megaspores of the same tetrad have been reported in *Oenothera pycnocarpa* and *O. lamarkiana*.

On the basis of the functional megaspore, Sundara Rajan (1972) reported the following variations in *Blainvillea rhomboidea* (Asteraceae): (i) Mostly the chalazal megaspore is functional; (ii) Two micropylar megaspores degenerate and two chalazal ones develop; (iii) The chalazal and micropylar megaspores enlarging, middle two degenerating; (iv) The second micropylar megaspore degenerating, and the remaining three showing signs of development.

In taxa where second meiotic division is not followed by wall formation, one of the dyad cells with two haploid megaspores nuclei takes part in the formation of the embryo sac, whereas the other dyad cell degenerates. This degeneration of the megaspores or dyad cells is accompanied by an increase in ribosomes, appearance of dictyosomes, lipid granules, vesicles autophagic vacuoles, loss of turgidity and cellular activity. In cases where no wall formation occurs throughout meiosis, all of the four nuclei of megaspore contribute towards the formation of the embryo sac (for a detailed account, see Chapter 7).

During megasporogenesis, a notable feature is the development of callus thickenings around the non-functional megaspores. In the Polygonum type of embryo sac, the callose appears first in the chalazal region of the megaspore mother cell, spread over its entire surface, and isolates from the circumjacent maternal tissue. Then the callose is retained only around the micropylar dyad, or around three upper megaspores, so that the functional chalazal megaspore does not have any callose.

In *Gasteria verrucosa* (Willemse and Bednara, 1979), a callose wall begins to appear at the micropylar end of the megaspore mother cell and thickens continuously (Figure 6.17). At diakinesis, the callose wall surrounds the entire cell, but it is thicker at the micropylar side. At the basal end of the cell, the wall is very thin or sometimes absent. During the dyad stage, the callose wall at the chalazal end disappears but envelops the micropylar cell completely. In the tetrad, all the cells are surrounded by a callose wall except the megaspore, which has only a cap of callose at the micropylar end (Figure 6.17). A close contact with the plasma membrane and the callose wall can be observed in the megasporocyte. It dissolves at tetrad stage, first around the tetrad; thereafter, the walls between the cells disappear.

In *Gasteria*, contact by means of plasmodesmata with the surrounding nucellar cells exists during megasporogenesis. At the chalazal end, more plasmodesmata are present. Depending on the formation and position of the callose wall,

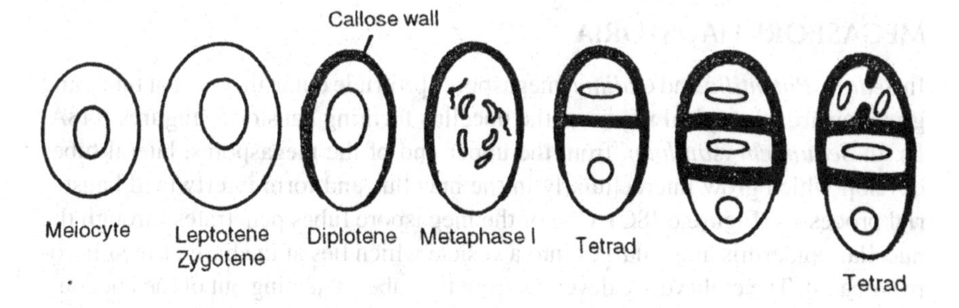


**FIGURE 6.17** *Gasteria.* Callose development during megasporogenesis (after Willemse and Bednara, 1979).

these plasmodesmata can be blocked and cut off. Some membrane-like remnants become visible against the plasma membrane. The megaspore is in contact with the chalazal nucellar cells only by means of plasmodesmata.

In *Rheum palmatum* (Yaozhi et al., 1992), callose first deposits on the micropylar end cell wall of the megasporocyte when it is at the pachytene, but it is not completely surrounded by callose. Besides the fluorescence at the micropylar end, callose fluorescence also appears on the cross walls of the megaspore dyad and tetrad. At the functional megaspore stage, fluorescence only exists on its micropylar end. Callose is never seen deposited on the cell wall of the chalazal end during meiosis. From megaspore tetrad onwards, callose begins to deposit on some nucellar cell walls at the chalazal region, and it does not disappear until the uninucleate embryo sac stage.

During the development of the nucellar tissue, a polarity is built up along the micropylar chalazal axis. Within the nucellus, the regulated development of the embryo sac is the result of several events, such as the position and migration of nuclei, orientation of the spindle, direction of cell elongation, selection of functional megaspore and the rate of differentiation, which determines the polarity of the megagametophyte during subsequent stages. Based on genetic factors, the development of the gametophyte is a programmed process resulting in a highly differentiated structure. The factors which govern polarity include mechanical, nutritive and organizing (hormonal control) nature.

Electron microscopic studies done on polarization of cell organelles during megasporogenesis in *Gasteria* (Willemse and Bednara, 1979) reveal that, in a young meiocyte, a few changes can be observed in the plastid, mitochondria and endoplasmic reticulum. At zygotene, vesicles appear in the cytoplasm probably pinched off from the outer nuclear membrane. In late diakinesis and prometaphase I, the micropylar end of the cell shows more dictyosomes. At the chalazal end, clear regions in the cytoplasm can be observed, in which parts of membrane are visible. An equal distribution of ribosome population is built up in the dyad. But the number of ribosomes per unit of cytoplasm differs among the megaspores in a tetrad. After the second meiotic division, the functional megaspore receives maximum ribosomes.

## MEGASPORE HAUSTORIA

In *Sedum*, *Potentilla* and *Galium*, megaspores protrude out in the form of tubes and grow towards the apical region of the nucellus forming haustoria (Figures 6.18A, B). In *Sedum chrysanthum*, from the upper end of the megaspores, lateral tubes develop which grow intercellularly in the nucellus and form intertwined haustorial processes (Figure 6.18C). One of the megaspore tubes penetrates through the nucellar epidermis and enlarges into a vesicle which lies at the base of the micropylar canal. The embryo sac develops from the tube projecting out of the nucellus. Occurrence of megaspore haustoria has been reported in *Sedum sempervivioides*, *Potentilla*, *Galium* and *Rosularia*.

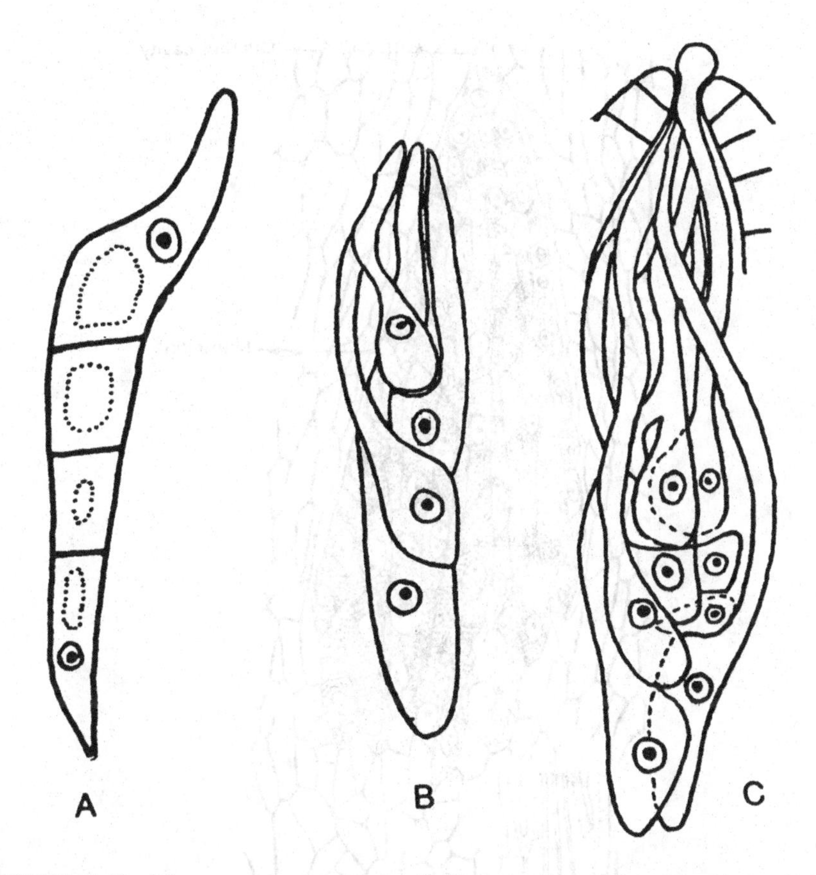


**FIGURE 6.18** Megaspore haustoria. A. *Gallium lucidum* (after Fagerlind, 1937); B. *Sedum sempervivoides* (after Mauritzon, 1933); C. *Sedum chrysanthum.* All the megaspores have given out tubes which are seen growing towards the micropyle (after Subramanyam, 1967).

## MAMELON

In the family Loranthaceae, instead of ovules, several members show a conical projection at the base of the ovary (Figure 6.19). Such projections are called *mamelon* (placenta). The morphological nature of mamelon is debatable. Coccuci (1983) does not accept the interpretation of mamelon as placenta. He refers to the ovules as "discrete" and "collective". Discrete ovules are equated to the lobes of mamelon, and unlobed mamelon with a massive sporogenous tissue to a collective ovule. According to Johri et al. (1992) the equivalence of mamelon to a placenta is fully justified.

In *Arceuthobium minutissimum* (Viscoideae, Loranthaceae) the central ovarian papilla (mamelon) develops from the base of the ovary and contains one or more archesporial cells. The true ovules, with nucellus and integuments, are absent. The archesporial cells directly function as megaspore mother cells.

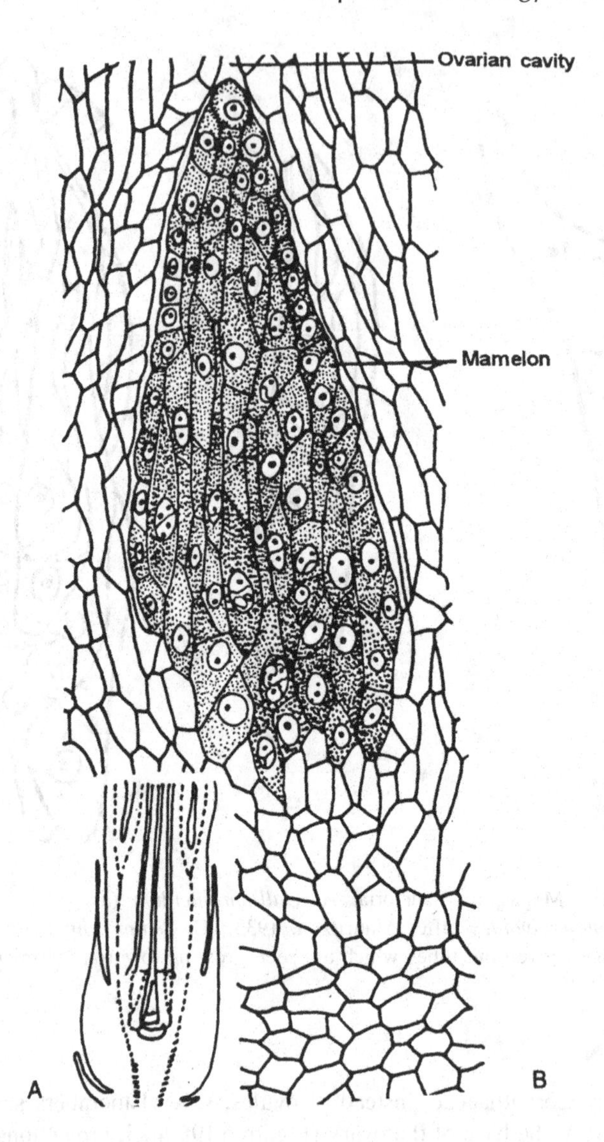


**FIGURE 6.19** *Helicanthes elastica*. A. Diagrammatic longitudinal section of the ovary to show the position of placenta or mamelon (stippled projection inside the block); B. Enlarged view of the placenta along with adjacent ovarian tissue, from A (after Johri et al., 1957).

The mamelon may be lobed (*Lysiana*, *Nuytsia*) or unlobed (*Helicanthes*). In lobed mamelon, the archesporium differentiates hypodermally in each lobe (e.g., *Amylotheca dictyophleba*, *Nuytsia floribunda*). In cases where mamelon is unlobed, the entire subepidermal tissue functions as the archesporium (e.g., *Amyema gravis*, *Tapinostemma acaciae*).

In *Dendrophthoe*, *Helixanthera* and *Moquiniella* there is no trace of any projection at the base of the ovary. The archesporial tissue differentiates in the bud stage as a compact tissue in the hypodermis at the base of the ovarian cavity. These archesporial cells directly function as megaspore mother cells.

## NAKED EMBRYO SAC

In the majority of angiosperms, the embryo sac is located inside the ovule surrounded by the nucellus and integument(s). In *Torenia fournieri* (Linderniaceae), the embryo sac protrudes from the micropyle. This feature has also been observed in *Gallium*, *Philadelphus*, *Thesium*, *Torenia*, *Utricularia* and *Vandellia*, in which the embryo sac protrudes out through the micropyle.

## PRACTICAL EXERCISES

### CLEARING TECHNIQUES

A. A simple and efficient method to study ovule and megagametophyte development is described following Herr (1971).

#### PROCEDURE

1. Fix excised pistils in $FPA_{50}$ and store in 70% ethyl alcohol.
2. Place fixed pistils in clearing fluid composed of lactic acid (85%), chloral hydrate, phenol, clove oil and xylene (2:2:2:2:1 by weight).
3. Dissect ovaries and place ovules with some of the fluid onto a slide, cover with a coverslip (coverslip is supported laterally by two permanently affixed coverslips).
4. Examine ovules under a phase contrast microscope.

B. Young et al. (1979) developed the following technique for detecting sexual (Figure 6.20) and aposporous embryo sacs.

#### PROCEDURE

1. For ovule clearing, fix pistils in formalin-acetic-alcohol (FAA: 70% ethanol: water: acetic acid: formaldehyde, 90:5:5 v/v).
2. After 24 hrs, excise fixed pistils and place in 70% alcohol in a 30 ml screw-cap.
3. Dehydrate pistils in an alcohol series: 70%; 85%; 100% (three changes); ethanol: methyl salicylate 1:1; ethanol: methyl salicylate 1:3; 100% methyl salicylate (two changes).
4. Mount cleared ovules in 100% methyl salicylate under an unsealed cover slip on a microscope slide.
5. Examine ovules under a differential interference contrast microscope.

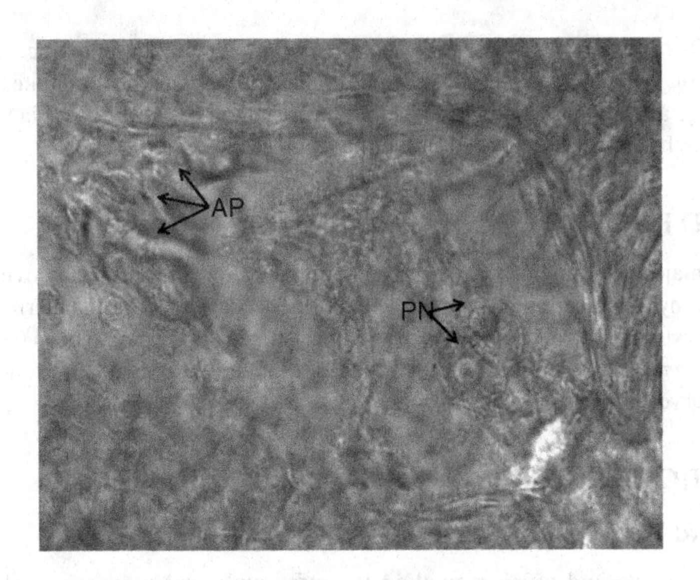


**FIGURE 6.20** *Cenchrus ciliaris.* Ovule after clearing protocol of Young et al. (1979). Embryo sac with two polars and three antipodals (PN=polar nuclei, AP=antipodals) (courtesy: Dr Vishnu Bhat, Delhi).

## SUMMARY

- The megasporangium (or ovule) consists of the nucellus which is surrounded by one or two integuments.
- Mature ovules may be of different types: orthotropus (atropous), anatropous, campylotropous, hemianatropous (hemitropous), amphitropous and circinotropous.
- The ovules may be unitegmic (having one integument) or bitegmic (having two integuments).
- In plants having unitegmic and tenuinucellate ovules, the inner epidermis elongates radially and differentiates into the endothelium.
- Obturators are specialized glandular structures involved in directing the pollen tube towards the micropyle.
- Ovules may be crassinucellate or tenuinucellate.
- Megaspore mother cell (megasporocyte or meiocyte) divides meiotically to form four megaspores.
- In some plants, nucellus persists in the mature seeds and serves as nutritive tissue. It is called *perisperm*.

## SUGGESTED READING

Archebald, E.E.A. 1939. The development of the ovule and seed of jointed cactus (*Opuntia aurantiaca* Lindley). *South Afr. J. Sci.* 36: 195–211.

Boesewinkel, F.D. 1977. Development of ovule and testa in Rutaceae: *Ruta*, *Zanthoxylum* and *Skimmia*. *Acta Bot. Neer.* 26 (3): 193–221.

Bouman, F. 1984. The ovule. In *Embryology of Angiosperms*, ed. B.M. Johri, 123–157. Berlin: Springer-Verlag.

Chandra, S., R.P. Singh and A.K. Pandey. 1980. Embryological studies in *Alectra parasitica* A. Rich. var. *chitrakutensis* Rau (Scrophulariaceae). *Flora* 169: 111–120.

Chikkanaiah, P.S. 1962. Morphological and embryological studies in the Commelinaceae. In *Plant Embryology*, 23–36. New Delhi: Symp. Council Sci. Ind. Res.

Chopra, R.N. and A.J. Mukkada. 1966. Gametogenesis and pseudoembryo sac in *Indotristicha ramosissima* (Weight) van Royen. *Phytomorphology* 16: 182–188.

Cocucci, A.E. 1983. New evidence from embryology in angiosperm classification. *Nord. J. Bot.* 3: 67–73.

Cresti, M., P. Gori and E. Pacini. 1988. *Sexual Reproduction in Higher Plants*. Berlin: Springer-Verlag.

Davis, G.L. 1966. *Systematic embryology of the angiosperms*. New York: John Wiley.

De Bore-de Jeu, M.J. 1978. Megasporogenesis, Meded. *Landbhogesch. Wageningen* 16: 1–127.

Fagerlind, F. 1937. Embryologische, zytologische und bestäubungsexperimentelle studien in der familie Rubiaceae nebst bemerkungen über einige polyploiditätsrobleme. *Acta Hortc. Bergiani* 11: 195–470.

Franssen-Verheijen, M.A.W. and M.T.M. Willemse. 1993. Micropylar exudate in *Gasteria* (Aloaceae) and its possible function in pollen tube growth. *Am. J. Bot.* 80 (9): 253–262.

Goebel, K. 1933. *Organographie der Pflanzen. Samenpflanzen*. 3rd ed. Jena.

Govindappa, D.A. and C.R. Nagendran. 1975. Is there a podostemum type of embryo sac in the genus Farmeria. *Caryologia* 28: 229–235.

Herr, J.M. Jr. 1971. A new clearing squash technique for the study of ovule development in angiosperms. *Am. J. Bot.* 58 (8): 785–790.

Johansen, D.A. 1928. The hypostase: Its presence and function in the ovule of the Onagraceae. *Proc. Natl. Acad. Sci. USA* 14 (9): 710–713.

Johri, B.M. 1992. Haustorial role of pollen tubes. *Ann. Bot.* 70 (5): 471–475.

Johri, B.M., J.S. Agrawal and S. Garg. 1957. Morphological and embryological studies in the family Loranthaceae I. *Heliacnthes elastica* (Desr.) Dans. *Phytomorphology* 7: 336–354.

Johri, B.M., K.B. Ambegaokar and P.S. Srivastava. 1992. *Comparative Embryology of Angiosperms*. Berlin: Springer.

Kapil, R.N. and S.C. Tiwari. 1978. The integumentary tapetum. *Bot. Rev.* 44: 457–490.

Kaur, H. 1969. Structure and development of seed in *Ipomoea obscura* Ker-Gawl. *J. Indian Bot. Soc.* 47: 346–351.

Lora, J., X. Yang and M.R. Tucker. 2019. Establishing a framework for female germline initiation in the plant ovule. *J. Exp. Bot.* 70 (11): 2937–2949.

Maheshwari, P. 1950. *An Introduction to Embryology of Angiosperms*. New York: McGraw Hill Book Co.

Mauritzon, J. 1933. *Studien über die embryologie der familien Crassulaceae und Saxifragaceae*. Lund: ÜDiss. Univ.

Malik, C.P. and S. Vermani. 1975. Physiology of sexual reproduction. I. A histochemical study of the embryo sac development in Zephyranthes rosea and Lagenaria vulgaris. *Acta Histochemica* 53 (2): 244–280.

Mohan Ram, H.Y. and B. Hari Gopal. 1981. Some observations on the flowering of bamboos in Mizoram. *Curr. Sci.* 50: 708–710.

Mohana Rao, P.R. and A. Kaur. 1979. Sporogenesis and gametophytes of *Polygonatum cirrhifolium*. *Phytomorphology* 29: 93–97.

Pandey, A.K. 1980. Structure and behaviour of endothelium in some Compositae. *Acta Botanica Indica* 8: 50–56.

Pandey, A.K. and S. Chopra. 1979. Development of seeds and fruits in *Gerbera jamisonii*. *Geophytology* 9: 171–174.

Pandey, A.K., T.G. Lammers and A. Jha. 1992. Sporogenesis and gametogenesis in Pacific Island *Lobelia* (Campanulaceae, Lobelioideae). *Phytomorphology* 42 (1): 63–69.

Prakash, N. and J.M. Herr Jr. 1979. Embryological study in *Glottidium vesicarium* through the use of clearing technique. *Phytomorphology* 29: 71–77.

Savchenko, M.I. 1973. *Morfologiya semyapochki pokrytosemwnnych rasteniy.* [*Ovule morphology of Angiosperms*] – Leningrad: Nauka. [In Russian]

Shamrov, I.I. 1991. The ovule of *Swertia iberica* (Gentianaceae): Structural and functional aspects. *Phytomorphology* 41: 213–229.

Singh, B. 1961. Studies in the family Malpighiaceae III. Development and structure of seed and fruit of *Malpighia glabra* Linn. *Hort. Adv. (Saharanpur)* 5: 145–155.

Singh, R.P. 1959. Structure and development of seed in *Phyllanthus niruri* L. *J. Indian Bot. Soc.* 51: 73–77.

Subramanyam, K. 1967. Some aspects on the embryology of *Sedum chrysanthum* (Boissier) Raymond-Hamlet with a discussion on its systematic position. *Phytomorphology (Panchanan Maheshwari Memm* 17: 240–247.

Sundara Rajan, S. 1972. Embryological studies in Compositae- III. A contribution to the embryology of *Blainvillea rhomboidea* Cass. (*B. latifolia* DC.). *Indian Acad. Sci.* 75 (4): 167–176.

Tilton, V.R. 1980a. Hypostase development in *Ornithogalum caudatum* (Liliaceae) and notes on other types of modifications in the chalaza of angiospermous ovules. *Can. J. Bot.* 58: 2059–2066.

Tilton, V.R. 1980b. The nucellar epiderrnis and micropyle of *Ornithogalum caudatum* (Liliaceae) with a review of these structures in other taxa. *Can. J. Bot.* 58: 1872–1884.

Van Tieghem, Ph. 1901. L,hypostase, sa structure et son role constants, sa position et sa forme variables. *Bul. Mus. Hist. Nat.* 7: 412–418.

Venkata Rao, C. 1953. Floral anatomy and embryology of two species of *Elaeocarpus*. *J. Indian Bot. Soc.* 32: 21–33.

Vijayraghavan, M.R. and N.N. Bhandari. 1970. Studies in the family Ranunculaceae: Embryology of *Thalictrum javanicum* Blume. *Flora* 159: 450–458.

Willemse, M.T.M. and J. Bednara. 1979. Polarity during megasporogenesis in *Gasteria verrucosa*. *Phytomorphology* 29: 156–165.

Yaozhi, W., C. Chao and D. Huibin. 1992. Development of ovule and female gametophyte in *Rheum palmatum*. *Phytomorphology* 42: 71–79.

Yeung, E.C., S.Y. Zee and X.L. Ye. 1994. Embryology of *Cymbidium sinense*: Ovule development. *Phytomorphology* 44: 55–63.

Young, B.A., R.T. Sherwood and E.C. Bashaw. 1979. Cleared-pistil and thick sectioning technique for detecting aposporous apomixis in grasses. *Can. J. Bot.* 57: 1668–1672.

# 7 Female Gametophyte

The organized female gametophyte, also referred to as the embryo sac or megagametophyte, consists of an egg apparatus (one egg and two synergids), two polar nuclei and three antipodals (Figure 7.1). The polar nuclei later fuse to form a secondary nucleus. The egg apparatus is situated towards the micropylar side, while the antipodals lie on the chalazal side.

## TYPES OF FEMALE GAMETOPHYTES

During the development of the female gametophyte, the megaspore enlarges considerably and undergoes three successive mitotic divisions resulting in the formation of, generally, an 8-nucleate embryo sac (Figure 7.2). However, many deviations from the typical manner of development of megaspore and the embryo sac have been observed. Maheshwari (1950) classified types of development of the angiosperm embryo sac based on the following features:

(i) The number of megaspores or megaspore nuclei that take part in the formation of the embryo sac.
(ii) The total number of the divisions that occur during the formation of the megaspore and the gametophyte.
(iii) The number and arrangement of the nuclei and their chromosome number in the mature embryo sac.

Based on the number of megaspore nuclei involved in the development, the female gametophyte of angiosperms may be classified into monosporic, bisporic or tetrasporic types.

## MONOSPORIC EMBRYO SACS

This type of embryo sac is derived from only one of the megaspores. Monosporic embryo sacs may be 8-nucleate or 4-nucleate, and on this basis, they may be distinguished into two categories:

### Polygonum Type

This type of embryo sac is the most common and was described by Strasburger (1879) in *Polygonum divaricatum*. In this type, out of megaspores, usually the chalazal one functions, and the remaining three degenerate. The functional megaspore enlarges, and its nucleus undergoes three successive mitotic divisions unaccompanied by wall formation to form 2, 4 and, ultimately, 8-nucleate female gametophytes showing an egg apparatus, two polar nuclei and three antipodal

DOI: 10.1201/9781003260097-7

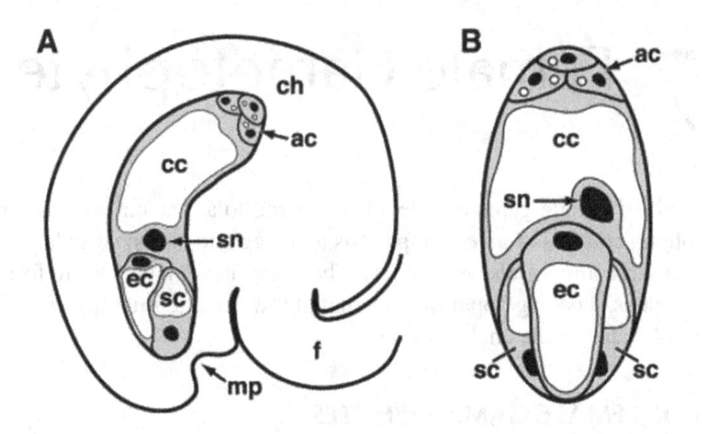


**FIGURE 7.1** *Arabidopsis* female gametophyte. A. Ovule showing embryo sac; B. Female gametophyte (ac, antipodal cells; cc, central cell; ch, chalazal region of the ovule; ec, egg cell; f, funiculus; mp, micropyle; sc, synergid cell; sn, secondary nucleus) (after Yadegari & Drews, 2004).

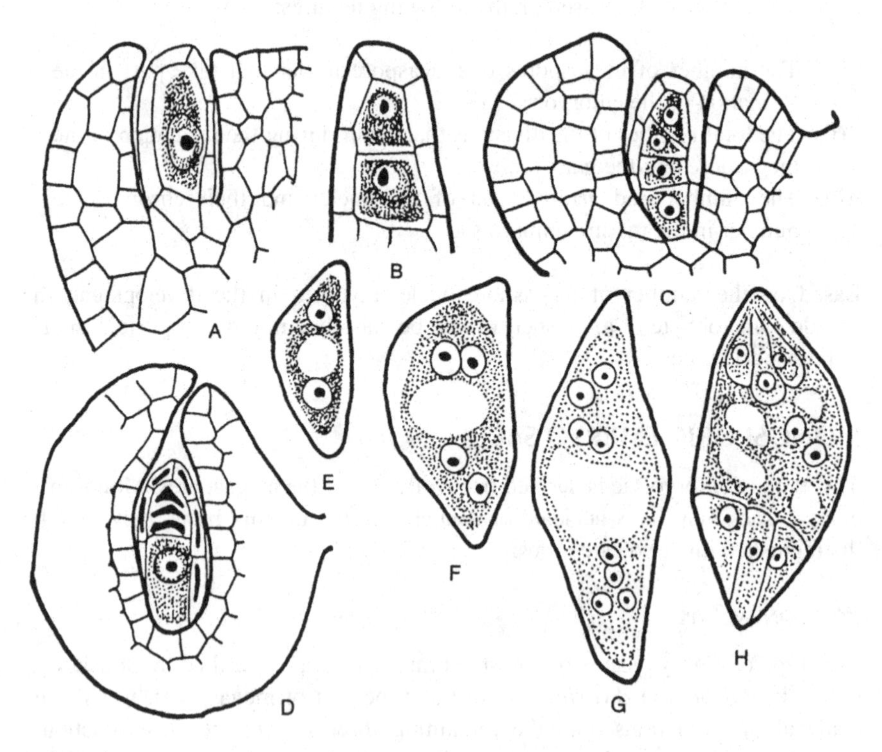


**FIGURE 7.2** Development of female gametophyte in *Cynoctonum oldenlandioides* (after Padhye and Jaolekar, 1995).

cells (Figure 7.3). Over 70% of the flowering plants share the polygonum type of embryo sac development (Zhou and Dresselhaus, 2019).

### Oenothera Type

In this type, the embryo sac is formed by the micropylar megaspore of the tetrad which undergoes only two nuclear divisions, forming four nuclei. These nuclei organize into an egg, two synergids and a single polar nucleus (Figure 7.3). Fertilization in this type results in the formation of a diploid endosperm nucleus – not a triploid one as in the Polygonum type. This type of female gametophyte is characteristic of the family Onagraceae. *Schisandra chinensis*, however, is the only example outside that family where such type of embryo sac occurs. The difference is that whereas in the Onagraceae the functional megaspore is the micropylar one, in *S. chinensis*, it is the chalazal megaspore that functions (Yoshida, 1962).

## BISPORIC EMBRYO SACS

In this type of embryo sac, the megaspore mother cell divides meiotically to produce two dyad cells. Only one of the dyad cells undergoes the second meiotic division, while the remaining one degenerates (Figure 7.4). Bisporic embryo sac may be of Allium type or Endymion type.

### Allium Type

This type of embryo sac was first described in *Allium fistulosum* (Strasburger, 1879). The first meiotic division in the megaspore mother cell results in the formation of a dyad. Of these two cells, the chalazal one becomes functional, while the micropylar megaspore degenerates. The nucleus of the functional megaspore

| | Megasporogenesis | | | Megagame-togenesis | | | |
|---|---|---|---|---|---|---|---|
| | Megaspore mother cell | Meiosis I | Meiosis II | Mitosis I | Mitosis II | Mitosis III | Mature embryo sac |
| Polygonum type | | | | | | | |
| Oenothera type | | | | | | | |

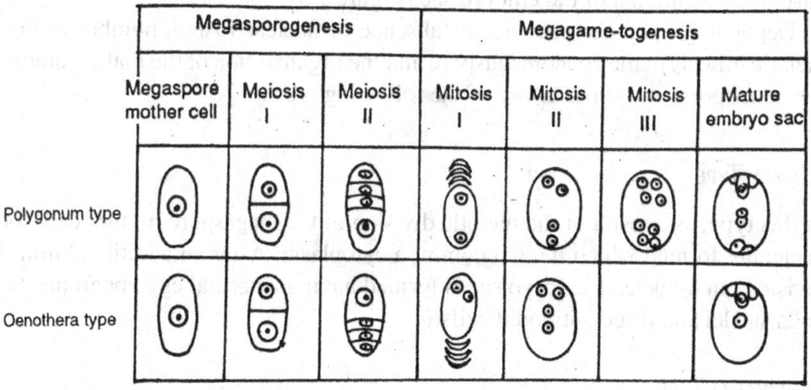

**FIGURE 7.3** Diagrammatic representation of different types of monosporic embryo sac development.

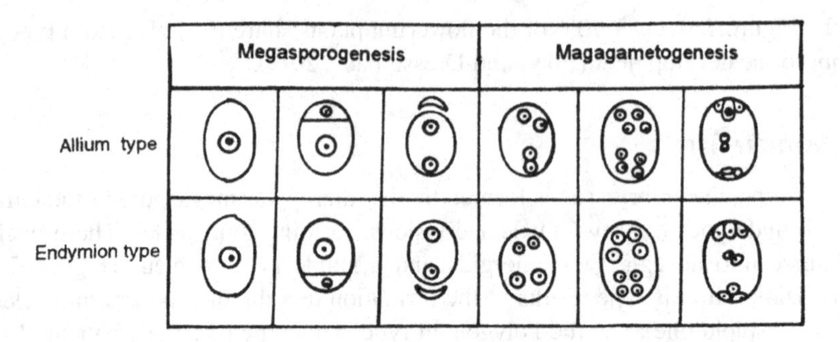


**FIGURE 7.4** Diagrammatic representation of different types of bisporic embryo sac development.

divides to form a 2, 4 and finally 8-nucleate embryo sac comprising of an egg, two synergids, two polars and three antipodals (Figure 7.4).

### Endymion Type

In this type of embryo sac, the 8-nucleate embryo sac is derived from the micropylar cell (Figure 7.4) of the dyad, and the chalazal one degenerates, e.g., *Scilla*. In *Schisandra chinensis*, some of the ovules show bisporic type of embryo sac development, but the final organization of the embryo sac is similar to *Oenothera*. This is the only authentic report of a bisporic, 4-nucleate embryo sac.

## TETRASPORIC EMBRYO SACS

In this type, the meiotic divisions in the megaspore mother cell are not accompanied by wall formation resulting in the formation of a coenomegaspore. The four haploid nuclei lie in a common cytoplasm. In the coenomegaspore, all nuclei take part in the formation of the embryo sac (Figure 7.5).

Depending upon the presence or absence of nuclear fusion, number of post-meiotic mitosis in the coenomegaspore and the organization of the mature embryo sac, tetrasporic embryo sacs are of the following types.

### Adoxa Type

In this type, as a result of the meiotic division of the megaspore mother cell, four nuclei are formed which lie in a common cytoplasm. After an additional mitotic division, an 8-nucleate embryo sac is formed having a normal egg apparatus, two polar nuclei and three antipodal cells.

### Plumbago Type

In this type, the 4-megaspore nuclei which are arranged in a crosswise manner, undergo further mitotic division forming free nuclei arranged in four pairs. The

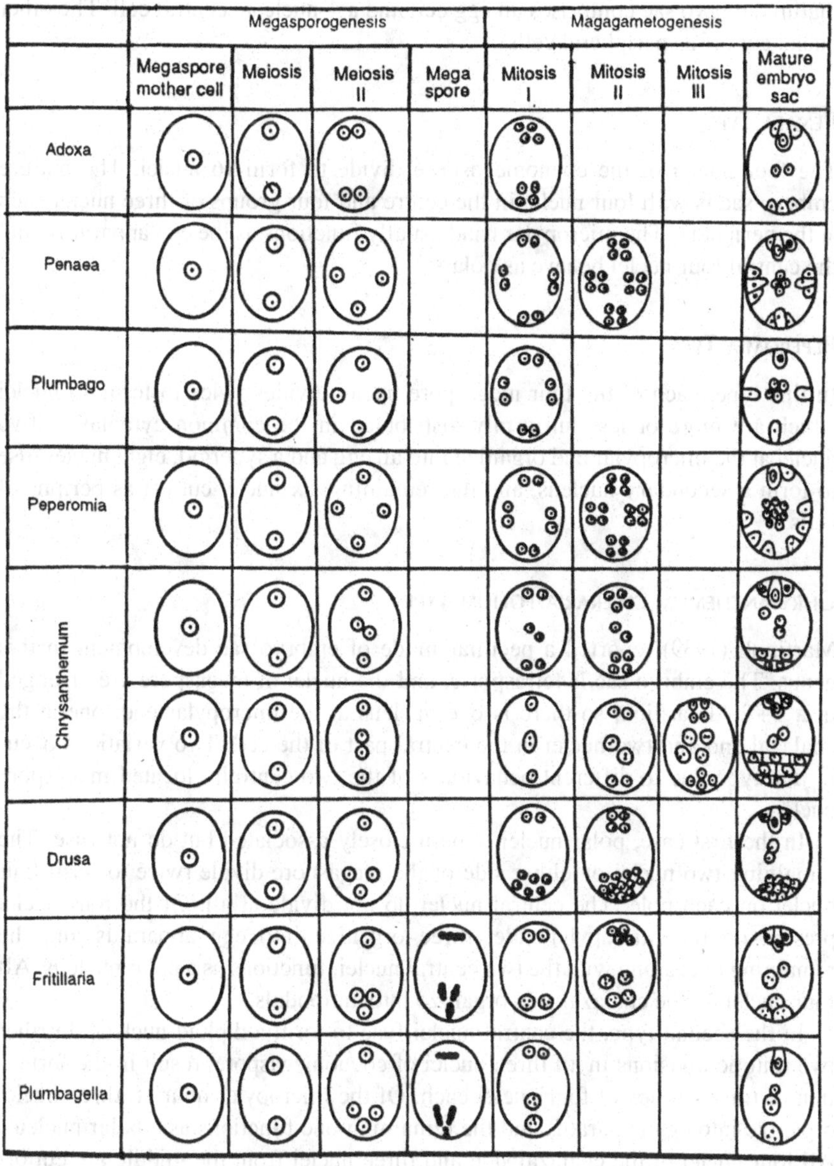

| | Megasporogenesis | | | | Magagametogenesis | | | |
|---|---|---|---|---|---|---|---|---|
| | Megaspore mother cell | Meiosis | Meiosis II | Mega spore | Mitosis I | Mitosis II | Mitosis III | Mature embryo sac |
| Adoxa | | | | | | | | |
| Penaea | | | | | | | | |
| Plumbago | | | | | | | | |
| Peperomia | | | | | | | | |
| Chrysanthemum | | | | | | | | |
| | | | | | | | | |
| Drusa | | | | | | | | |
| Fritillaria | | | | | | | | |
| Plumbagella | | | | | | | | |

**FIGURE 7.5** Diagrammatic representation of various types of tetrasporic embryo sac development.

mature embryo sac comprises an egg cell and a 4-nucleate central cell. The other nuclei cut off as peripheral cells.

### Penaea Type

The four nuclei of the coenomegaspore divide to form 16 nuclei. The mature embryo sac is with four nuclei in the centre and four groups of three nuclei each at the periphery. The micropylar triad usually functions as the egg apparatus, and the central four nuclei behave as polars.

### Peperomia Type

In this type, each of the four megaspore nuclei divides twice to form 16 nuclei which are more or less uniformly distributed in the common cytoplasm. Two nuclei at the micropylar end organize into an egg and a synergid, eight nuclei fuse to form a secondary nucleus, and the remaining six nuclei cut off as peripheral cells.

### Chrysanthemum cinerariaefolium Type

Martinoli (1939) reported a peculiar mode of embryo sac development in this genus. The embryo sac is tetrasporic, and the nuclei of megaspore are arranged in a 1+2+1 fashion; so there is one nucleus at the micropylar end, one at the chalazal end and two nuclei in the central part of the cell. Two variations occur in this type due to different behaviours of the two centrally located megaspore nuclei.

In the first case, polar nuclei remain closely associated but do not fuse. The remaining two nuclei on either side of the megaspore divide twice to form four nuclei on each pole. The central nuclei do not divide at all. Of the four nuclei present on the micropylar side, three organize into egg apparatus and the remaining one, along with the two central nuclei, functions as a polar nucleus. All four nuclei on the chalazal side organize into antipodals.

In the second type, the central nuclei fuse to form a diploid nucleus. Further two mitotic divisions in all three nuclei of coenomegaspore, result in the formation of three groups of four nuclei each. Of the micropylar quartet, three nuclei organize into egg apparatus and the remaining one functions as a polar nucleus. All four nuclei of the chalazal side and three nuclei from the middle are cut off as antipodal cells. The central cell consists of a diploid nucleus from the middle quartet and a haploid nucleus from the micropylar quartet.

### Drusa Type

This type of embryo sac is also 16-nucleate. After meiotic division in one megaspore mother cell, the four nuclei are arranged in a 1+3 fashion. Further meiotic division results in a 2+6, followed by a 4+12, stage. The four micropylar nuclei organize into egg apparatus and upper polar nucleus. The remaining 12 chalazal nuclei give rise to a lower polar nucleus and 11 antipodal cells.

### Fritillaria Type

Here three of four nuclei, formed after the meiotic divisions in the megaspore mother cell, move towards the chalazal side, and a fourth remains at the micropylar end. The latter divides in the usual manner, while the other three nuclei fuse to form a single triploid nucleus, which immediately divides to form two nuclei. Further divisions on both sides give rise to an 8-nucleate embryo sac. The mature embryo sac has a normal haploid egg apparatus, three triploid antipodal cells and tetraploid secondary nucleus, which results from the fusion of a haploid and a triploid polar nucleus.

### Plumbagella Type

In this type, the four megaspore nuclei take up a 1 + 3 arrangement, one being on the micropylar side and three towards the chalazal end. The latter fuse to form a single triploid nucleus. Further division results in the formation of four nuclei, two on each side. The nucleus lying on the micropylar side organize into an egg, one of the triploid nuclei on the chalazal end as antipodal cell, and the remaining two nuclei, one haploid and the other triploid, fuse to form a tetraploid secondary nucleus.

## MATURE EMBRYO SAC

In the majority of angiosperms, the female gametophyte consists of seven cells: the egg cell, two synergids, the central cell and three antipodal cells (Figure 7.6). This type of female gametophyte is referred to as normal type.

### Egg

Among the components of the female gametophyte, the egg cell is very important because it produces the embryo after fertilization. The egg is large and highly vacuolate. The nucleus is situated in the lower end and, in the centre, there is a large vacuole. The cytoplasm is accumulated in the chalazal end of the cell (Figure 7.7). In *Epidendrum* (Cocucci and Jensen, 1969), however, the nucleus is located in the centre. Usually, the egg cell is only partially surrounded by the cell wall. The wall is complete in the upper region and thus appears in the chalazal region. In this portion, the cell is surrounded only by a plasma membrane.

In *Capsella*, the egg cell in its chalazal region is surrounded by a thin and irregular PAS-positive cell wall. The egg cell wall is thickest at the micropylar region of the cell and gradually thins toward the chalazal region.

The egg cytoplasm contains varying amounts of cell organelles. In *Gossypium*, the egg cell contains a large amount of endoplasmic reticulum (ER), numerous

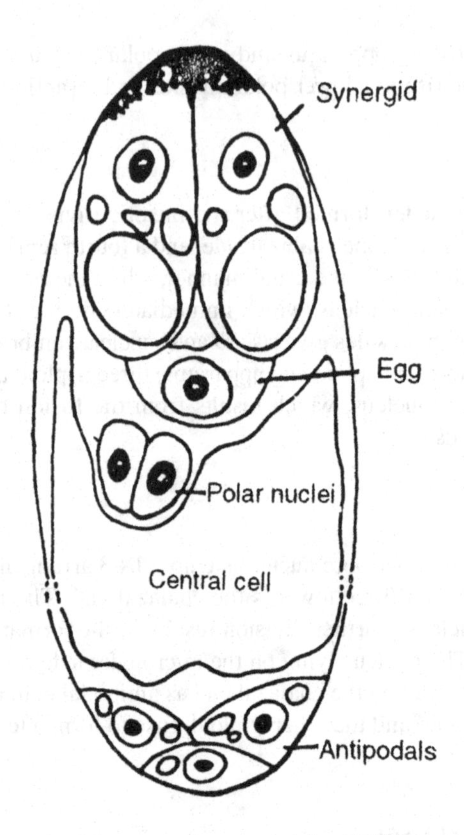


**FIGURE 7.6** An organized female gametophyte (after Willemse and Went, 1984).

ribosomes, numerous mitochondria with only few cristae and only a few dictyosomes and plastids (Jensen, 1965a, b). In *Capsella* (Schulz and Jensen, 1968), mitochondria are concentrated in the chalazal end. The plastids are seen in a shell around the nucleus. There are a few dictyosomes and lipid bodies. In *Petunia*, the egg contains only a few mitochondria, plastids and dictyosomes.

In *Plumbago*, where synergids are absent, filiform apparatus is present in the micropylar part of the egg (Cass, 1972). The filiform apparatus is PAS-positive and resembles the filiform apparatus of the synergids. The egg cell contains one large vacuole, which occupies the major portion of the cell volume. In *Plumbago*, the function of the synergid is taken over by the egg, which is probably involved in both nutrition and pollen tube entry.

## Synergids

The synergids are pyriform and "hooked", or the tip is prolonged into a "beak". There is a common wall between the synergid and the egg and the other synergid. Ultrastructural studies have revealed that the wall is absent at the chalazal end of the synergids in cotton, *Torenia* and maize. In *Capsella* and *Aquilegia*, however,

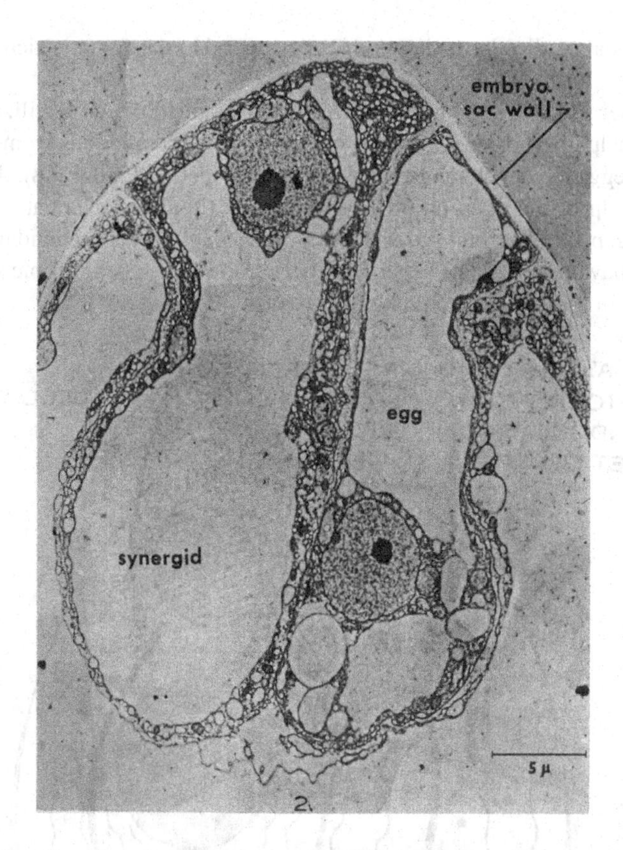


**FIGURE 7.7** Electron micrograph of a longitudinal section through the egg of an unfertilized embryo sac of *Torenia fournieri* (from Linskens, 1969).

a cell wall surrounds each synergid. The wall is thickest in the micropylar region of the synergid, but it gradually gets thin and irregular over the chalazal end. In *Epidendrum* the persistent synergid has a wall continuous over the cell. In the degenerated synergid, on the other hand, the wall is incomplete at the chalazal end.

The synergid usually possesses a vacuole in the chalazal region, and the nucleus in the middle region. A peculiar feature of the synergid is the filiform apparatus (FA). It usually consists of a mass of wall projections extending deep into the cytoplasm, and vice versa cytoplasmic channels penetrating deep into the FA. The filiform apparatus is PAS-positive, thus indicating its wall nature. It gives negative reaction for nucleic acids and proteins.

Most of the cytoplasm of the synergids is located in the micropylar half of the cell. The synergids are rich in plastids, mitochondria, dictyosomes, masses of ER and ribosomes. These organelles are associated with the filiform apparatus in large amounts. The synergid vacuoles are rich in inorganic compounds.

The synergids are considered as highly active cells. The abundance of mitochondria, plastids, dictyosomes and rough ER indicate that these are active cells.

Histochemical studies have shown that the synergid cytoplasm is rich in protein and RNA.

The synergids play an important role both in nutrition and fertilization. The synergids help in absorption, storage and transport of compounds from the nucellus to the egg, for developing embryo and endosperm (Figure 7.8). The synergids also help in the opening of the pollen tube tip and placement of the male gametes deep into the embryo sac in between the egg and secondary nucleus (Vijayaraghavan, 1972). The synergids also secrete some chemotropic substances which help in directing the pollen tube growth.

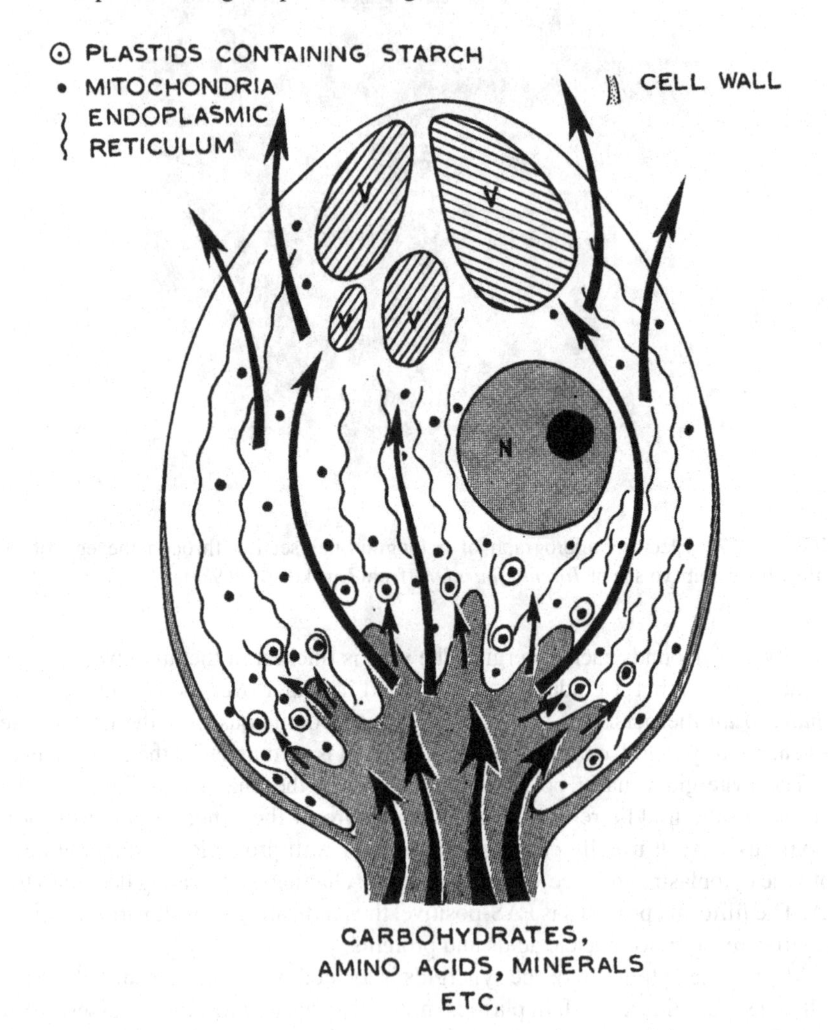


**FIGURE 7.8** Diagrammatic representation of the proposed function of synergid in the absorption, storage and transport of metabolites (from Jensen, 1965).

## Central Cell

The central cell is large and highly vacuolate. It usually contains two polar nuclei. The cytoplasm of the central cell is confined to a thin layer along the embryo sac wall and accumulates near the egg apparatus and the antipodal cells. The cytoplasm of the central cell is rich in ER which may be smooth or rough. Plastids are common and may be filled with starch. There are numerous mitochondria with well-developed cristae. There are large amounts of RNA and proteins in the cytoplasm and the nucleoli.

Like the cells of the egg apparatus, the central cell is surrounded only by a partial cell wall. The wall against the nucellus or integuments is usually thick and regular. The wall stains strongly for PAS and is rich in pectic substances. The central cell is separated from the egg cell and synergids by plasma membranes, and from the antipodals by thin walls often transverse by plasmodesmata.

The cell wall projections in the micropylar and chalazal regions of the central cell are observed in *Capsella*, *Helianthus* and *Euphorbia*. The wall projections strongly resemble the filiform apparatus. According to Newcomb and Steeves (1971), the presence of embryo sac wall projections suggest that the central cell plays an important role in the absorption of metabolites from the surrounding tissue.

The polar nuclei are quite large with a single nucleolus in each. In *Epidendrum* the polar nuclei show a lobed outline. The nuclei show some large fingerlike structures. The two polar nuclei fuse, before or during fertilization, to form the secondary nucleus.

## Antipodals

Antipodals vary widely in number, size, ploidy level and may be ephemeral or persistent. Normally, there are three cells, but in several taxa, they undergo secondary multiplication. In grasses, antipodal cells may proliferate into a multicellular tissue consisting of up to 100 cells. In *Lilium* (Vazart, 1969), the antipodal cells are only partly separated by walls. These incomplete walls subsequently dissolve, and a syncytium is formed.

In *Capsella* and *Helianthus*, the antipodal cells are surrounded by cell walls of uniform thickness, invariably containing plasmodesmata. These cells are connected to the nucellus and central cell by means of plasmodesmata. The antipodals are also interconnected. In *Zea mays*, the inner face of antipodals walls, adjacent to the nucellus, is papillate.

The antipodals do not organize into cells in *Cassytha*, *Hydnora*, *Scilla*, *Thismia* and *Xyris*. In *Helianthus*, instead of three antipodals, only two antipodals are present (Newcomb, 1973). One of the antipodals cells is mononucleate, the second antipodals cell contains two nuclei.

The antipodal nuclei stain intensely for DNA. The antipodal cells become polyploid, due to endomitosis (or polyteny). They may be multinucleate and multinucleolate.

In *Zephyranthes* and *Lagenaria*, histochemical studies have shown that antipodal cells contain high amounts of protein, polysaccharides, lipids and RNA (Malik and Vermani, 1975). In *Stellaria* (Pritchard, 1964), antipodal cells show rich amounts of DNA and proteins. There is a lower content of RNA, and there is no reaction for PAS. *Dendrobium* shows high concentration of peroxidases, cytochrome oxidase, ascorbic acid and SH compounds. Generally, three main functions have been attributed to the antipodals cells.

1. They may be involved in the transfer of nutrients and serve as a pathway for metabolites from the nucellus to the central cell.
2. The antipodal cells may store large quantities of starch, lipids and proteins which are utilized by the developing endosperm and embryo.
3. The antipodal cells may have a secretory function. They possibly secrete growth-controlling substances which regulate the development of the adjacent endosperm.

## Ultrastructure

To illustrate the ultrastructure of organized embryo sac, the work of Tilton and Mogensen (1979) on *Agave parryi* is described.

### Egg

The area of egg which contains the nucleus is large and generally spherical in shape. A large vacuole occurs in its micropylar end (Figures 7.7, 7.9), and smaller vacuoles are adjacent to the nucleus. Egg cytoplasm is dense, and within it, some ribosomes occur in clustered groups or as free monosomes. Others are found associated with the nuclear membrane and plasmalemma. Mitochondria are few and although plastids are not abundant either, they may become quite large and tend to be perinuclear in distribution. Plastids containing starch were not seen in the egg, but a few small lipid bodies do occur. Dictyosomes are fairly common and are unevenly distributed. Endoplasmic reticulum is rare and is found almost entirely as short segments of the smooth lamellate type. Groups of small vesicles included within a double membrane structure were found in some of the small vacuoles.

The egg nucleus is very large and occupies much of the cell's volume (Figures 7.8, 7.9). It is spherical with a single distinct nucleolus. Clumps of the condensed chromatin appear scattered throughout the nucleoplasm.

### Synergids

In the longitudinal section, each synergid appears to have a hook which extends laterally for a short distance along the nucellar cap-megagametophyte boundary (Figures 7.8, 7.9). But these hooks are actually part of an umbrella-like cap of the egg apparatus formed by the micropylar end of the synergids.

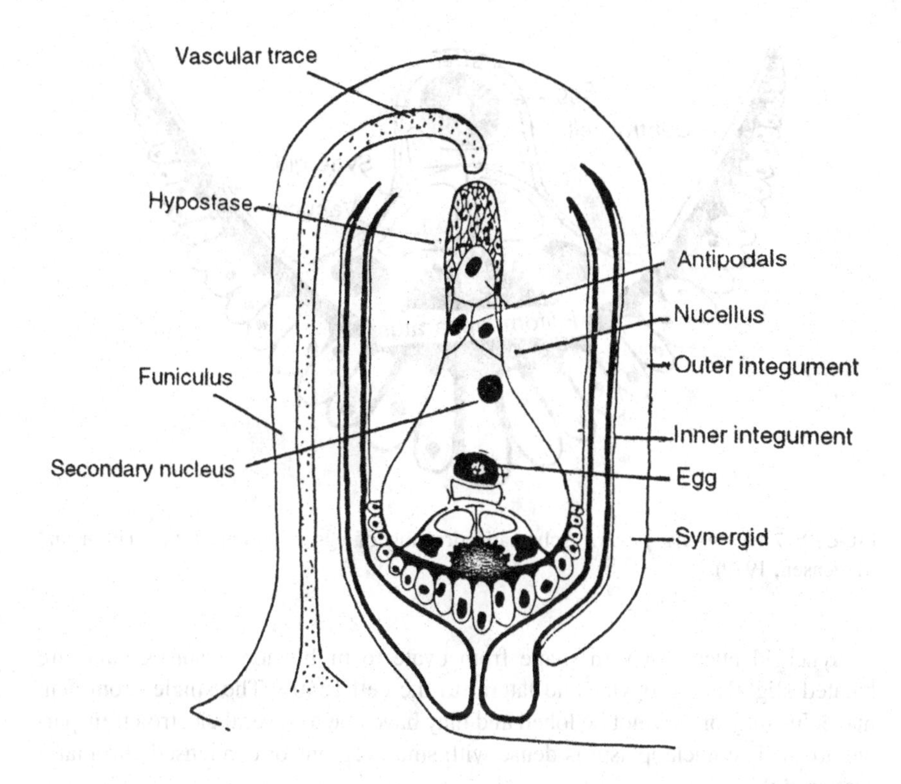


**FIGURE 7.9** *Agave parryi.* Schematic drawing showing salient features of ovule after (Tilton and Mogensen, 1979).

The micropylar end wall of the synergids is thick and fibrous. The cell wall between sister synergids thins as it approaches the egg, as do the lateral walls (Figure 7.10). The wall between egg and synergids is discontinuous in some areas, and thus in these places, the cells are separated only by their plasmalemma. Plasmodesmata occur in the wall between sister synergids and the central cell, and between synergids and egg, but there are no plasmodesmata between synergids and the nucellus. In the egg apparatus, a middle lamella appears to be present only between the two halves of the filiform apparatus.

After the mature embryo sac has expended and is almost mature, the filiform apparatus is formed by the proliferation of the micropylar end wall of each synergid. The filiform apparatus is hemispheric in its general outline, but a detailed view reveals slender projections extending into the synergid cytoplasm (Figure 7.10). The projections taper distally and may have a knobbed terminus. Narrow channels of cytoplasm are present in the filiform apparatus, and the plasmalemma is appressed closely against it.

Synergid cytoplasm is very dense and is rich in organelles which tend to be peripheral in distribution. Plastids and mitochondria are most numerous in the chalazal end of the cells, whereas dictyosomes and ER occur more commonly in the micropylar end.

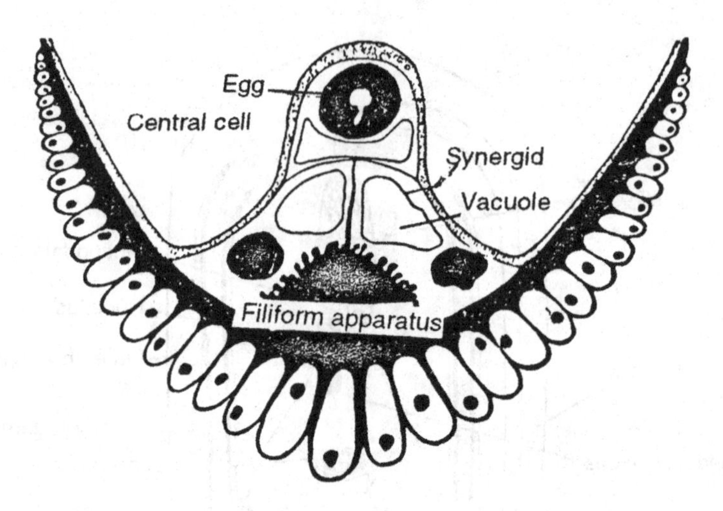


**FIGURE 7.10** *Agave parryi.* Schematic drawing of egg apparatus (after Tilton and Mogensen, 1979).

Synergid nuclei vary in shape from ovate to highly lobed spheres and are located slightly micropylar and lateral to the cell centre. The single prominent nucleolus may or may not be lobed and may have one to several electron transparent areas. The nucleoplasm is dense with small regions of condensed chromatin scattered throughout.

Ribosomes are quite numerous, but only a few are aggregated into the polysomes. The free ribosomes are evenly distributed and give the synergid cytoplasm its dense, homogeneously granular appearance. Plastids are more numerous in the synergids than in any other cells of the unfertilized ovule. Many small vesicles are associated with most of the dictyosomes, which are numerous. A large number of vesicles have contents which are relatively electron dense. The endoplasmic reticulum is fairly abundant.

## Central Cell

The central cell borders on the antipodals at its chalazal end, and at its micropylar end, it partially surrounds the egg apparatus. The central cell conforms laterally to the shape of the nucellar cells which bound it. The secondary nucleus remains close to the antipodals in the chalazal end of the cell (Figure 7.9). Although the central cell wall is very thin laterally, there are no plasmodesmata between the central cell and nucellus. Plasmodesmata do occur, however, between the central cell and egg apparatus, particularly the synergids. A thin wall is present towards the micropylar end of the egg with plasmodesmata that connect egg and the central cell cytoplasm. For the most part, however, the central cell and egg are separated by their plasmalemmas only (Figure 7.8, 7.9), but a few small disjointed deposits of cell wall material are scattered between the two cells.

Dictyosomes were not seen in the lateral cytoplasm but some are present near the egg apparatus. Many vesicles are associated with dictyosomes close to the egg, most of which are large and have transparent contents. The ER is scarce in the lateral cytoplasm. Plastids in the central cell may become fairly large, but only few of the largest may contain starch. Some small lipid bodies are found in the central cell also.

### ANTIPODALS

The antipodal cells degenerate soon after the gametophyte is fully mature. They contain abundant mitochondria, plastids and multicisternal dictyosomes. The cytoplasm is full of small vesicles derived from the ER or dictyosomes.

## NUTRITION OF THE EMBRYO SAC

The ovule gets its nutritional requirements through the vascular supply which transverses the funiculus and usually terminates in the chalaza. The morphology of the ovule suggests that the main pathway of nutrients into the embryo sac is the chalazal end. According to Mogensen (1973), in *Quercus gambelii*, the most likely pathway of food transport within the ovule is from the outer integument to the chalaza (facilitated due to occurrence of a group of persistent nucellar cells, called postament) and through the postament to the embryo sac. Nucellus is the obvious pathway for nutrients to enter the embryo sac because it surrounds the latter on all sides.

### EXTENSIONS OF THE EMBRYO SAC

The entire embryo sac may grow beyond the ovular tissue thus becoming haustorial in nature. In *Santalum*, the tip of the embryo sac becomes extraovular and reaches up to the base of the style. The chalazal end forms a long caecum (Figure 7.11).

In *Exocarpos menziesii*, there is no ovule, and the 4-nucleate embryo sac lies within the placental tissue. Several tubular structures arise from the apex of the gametophyte, which break through the placenta. In *Exocarpos strictus*, numerous fingerlike outgrowths arise from the middle of the embryo sac (Figure 7.12). According to Ram (1959), these outgrowths provide an additional absorptive surface.

The elongation of the embryo sac in *Macrosolen* and *Atkinsonia* (Loranthaceae) begins at the 4-nucleate stage, and the tip extends only up to the base of the style. In *Lepeostegeres*, the tip of the embryo sac reaches up to one-fifth the length of the style.

### SYNERGIDS

The upper part of the synergids is beaked or hooked. The prolongation of the beaks of synergids beyond the micropyle has been reported in *Mutisia candolleana* and

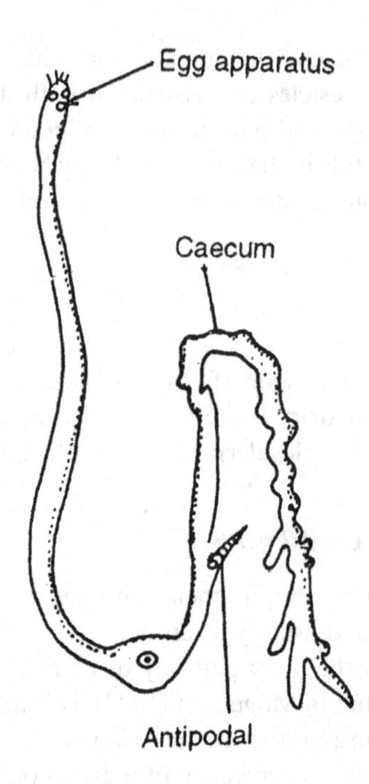


**FIGURE 7.11** A mature embryo sac of *Santalaum album*. The chalazal end of the embryo sac has grown into an extensive caecum leaving the antipodals *in situ* (after Paliwal, 1956).

*Ursinia anthemoides*. In *Quinchamalium chilense* (Santalaceae), after fertilization, the tips of the synergids elongate (Figure 7.13) and travel along the surface of the conical placenta finally entering into the basal part of the stylar tissue (Agarwal, 1962). Kallarackal and Bhatnagar (1980) have suggested that before fertilization synergids are involved in the nutrition of the egg.

In *Cortaderia jubata* (Philipson, 1978), the mature gametophyte shows a well-developed synergid haustorium (the other synergid collapses), which penetrates the micropyle and extends between the outer integument and ovary wall. Synergid haustoria are common in the family Asteraceae.

The filiform apparatus in the synergids performs the function of absorbing nutritive substances from the surrounding tissue. These substances pass from synergids to egg cell and central cell (Figure 7.7). In crassinucellate ovules, the FA absorbs nutrition from surrounding parietal tissue or long persistent nucellus. Such a function in the embryo sac before fertilization, in the tenuinucellate ovules with ephemeral nucellus, becomes possible only with the formation of an intervening cuticle at the site of contact of synergids and integument.

FA serves as a storehouse of reserve substances for supplying nutrition to the embryo at the very early stages of its development. The connection between the synergids and the central cell is important in providing nutrition to the embryo sac. According to Wang and Wang (1991), in *Vicia faba*, a connection is formed between the synergids and central cell.

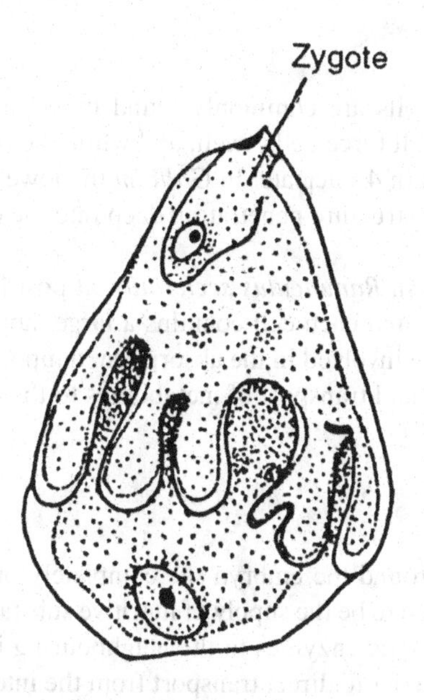


**FIGURE 7.12** Fertilized embryo sac of *Exocarpos strictus* (whole mount), showing a fringe of fingerlike processes (arrow-marked) (after Ram, 1959).

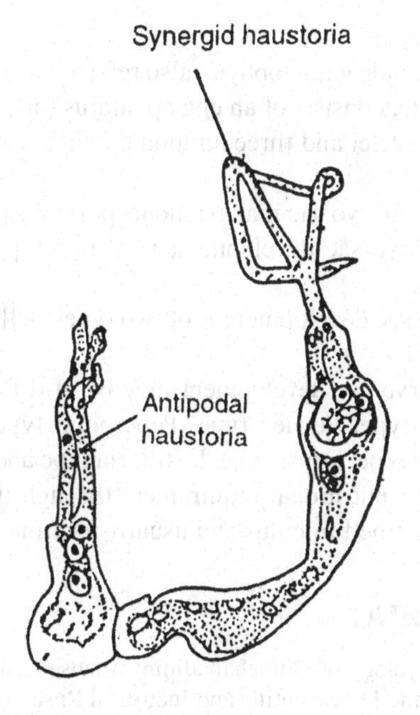


**FIGURE 7.13** Mature embryo sac of *Quinchamalium chilense* showing synergid and antipodal haustoria (after Johri and Agarwal, 1965).

### Antipodals

Haustorial antipodal cells are commonly found in the members of the family Rubiaceae. In *Phyllis*, all three cells swell, and while the basal becomes 8-nucleate, the other two remain 4-nucleate. In *Gallium* the lowest antipodal cell develops into a long tubular structure penetrating deep into the chalazal portion of the ovule.

The antipodal cells in *Ranunculus sceleratus*, at post fertilization stages, are considerably enlarged, and each cell contains a large, apparently endopolyploid nucleus. These cells are involved in the absorption, temporary stage, biochemical conversions, synthesis and transport of metabolites to the developing endosperm and embryo.

### Endothelium

Endothelium may surround the embryo sac completely or partially. Its possible functions are considered to be the supply of nutritive substances to the embryo sac and secretion of hydrolytic enzymes to the neighbouring integumentary cells. It is also considered a barrier to direct transport from the integument to the embryo sac or endosperm. In Scrophulariaceae and Orobancaceae, it functions as a metabolic regulator between integument and endosperm (Shamrov, 1991).

## SUMMARY

- The organized female gametophyte, also referred to as the embryo sac or megagametophyte, consists of an egg apparatus (one egg and two synergids), two polar nuclei and three antipodals. Filiform apparatus is present in synergids.
- Development of embryo sac may be monosporic, bisporic or tetrasporic.
- Monosporic embryo sac development is of two types: Polygonum and Oenothera types.
- Bisporic embryo sac development is of two types: Allium and Endymion types.
- Tetrasporic embryo sac development may be of different types: Adoxa type, Plumbago type, Penaea type, Peperomia type, *Chrysanthemum cinerariaefolium* type, Drusa type, Fritillaria type and Plumbagella type.
- The ovule gets its nutritional requirement through the vascular supply, which transverses the funiculus and usually terminates in the chalaza.

## SUGGESTED READING

Agarwal, S. 1962. Embryology of Quinehamalium chilense Lam. In *Plant Embryology: A Symposium*. Council of Scientific and Industrial Research, New Dehli, 162–169.

Cass, D.D. 1972. Occurrence and development of a filiform apparatus in the egg of *Plumbago capensis*. *Am. J. Bot.* 59: 279–283.

Chitralekha, P. and N.N. Bhandari. 1991. Post-fertilization development of antipodal cells in *Ranunculus sceleratus. Phytomorphology* 41: 200–212.

Cocucci, A.E. and W.A. Jensen. 1969. Orchid embryology: The mature megagametophyte of *Epidendrum scutella. Kurtziana* 5: 23–38.

Diboll, A.G. and D.A. Larson. 1966. An electron microscopic study of the mature megagametophyte in *Zea mays. Am. J. Bot.* 53: 391–402.

Jensen, W.A. 1965a. The ultrastructure and histochemistry of the synergids of cotton. *Am. J. Bot.* 52: 238–256.

Jensen, W.A. 1965b. The ultrastructure and composition of the egg and central cell of cotton. *Am. J. Bot.* 52: 781–797.

Johri, B.M. and S. Agarwal. 1965. Morphological and embryological studies in the family Santalaceae 8. *Quinchamalium chilense* Lam. *Phytomorphology* 15: 360–372.

Kallarackal, J. and S.P. Bhatnagar. 1980. Cytochemical studies on the developing female gametophyte of *Linaria bipartita. Acta Botanica Indica* 8 (1): 11–22.

Kapil, R.N. and A.K. Bhatnagar. 1981. Ultrastructure and biology of female gametophyte in flowering plants. *Int. Rev. Cytol.* 53: 291–341.

Linskens, H.F. 1969. Fertilization mechanisms in higher plants. In *Fertilization: comparative morphology, biochemistry, and immunology*, eds. C.B. Metz and A. Monroy, Vol. II, 189–253. New York: Academic Press.

Maheshwari, P. 1950. *An Introduction to the Embryology of Angiosperms.* New York: MacGraw Hill.

Malik, C.P. and S. Vermani. 1975. Physiology of sexual reproduction I. Ahistochemical study of the embryo sac development in *Zephyranthes rosea* and *Lagenaria vulgaris. Acta Histochemica* 53 (2): 244–280.

Martinoli, G. 1939. Contributo all'embriologia delle Asteraceae. I-III. *Nuovo Gior. Bot. Ital. N.S.* 46: 259–298.

Mogensen, H.L. 1973. Some histochemical, ultrastructural and nutritional aspects of ovule of Quercus gambelii. *Am. J. Bot.* 60: 48–54.

Mogensen, H.L. 1992. The male germ unit: Concept, composition and significance. *Int. Rev. Cytol.* 140: 129–147.

Newcomb, W. 1973. The development of the embryo sac of sunflower *Helianthus annuus* before fertilization. *Can. J. Bot.* 51: 863–878.

Paliwal, R.L. 1956. Morphological and embryological studies in some Santalaceae. *Agra Univ. J. Res. Sci.* 5: 191–284.

Philipson, M.N. 1977. Haustorial synergids in *Cortaderia* (Gramineae). *New Zealand J. Bot.* 15 (4): 777–778. https://doi.org/10.1080/0028825X.1977.10429647.

Pritchard, H.A. 1964. A cytochemical study of embryo sac development in *Stellaria media. Am. J. Bot.* 51: 371–378.

Ram, M. 1959. Morphological and embryological studies in the family Santalaceae 2. *Exocarpus* with a discussion on its systematic position. *Phytomorphology* 9: 4–19.

Schulz, S.R. and W.A. Jensen. 1968. Capsella embryogenesis: The synergids before and after fertilization. *Am. J. Bot.* 55: 541–552.

Shamrov, I.I. 1991. The ovule of *Swertia iberica* (Gentianaceae): structural and functional aspects. *Phytomorphology* 41 (3–4): 213–229.

Strasburger, E. 1879. *Die Angiospermen und die gymnospermen.* Jena.

Tilton, V.R. and H.L. Mogensen. 1979. Ultrastructural aspects of the ovule of *Agave parryi* before fertilization. *Phytomorphology* 29: 338–350.

Vazart, B. 1969. Structure et evolution de la cellule generatrice du Lin (*Linum usitatissimum* L.) au cours de premier stades de la maturation du pollen. *Rev. Cytol. Biol. Veg.* 12: 101–114.

Vijayaraghavan, M.R., W.A. Jensen and M.E. Ashton. 1972. Synergids of *Aquilegia formosa*: Their histochemistry and ultrastructure. *Phytomorphology* 22 (2): 144–159.

Willemse, M.T.M. and J.L. Van Went. 1984. The female gametophyte. In *Embryology of Angiosperms*, ed. Johri, B.M., 159–196. Berlin: Springer-Verlag.

Wilms, H.J. 1981. Ultrastructure of the developing embryo sac of spinach. *Acta Bot. Neerl.* 30: 75–99.

Yadegari, R. and G.N. Drews. 2004. Female gametophyte development. *Plant Cell* 16 Supplement: 133–141.

Yoshida, O. 1962. Embryologische studien u¨ber *Schisandra chinensis* Bailey. *J. Coll. Arts Sci. Chiba Univ.* 3: 459–462.

Zhou, L.Z. and T. Dresselhaus. 2019. Friend or foe: Signaling mechanisms during double fertilization in flowering seed plants. *Curr. Top Dev. Biol.* 131: 453–496. https://doi.org/10.1016/bs.ctdb.2018.11.013.

# 8 Pollination

Flowers have to be pollinated before their ovules can be fertilized and seeds produced. The term pollination refers to the transfer and deposition of pollen on the stigmatic surface of the flower. Dehiscence of the anther is a prerequisite for pollination.

Pollination biology provides insight into the relationship between plants and their pollinators, foraging strategies of pollinators, community structure, reproductive strategy and co-adaptation of plants and pollinators. An understanding of pollination and breeding system is highly essential for the success of any conservation and restoration programme.

## ANTHER DEHISCENCE

Dehiscence of anther involves the splitting open of the pollen sacs at maturity (Figure 8.1). The pollen grains are exposed, which are carried by various agencies and later effect pollination.

Dehiscence of an anther is described as introrse if it is towards the centre of the flower (e.g., Fabaceae, Bromeliaceae) and extrorse if it is away from the centre (e.g., Papaveraceae). In some families (e.g., Butomaceae) the anther opens at the side, and this is known as lateral dehiscence (latrorse).

The general mode of the dehiscence of an anther is by longitudinal splitting (Figure 8.2A). In this case, the opening occurs between each pair of pollen sacs, and pollen from both sacs of an anther lobe is released from the same split. In poricidal dehiscence, the pollen grains are shed through terminal pores (Figure 8.2B). This pattern of anther dehiscence is common in members of the family Ericaceae. When anthers open like valves, this type of dehiscence is known as valvate. In Lauraceae, a flap-like valve opens upwards to liberate the pollen (Figure 8.2C).

At maturity, endothecial cells develop fibrous thickenings all around the anther locule except along the line of dehiscence. This region, which is characterized by the presence of thin-walled cells, is called stomium. The hygroscopic nature and differential expansion of the tangential walls of the endothecium are chiefly responsible for anther dehiscence. In some taxa (e.g., *Cassia*) fibrous thickenings are absent in endothecium. In such cases, the anther dehisces due to unequal thickening of outer, inner and radial walls. The differential shrinkage of these walls develops a mechanical pressure causing dehiscence of the anther.

DOI: 10.1201/9781003260097-8

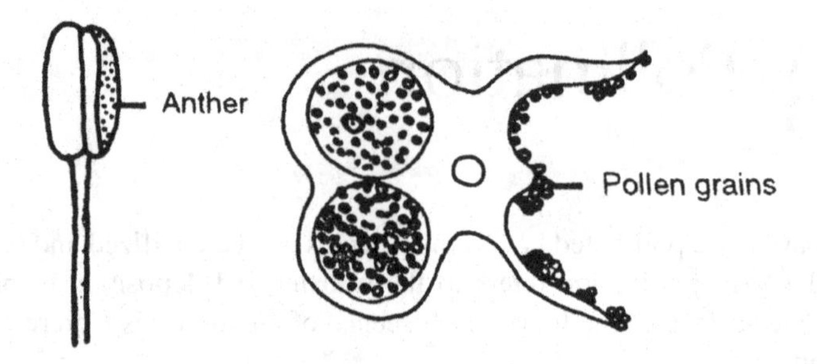


**FIGURE 8.1** Diagram showing how the anthers split and expose the pollen.

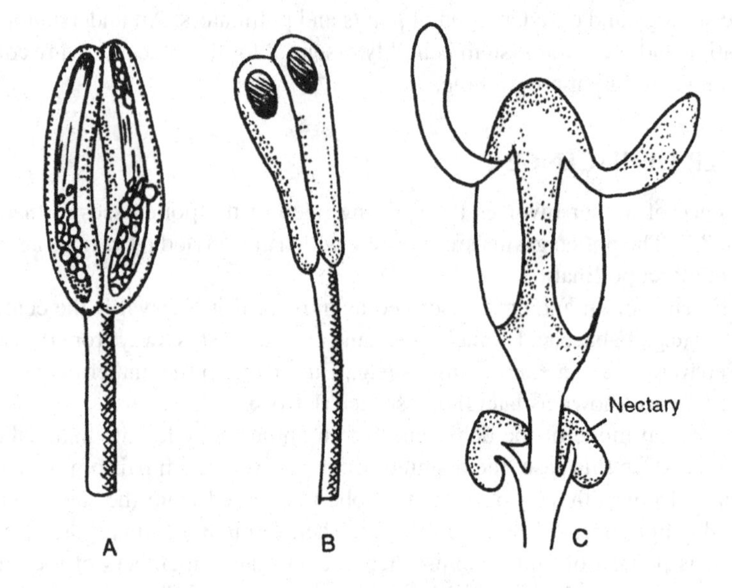


**FIGURE 8.2** Types of anther dehiscence. A. Longitudinal dehiscence; B. Poricidal dehiscence; C. Anther of *Laurus nobilis* showing dehiscence by lateral valves.

## TYPES OF POLLINATION

Pollination is categorized into two broad groups: self-pollination (autogamy) and cross-pollination (allogamy). The different types of effective pollen transfer are shown in Figure 8.3. Asexually reproducing clones are called ramets, whereas genetically different individuals are referred to as genets.

## SELF-POLLINATION

Transfer of pollen grains from the anther of a flower to the stigma of the same flower is termed self-pollination. No external agencies are involved in achieving autogamy. But the flower structure itself is provided with a mechanism which

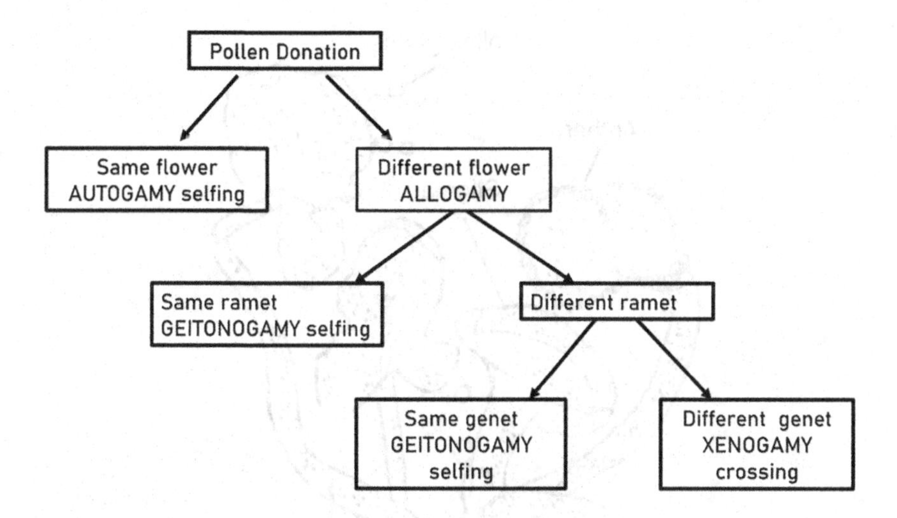


**FIGURE 8.3** Modes of transfer of pollen within and between flowers and plants.

encourages or enforces self-pollination. Autogamous flowers are always bisexual. Autogamy occurs naturally in almost all the legumes (e.g., pea, beans) and many cereals (e.g., wheat, rice, maize, etc.).

Most of the flowering plants bear chasmogamous flowers (the flowers which open normally), but in some plants cleistogamous flowers develop (e.g., *Viola*, *Oxalis*). Pollination and fertilization in an unopened flower bud are called cleistogamy. In cleistogamous flowers, the stigma and the anthers mature at the same time. This condition is known as homogamy. Cleistogamy is a remarkable way of avoiding cross-pollination.

In *Viola odorata*, the pollen grains germinate while they are still present inside the undehisced anthers. The pollen tubes reach up to the stigma by penetrating the anther walls.

Most of the plants which bear cleistogamous flowers also bear chasmogamous flowers. In *Commelina benghalensis*, and some species of *Cardamine*, cleistogamous flowers produced on subterranean or underground rhizomes are small and dull; and chasmogamous flowers borne on aerial branches are usually brightly coloured and attractive.

*Boerhavia diffusa* is usually pollinated by insects, but in the rainy season when insects do not visit flowers, the flowers are pollinated by their own pollen grains due to coiling of style and stigmas (Figure 8.4). Autogamy is also reported in *Tylophora* (Kunze, 1991).

In *Volulopsis nummularium* (Convolvulaceae), the flowers open partly under lower temperature on rainy days, providing an opportunity for the terrestrial snails (*Lamellaxis gracile*) to pollinate rather than bees (*Apis cerana indica*), which are more effective on bright sunny days (Sarma et al., 2007).

Continuous inbreeding results in progressive loss of genetic variability until each individual has become homozygous for all or almost all the genes

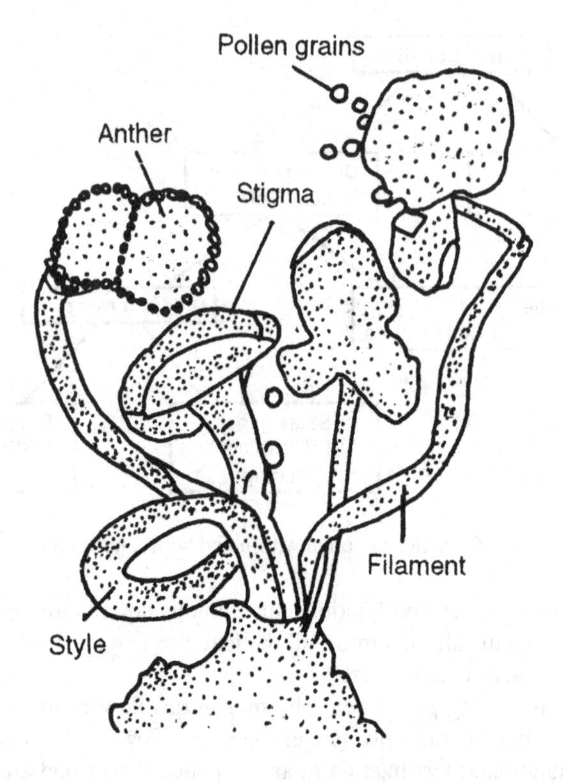


**FIGURE 8.4** *Boerhavia diffusa*. Diagram showing coiling of style and a dehisced anther, calyx and corolla removed (after Chaturvedi, 1989).

(characters). In other words, it leads to the fixation of alleles and, hence, the fixation of genotypes.

## CROSS-POLLINATION

Cross-pollination is the transfer of pollen grains from one flower to another born on the same or different plant of the same or allied species regardless of whether the flower is bisexual or unisexual. When pollination takes place between two flowers on the same plant, the process is known as geitonogamy. When pollination occurs between two flowers on different plants, the process is known as xenogamy.

Cross-pollination is accomplished chiefly by external agencies since floral structure and physiological factors in this group of plants pose a great barrier to the occurrence of self-pollination.

Frequent cross-pollination tends to create and maintain a tremendous amount of genetic variability chiefly because of the high degree of heterozygosity in populations. The occurrence of heterozygosity due to allogamy prevents genetic stagnation.

## AGENCIES OF CROSS-POLLINATION

Cross-pollination requires an abiotic or biotic agent (Table 8.1). Pollination by abiotic agents is not very efficient, as pollen movement is not directed to the stigma. Pollen grains are released into the air or water passively; if they happen to come in contact with the stigma during their movement, then pollination is brought about. Thus there is lot of wastage of pollen grains. In order to compensate such wastage, wind and water-pollinated plants produce enormous amounts of pollen for which the plant had to spend much more resources when compared to animal-pollinated plants.

### ANEMOPHILY

Flowers which are wind-pollinated include grasses, date palm, maize, *Cannabis*, etc. Wind-pollinated flowers have the following characteristics, most of which help to increase the chances of pollen reaching the stigmas:

1. They are small, inconspicuous and produce no nectar.
2. They produce large amounts of pollen grains that are so small and light that they can be carried for considerable distances on air currents.
3. They occur in a position where they are easily blown.
4. They have stamens with large anthers born on long filaments which hang or project well outside the flower. The anthers shake out their pollen when disturbed by the slightest breeze.
5. The stigmas are large and feathery, which have a greater chance of receiving the blown pollen.

During the transfer of pollen through air currents, a considerable amount of pollen is lost because it never reaches a proper stigma. As there is much wastage, pollen is produced in enormous quantities. For example, a single flower of *Cannabis* produces 5,000,000 pollen grains.

In Urticalean rosides (e.g., *Morus nigra*, *Urticaria dioica*) an explosive mechanism of pollen dispersal involving sepals and inflexed stamens has been observed.

**TABLE 8.1**
**Agencies and Types of Cross-Pollination**

| Agencies | | Types of Pollination |
|---|---|---|
| Abiotic agencies | Wind | Anemophily |
| | Water | Hydrophily |
| Biotic agencies | Insects | Entomophily |
| | Birds | Ornithophily |
| | Bats | Chiropterophily |
| | Ants | Myrmecophily |
| | Snails and slugs | Malacophily |

The anthers and filaments dehydrate during the day when the temperature is higher and relative humidity is lower, causing the filaments to distend and anthers to open, which in a rapid movement explosively releases large amounts of pollen. These grains are then transported by the wind currents (Pedersoli et al., 2019).

## Hydrophily

The transfer of pollen grains from anther to stigma through the agency of water is called hydrophily. Like anemophilous flowers, hydrophilous flowers are also small and inconspicuous. In aquatic habitats, plants may use alternative environmental agents to pollinate their flowers. The peculiar structure of pollen grains, with well-developed intine, strongly reduced exine and practically no sporopollenin, increases their floating ability and enhances pollination success. The grass-wrack pondweed (*Zostera*), for example, has long threadlike pollen grains (see Chapter 4, Figure 4.12), which have a density very similar to that of surrounding seawater. Consequently, they are dispersed by water currents and can be trapped by submerged feathery stigmas. Hydrophily is further categorized into two groups: Ephydrophily and hyphydrophily.

**Ephydrophily**: In this category, pollination takes place on the surface of the water. *Vallisnaria spiralis*, a dioecious plant, offers a common example of ephydrophily. The male flowers are minute and are born in groups. The male flowers are released to the surface of the water, where they attach themselves to the stigma of floating female flowers held afloat by long stalks. After pollination, the female flower is withdrawn inside the water by coiling of the stalk. The fruits mature in a submerged condition. A similar mechanism is also reported in *Hydrilla* and *Lemna trisulca*.

In *Willisia arekaliana*, flowering starts under submerged conditions. Pollination occurs at the air–water interface soon after the flower is partly exposed to air by a small opening in the spathella. Concurrence of stigma receptivity with anther dehiscence and apposition of the two organs facilitates autopollination (Khanduri et al., 2014).

**Hyphydrophily**: In this category, pollination takes place under the surface of water, (e.g., *Ceratophyllum*, *Najas*, *Zostera*, etc.). In *Ceratophyllum*, a monoecious plant, the male flowers bear a large number of spirally arranged stamens, each with an apical float. At maturity, stamens abscise and rise to the surface of water by the buoyancy of the floats. A few pollen grains germinate *in situ* within the anthers of the intact stamen. Following abscission of the stamens, more pollen grains germinate *in situ*. The stamens dehisce after floating on the surface of the medium for a day and liberate both germinated and ungerminated pollen grains. The pollen grains gradually sink and come in contact with the stigma of submerged female flowers.

## Entomophily

Insect pollination is one of the most reliable methods of transferring pollen. Bees, beetles, flies, butterflies, wasps, moths and thrips are the major groups involved

**TABLE 8.2**
**Forms of Entomophily and Pollinating Agents**

| Forms of Entomophily | Pollinating Agent |
|---|---|
| Canthatrophily | Beetles |
| Melittophily | Bees, bumblebees, wasps |
| Psychophily | Butterflies |
| Phalaenophily | Moths |
| Myophily | Flies, mosquitoes, flower flies |
| Myrmecophily | Ants |

in the process of pollination (Table 8.2). Moths perform this function during the night (nocturnal) while all other insects are day-pollinators (diurnal). Flowers pollinated by insects (entomophilous flowers) have certain characteristics which make the process effective:

1. Individual flowers are large and brightly coloured, while small flowers are usually found together in a group, or inflorescence, which presents a broad expanse of colour.
2. The flowers are usually scented and provide pollen and nectar as food for insects.
3. The surface of the stigma is coated with a sticky secretion which taps pollen grains, and the flower parts are arranged so as to make contact with the bodies of visiting insects.
4. The pollen grains are large and thick-walled with projecting spikes. This tends to make the pollen clump together and cling better to the hairy legs and bodies of insects.

The pollination mechanism in some entomophilous flowers is described below.

In *Cucurbita pepo* (Cucurbitaceae) pollination is almost exclusively performed by bees (*Apis mellifera*). Bee visits soon after the flowers open. Female flowers are visited more than male flowers. The bees follow a different path in the two types of flower because access to the nectar is different (Figure 8.5). In male flowers the insect lands near the edge of the corolla and goes towards one of the three nectary pores, through which it sucks the nectar. While in a vertical position, it returns to its landing position and then flies on to the next flower. During this operation, the hairs on the bee's back brush the anthers, and pollen adheres to them.

In the female flower, the insect usually lands on the edge of the corolla, sometimes even on the tip of the stigma. From one of these two positions, it heads for the base of the corolla where it moves around the annulus, collecting nectar. Pollen is mostly left on the stigma during descent into the flower, but may also be deposited during nectar collection.

In the dandelion (*Taraxacum*), individual florets are small, but the whole inflorescence is very conspicuous, highly scented and produces much nectar. The

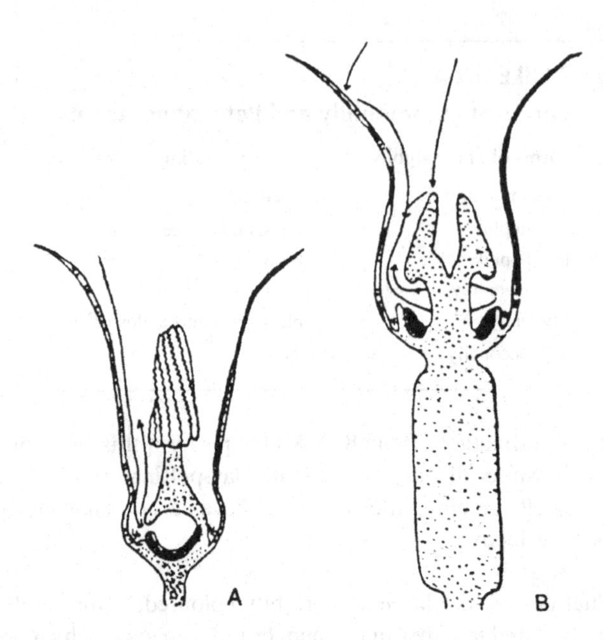


**FIGURE 8.5** *Cucurbita pepo.* Male and female flowers showing path of insect visit (after Nepi and Pacini, 1993).

florets are pollinated mainly by bees and butterflies. The pollination mechanism is adapted for cross-pollination, but self-pollination can occur if cross-pollination fails.

In young florets, the stamens shed their pollen into the stamen sheath, and as the style grows, it acts like a piston and forces the pollen out of the top. At this stage the two stigmas are pressed closely together and cannot be self-pollinated. Later the stigmas expand and can receive pollen from another floret if an insect comes in contact with them. If this is not effected, then, later still, the stigmas curl around and touch the hairy style to which the floret's own pollen adheres and self-pollination may occur (Figure 8.6).

The genus *Passiflora* exhibits numerous flower shapes, ranging from small to large, open, dish-like forms to campanulate types to long cylindrical tubes. Much of this diversity is the result of different degrees of expansion of the hypanthium. In *Passiflora caerulea* (Figure 8.7) the corona has the appearance of differently coloured concentric rings which act as nectar guides to bees and lead them to the nectar secreted by a ring within the operculum. In the first stage of anthesis, the insect probing for nectar receives pollen on its back from the downward curving anthers. Later the styles bend down even further than the anthers and the now-receptive stigmas are able to collect pollen from other insect visitors.

Insect pollination provides many examples of co-adaptation in which different species evolve to become dependent on each other for their continued survival. Bees benefit by obtaining food, while the flowers benefit by the transport of pollen from one flower to another.

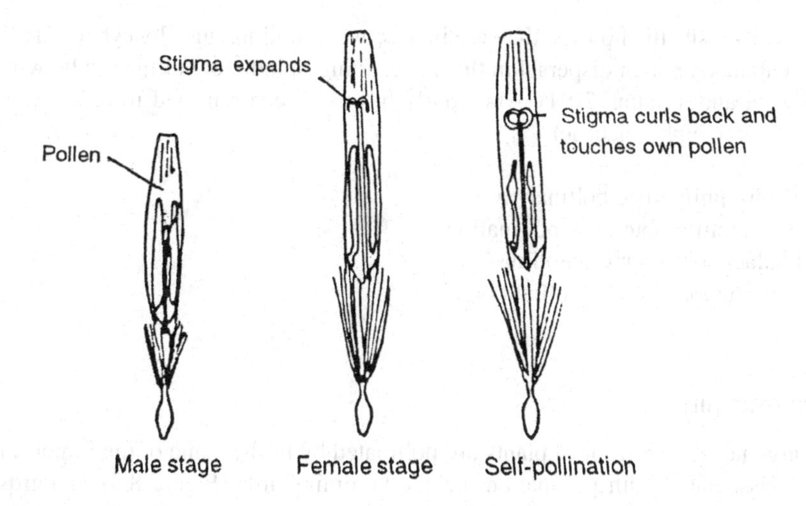


**FIGURE 8.6** Pollination mechanism in *Taraxacum*.

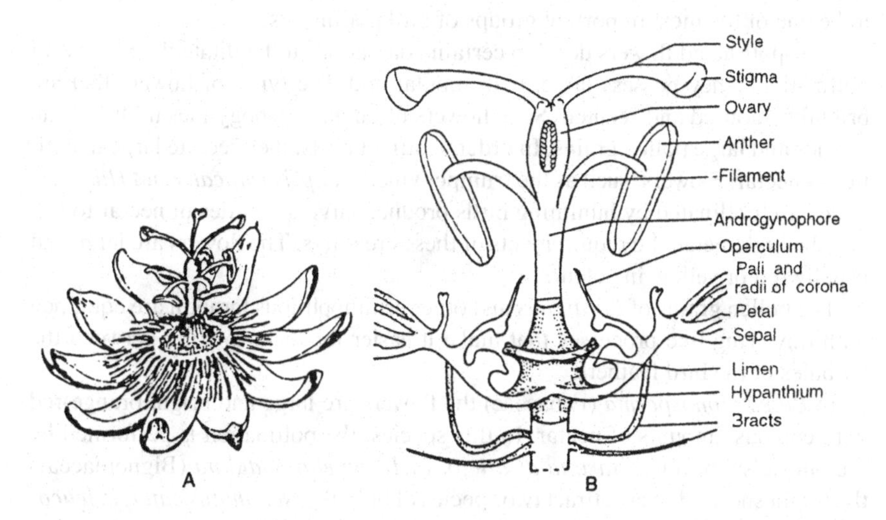


**FIGURE 8.7** *Passiflora caerulea*. A. Anterior view of flower; B. Longitudinal section of flower.

In *Acacia senegal*, flowers emit a mild fragrance, produce a minute quantity of nectar and pollination is effected by the Asian honeybee (*Apis dorsata*) (Tandon et al., 2001).

*Ophrys apifera*, an orchid, achieves pollination by sexual impersonation, producing flowers which resemble the female mating partners of bees or wasps in size, colour and, sometimes, in odour. As the male insect attempts to copulate with the flower, pollen is transferred.

*Crateva adansonii* (Capparaceae) exhibits both anemophily and entomophily indicating a mixed system of pollination known as ambophily. Although a variety of insects visits the flowers, they are ineffective in pollinating. Honeybees facilitate enhanced pollen dispersal in the air resulting in indirect pollination by wind (Mangla and Tandon, 2011). Ambophily has also been reported in *Excoecaria agallocha* (Euphorbiaceae).

Melittophily = bee pollination
Psychophily = butterfly pollination
Phalaenophily = moth pollination
Sapromyiophily = fly pollination

## Ornithophily

A large number of tropical plants are pollinated by birds. Some of the important birds associated with pollination include hummingbirds (Figure 8.8), sunbirds, leaf birds and flower peckers. Among Indian birds, sunbirds (members of the family Nectariniidae) frequent up to 58 species of flowering plants and continue to be one of the most important groups of bird pollinators.

Bird-pollinated flowers develop certain adaptations to facilitate the process of pollination. They possess large, both tubular and disc types of flowers that are brightly coloured and scented. Such flowers often have hypogynous multiovulate ovaries and large pollen grains. In order to attract birds, they secrete large quantities of nectar. Flowers, such as the trumpet vine, *Campsis radicans* and *Hibiscus*, which are pollinated by humming birds produce large quantities of nectar to satisfy the high metabolic requirements of these creatures. The flowers are large and usually red or yellow in colour.

The pollen grains of *Loranthus* and other ornithophilous loranthi are equipped with tiny wing-like processes that make it easier for them to cling between the barbules of the bird feathers.

In *Butea monosperma* (Fabaceae) the flowers are large and bright orange-red with copious amounts of nectar. In this species, the pollination is performed by the purple sunbird (*Nectarinia asiatica*). In *Tecomella undulata* (Bignoniaceae) the bright showy flowers attract two species of bulbul (*Pycnonotus cafer*, *P. leucotis*) and one sunbird (*Nectarinia asiatica*) to bring about pollination in the species.

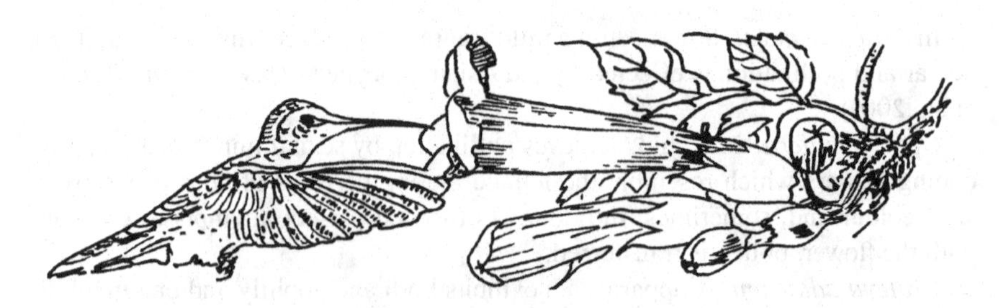

**FIGURE 8.8** Hummingbird hovering near a trumpet vine (*Campsis radicans*) flower.

The body size and beak characteristics of nectarivorous birds vary considerably. Among all the nectar feeder families, Dicaeidae (e.g., *Dicaeum* species) and Nectariniidae (e.g., *Nectarinia, Arachnothera*) show higher ratios of beak length to body size.

The members of families Malvaceae, Fabaceae, Myrtaceae, Bignoniaceae and Verbenaceae are the most ornithophilous plants of India. Certain plant species namely *Bombax ceiba, B. insigne, Erythrina variegata* and *E. stricta* are visited by nearly 58 different bird species for nectar (Subramanya and Radhamani, 1993).

### CHEIROPTEROPHILY

Bats are important for pollination of some trees in the tropics. Bat-pollinated (cheiropterophilous) flowers are characterized by a wide-throated corolla, foetid odour and production of copious amounts of pollen and nectar. Such flowers bloom in the nighttime. In *Oroxylum indicum* (Bignoniaceae), a bat-pollinated tree, the flowers are short-lived and open acropetally in the long protruding and exposed inflorescence. Bats are most suitable for pollination success as their trapline foraging behaviour helps in ensuring outcrossing (xenogamy) (Vikas et al., 2009).

In some species, the flowers are carried on ropelike branches which dangle down from the canopy and open at night to emit a strong fruity odour. The pollen produced contains much more protein than that of most other types of flower. According to Walker (1975), bats that frequent flowers feed mainly on pollen and nectar, and such species usually possess long pointed heads and long tongues with brush-like tips to aid in food gathering.

The mechanism of pollen transfer in bats is similar to that observed in birds. While lapping nectar, pollen grains adhere to the faces of birds and get transferred to other flowers. To aid the visitation of bats, the flowers open in the evening or night. For example, in *Bombax ceiba*, flowers open from 17:00 to 19:00 hr when nectar production is at its peak. McCann (1938) observed that the flowering seasons of bat-pollinated trees coincide with the breeding season of bats, a time when the requirement for food is greater.

Some of the plant species which are pollinated by bats are: *Adansonia digitata, Bombax ceiba, Ceiba, Kigelia pinnata, Anacardium occidentale* and *Careya arborea.*

**Pollination syndrome:** Suites of convergent floral traits among unrelated taxa adapted to attract a particular functional group of pollinators. The biotic and abiotic pollination syndromes include wind, water pollination, and animal pollination.

## BREEDING SYSTEMS

Breeding system refers to whether pollen from a plant can fertilize an egg from the same plant (self-compatible), or the pollen and egg donors must come from different plants (self-incompatible).

Plants can be predominantly outbreeding, inbreeding, or some mixture of the two. Outbreeding is the transfer of gametes from one individual to another individual which is genetically different. Outbreeding is also called outcrossing, allogamy, or xenogamy. Inbreeding, also called selfing, autogamy, or geitonogamy, is the union of gametes derived from a single individual. Some plant species have both outcrossing and selfing flowers, a breeding system known as alloautogamy (e.g., species of *Viola* and *Clarkia*). Inbreeders normally produce fewer pollen grains than outbreeders.

The family Apiaceae exhibits a variety of sex expression including hermaphroditism, andromonoecy, gynodioecy and dioecy. For example, fennel (*Foeniculum vulgare*) is categorized as a hermaphrodite type that exhibits protandrous, dichogamy which promotes cross-pollination (Koul et al., 1986).

In *Griffithella hookeriana* (Podostemaceae) the pollination is typically cleistogamous, and the species is an obligate inbreeder (Khosla et al., 2001). *Commelina benghalensis*, an andromonoecious weed, bears both staminate and bisexual flowers. The bisexual flowers bear two types of flowers: chasmogamous and cleistogamous (Kaul and Koul, 2009). *Valeriana wallichii* displays a mixed mating system and shows three types of sex expression: hermaphrodite, female and gynomonoecious (Verma et al., 2019). In *Oxystelma esculentum* (Asclepiaoideae, Apocynaceae) the breeding behaviours include autogamy, geitonogamy and xenogamy (Pal and Mondal, 2019). *Vitex negundo* exhibits a mixed mating system, and pollination is insect-dependent. Floral structure and placement of reproductive apparatus reveal that *V. negundo* exhibits hercogamy and protogyny (Khan et al., 2021).

*Hyptis suaveolens*, a weed, appears as an outbreeder on the basis of its floral features, but details of the floral phenology confirm it to be an inbreeder. This shift of breeding system from outcrossing to inbreeding helps the species for the successful invasion and colonization of new areas (Sharma and Sharma, 2019).

Both abiotic and biotic agencies are responsible for the transfer of pollen grains, but breeding systems control whether the pollen grain may germinate on a receptive stigma, penetrate the style and bring about fertilization. DNA data and phylogenetic analyses have allowed character mapping to reveal more clearly the evolution of breeding systems in plants.

> Mating system refers to how genes are transmitted from one generation to the next through sexual reproduction.

## ADAPTATIONS FOR CROSS-POLLINATION

In general, cross-pollination is better than self-pollination, as it provides greater variation within the species. Therefore, the majority of flowers have some device which makes self-pollination less likely or impossible.

Plants are equipped with devices and mechanisms to promote outcrossing. These devices (barriers to selfing) can be morphological, physiological or genetic in nature.

## Self-sterility

It is a condition where pollen grains of a flower do not germinate (e.g., *Malva*) or grow very slowly if they land on the stigma of the same flower (e.g., *Petunia*). Such growth inhibition is genetically governed.

## Dichogamy

The maturation of stigma and anthers at different intervals in bisexual flowers is known as dichogamy. In some dichogamous flowers, such as *Epilobium*, *Impatiens*, *Salvia* and *Saxifraga*, the anthers mature before the stigma becomes receptive. This condition is called protandry. On the other hand, in some flowers the stigma loses receptivity by the time anthers dehisce (e.g., *Magnolia*, *Aristolochia*, *Scrophularia*). This condition is known as protogyny. Both protandry and protogyny promote outcrossing when flowers of different individuals mature at slightly different rates. The pollen from one flower will normally not pollinate the same flower, but can pollinate a different flower in which the stigma is receptive.

## Hercogamy

Outcrossing has also been promoted by evolutionary changes in floral structure, particularly the spatial separation of anthers and stigmas, a phenomenon known as hercogamy (also spelled "herkogamy"). In some bisexual flowers, the structure of anthers and styles is such that self-pollination becomes impossible. In members of family Caryophyllaceae, the style is much longer compared to the stamens. This prevents pollen grains from falling on the stigma. In *Gloriosa*, the anthers dehisce at a distance from the stigma thus avoiding self-pollination.

**Enantiostyly:** The curvature of style to either the left or the right is known as enantiostyly. This style curvature usually corresponds with a curvature of at least one stamen to the side opposite the style. Enantiostyly results in the preferential deposition of pollen on one side of the body of the insect pollinators.

## Heterostyly

Heterostyly is a mechanism against self-pollination, and is characterized by the presence of different style lengths in different individuals of the same species. The phenomenon is reported in many families (e.g., Primulaceae, Boraginaceae, Lamiaceae, Plumbaginaceae, Solanaceae, etc.).

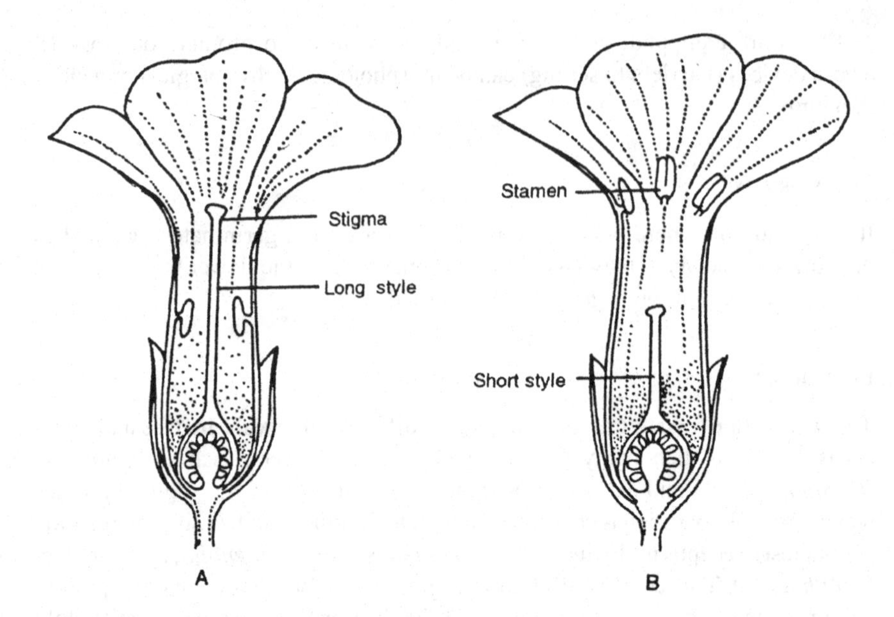


**FIGURE 8.9** *Primula vulgaris* showing heterostyly. A. long-styled flower; B. short-styled flower.

Heterostylous plants have flowers of two types viz., long styled (LS) and short styled (SS). Besides variation in their style length, the LS and SS flowers also differ in their size, stamen length, size of pollen grains, ornamentation of the pollen grains and the size of the stigmatic papillae.

The classic example of heterostyly is *Primula vulgaris* (Primrose). It has two types of flowers: (i) long styled (pin-eyed) and (ii) short-styled (thrum-eyed). In the former, flowers have long style and short stamens, while in the latter, they possess short style and long stamens (Figure 8.9). The long and short-styled flowers develop on different plants and show a specialized type of sexual segregation. Pollination from another individual of the same kind is relatively ineffective and results in poor fruits and few seeds. Such a less fertile pollination is called illegitimate. On the other hand, pollination of a long-styled and short styled flower or vice versa gives more than twice the number of good fruits and seeds. Such a fully fertile pollination is known as legitimate (Figure 8.10).

**Hermaphrodite:** A hermaphroditic form is one in which both pistils and stamens are borne in the same flower (perfect flower), and in which all the flowers on a plant show the same arrangement of parts. Example: *Solanum, Lilium.*

**Monoecious:** A monoecious form is one in which the pistillate and staminate inflorescences are borne separately on the same plant. Example: *Begonia, Zea mays.*

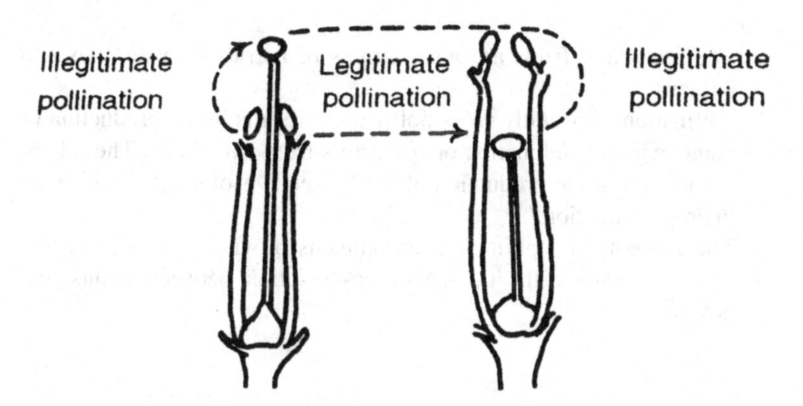


**FIGURE 8.10** Scheme of pollination in long and short-styled flowers of *Primula*.

**Andromoecious:** An andromonoecious form is one that bears perfect flowers and male flowers on the same plant. Example: many Apiaceae.

**Gynomonoecious:** A gynomonoecious form is one that bears perfect flowers and female flowers on the same plant. Example: *Atriplex* and many Asteraceae.

**Trimonoecious:** A trimonoecious form is one that has three distinct types of flowers, male, female, and hermaphroditic. Example: *Acer campestre*.

**Dioecy:** A dioecious form is one in which the sexes are completely separated so that one plant bears male flowers (staminate) only and the other plant female flowers (pistillate) only. Example: *Elodea canadensis*. A modified type of dioecy occurs in some flowering plants in which some individuals have flowers of one sex but others have bisexual flowers. These include gynodioecy, androdioecy and trioecy.

**Gynodioecy:** A gynodioecious form is one that has both female and bisexual (perfect) flowers as in *Plantago lanceolata*.

**Androdioecy:** An androdioecious form is one that has both male and bisexual (perfect) flowers as in *Caltha palustris*.

**Trioecy:** In a trioecious form, some individuals with staminate flowers only, some with pistillate flowers only, and some with perfect flowers.

## SIGNIFICANCE OF POLLINATION

1. The process of pollination leads to fertilization, resulting in the formation of seeds and fruits. It ensures the continuity of plant life.

2. The seeds and fruits are also sources of nutrition for animals and humans.
3. Pollination, especially cross-pollination, results in the production of plants with a combination of characters from two plants. The role of pollination in the production of hybrid seeds is of great significance in crop production.
4. The diversity of pollination mechanisms provides excellent opportunities to study the degree of co-evolution between plants and pollinators.

## POLLINATION CONSTRAINTS IN FLOWERING PLANTS

Pollination has been a major driving force in the evolution and the diversification of spermatophytes. The angiosperms and their pollinators have co-evolved since the Cretaceous. Flowering plants have evolved in terms of inflorescence, flower structure, colour and smell for insect attraction, secretion of nectar and pollen as reward to maximize their pollination chiefly by biotic agents like insects, birds and bats. In order to ensure effective pollen dispersal and capture, there have been reversals from insect pollination to wind pollination where there is limited biotic pollination. In recent decades, plant–pollinator interactions have been adversely affected by the decline of insect populations due to loss of habitat overexploitation, biological invasions, forest fire, pollution, use of pesticides and other anthropogenic activities. Due to pollen limitation, several species have either vanished or are on their way to the "Red List". Plants that are dioecious or self-incompatible and are dependent on specialists for their pollination have been found to be more vulnerable. Insect-mediated pollination services are also important for the sustenance of agro-ecosystems that provide food security to people across the globe (Thakur and Bhatnagar, 2013; Koul et al., 2018).

## CONTROLLED POLLINATIONS

Effective pollen transfer to the stigma and control of its germination and growth are most important for reproductive success of each species. Vectors are responsible for the transfer of pollen, but controlled pollinations determine if the pollen grain can germinate on a receptive stigma, penetrate the style and effect fertilization.

Control of pollination can be achieved by adopting different systems of artificial mating in crop plants. The purpose of deliberate crossing is generally to isolate superior recombinations in segregating generations rather than to develop better $F_1$ hybrids. Controlled pollination can be conducted on intact flowers as well as on excised pistils maintained under suitable conditions. Experimental studies on pollen-pistil interaction and incompatibility involve controlled pollinations and subsequent observations on pollen germination, pollen tube growth and the development of seed and fruit.

## POLLEN STORAGE

Hybridization is the most important plant improvement technique. Often, the two parents are used for flower hybridization at different times of the year, or grow in different ecological regions. The storage of pollen is one answer to this problem and, hence, it becomes important to store pollen in a viable condition for prolonged periods. The ultimate aim of the plant breeder is, in fact, to establish "pollen banks" through which viable pollen of any desired species can be obtained, at any time, at any place.

Pollen grains remain viable after dispersal for varying periods ranging from as low as 30 min to several weeks. For example, in the members of Poaceae and Asteraceae the viability of pollen grains is lost very rapidly, often within minutes after shedding. In many fruit trees the viability of pollen grains is maintained for months.

Among various factors, the flowering time, relative humidity (RH) and temperature are important in determining the viability of the pollen. In almost all taxa, storage at lower temperatures (upto –196°C) prolongs viability, presumably by reducing the metabolic activity of pollen. The optimal temperature and humidity suitable for pollen storage can vary from species to species.

### Effect of Temperature and Humidity

Lower temperature and humidity prolong the viability period, while higher temperature and humidity reduce it. Pollen grains of many species can withstand high temperatures. For example, pollen grains of *Nicotiana* and *Petunia* can withstand even 75°C for up to 12 h and set seeds. On the other hand, pollen grains of *Brassica* fail to set seeds after exposure to 75°C (Shivanna and Tandon, 2020). Generally, near freezing temperatures and 25–50% relative humidity are suitable for prolonging longevity. Low temperature range of –190° to 10° C has been found suitable for pollen storage. The pollen of deciduous trees can be preserved easily in desiccators containing $H_2SO_4$ at 2–8° C at 50% moisture. The maximum viability under these conditions have been shown by apple, followed by pears and plums. The optimum condition recorded for most of the deciduous fruit trees is 36°C and 28% moisture.

Graminaceous pollen remains viable for one to three weeks when stored at 0–10°C and 80–100% RH. The longevity of 336 days is reported for the pollen of *Typha laticifolia* when stored at 17–20°C at 8% RH.

In *Solanum tuberosum*, if the relative humidity fluctuates abruptly, the viability of pollen is soon lost. In mango (*Mangifera indica*), pollen remains viable for eight days, but at temperatures of 4.5–9°C and 10, 25 or 50% RH, they maintain viability for about 5 months; and at –23°C and 0% RH, they remain viable up to 14 months.

### Storage by Freezing and Dehydration

King (1959) standardized a method known as "freeze-drying method" for storage of pollen. Freeze-drying has been very successful in a large number of taxa. Dry ice can be used for the drying. At 15°C, apple pollen gave a high percentage of germination after nine months storage in dry ice. In the freeze-drying method, water is removed from the pollen after freezing it and sealing them in vacuo or in an inert gas like helium. The advantage of this method is that no humidity control is required. Freeze-drying technique is expensive and involves elaborate equipment.

### Cryogenic Storage

Long-term pollen storage (a few months to years) has been achieved largely through lyophilization and cryopreservation (i.e., storage under ultra-low temperatures). Application of cryogenic technique to preserve pollen viability for prolonged duration has been found suitable for a large number of economically important plants. In this technique, pollen grains are preserved in liquid nitrogen at –196°C. The pollen samples are first dried to a standard water content. Pollen grains of grape, papaya and onion, stored in liquid nitrogen, remained viable for 485, 448 and 360 days, respectively. Pollen grains of oil palm (*Elaeis guineensis*), cryopreserved in liquid nitrogen for up to eight years retained as high as 54% viability.

### Use of Diluents During Storage

A variety of diluents are used during pollen storage and transportation in order to minimize and deterioration due to desiccation. Out of over 30 diluents, powdered *Lycopodium* spores, egg albumen and milk powder have been found to be more effective.

## ORGANIC SOLVENTS

Iwanami (1972) was the first to demonstrate feasibility of storing pollen grains in organic solvents. Pollen storage in organic solvents is one of the simplest methods. Pollen grains kept in organic solvents like acetone, benzene, ether and chloroform germinated *in vitro*. Fertilization was successfully achieved by using such pollen grains. In *Chrysanthemum pacificum* the pollen grains stored for 30 minutes under 25°C over silica gel were soaked in diethyl ether for as long as 20 days. They showed 14% germination. Similarly, *Lilium* pollen grains, stored in organic solvents, did not show any loss of viability even after ten years.

Pollen grains of *Crotalaria retusa* stored under different organic solvents substantiated the role of membrane phospholipids in maintaining pollen viability. Pollen grains stored in organic solvents, which maintained viability, showed very little leaching of sugars, free amino acids and phospholipids into the solvents.

## SUMMARY


- The transfer and deposition of pollen on the stigmatic surface of the flower is known as pollination.
- Pollination biology provides insight into the relationship between plants and their pollinators, foraging strategies of pollinators, community structure, reproductive strategy and co-adaptation of plants and pollinators.
- Pollination is of two types: self-pollination (autogamy) and cross-pollination (allogamy).
- Pollination and fertilization in an unopened flower bud is known as cleistogamy, and such flowers are called cleistogamous.
- When pollination takes place between two flowers of the same plant, the process is known as geitonogamy.
- When pollination occurs between two flowers on different plants, the process is known as xenogamy.
- The agencies and types of pollination may vary from anemophily (wind), hydrophily (water), entomophily (insects), ornithophily (birds), cheiropterophily (bats), myrmecophily (ants) and malacophily (snails and slugs).
- Breeding system refers to whether pollen from a plant can fertilize an egg from the same plant (self-compatible), or the pollen and egg donors must come from different plants (self-incompatible).
- Long-term pollen storage has been achieved through lyophilization and cryopreservation.


## SUGGESTED READING


Barth, F.G. 1991. *Insects and Flowers*. Princeton, NJ: Princeton University Press.

Chaturvedi, S.K. 1989. A new devise of self-pollination in *Boerhavia diffusa* L. (Nyctaginaceae). *Betrag. Biol. Pflanzen* 64: 55–58.

Chauhan, Y.S. 1995. Stylar heteromorphism in *Solanum* (Solanaceae): An expression of andromonoecy. In *Advances in Plant Reproductive Biology*, eds. Y.S. Chauhan and A.K. Pandey, 79–90. Delhi: Narendra Publishing House.

Faegri, K. and L. van der Pijl. 1979. *The Principles of Pollination Ecology*. Oxford: Pergmon Press.

Heslop-Harrison, J., Y. Heslop-Harrison and K.R. Shivanna. 1984. The evaluation of pollen quality and a further appraisal of the fluorochloromatic (FCR) test procedure. *Theoretical and Applied Genetics* 67: 367–375.

Howell, G.J., A.T. Slater and R.B. Knox. 1993. Secondary pollen presentation in angiosperms and its biological significance. *Aust. J. Bot.* 41: 417–438.

Kaul, V. and A.K. Koul. 2008. Floral phenology in relation to pollination and reproductive output in *Commelina caroliniana* (Commelinaceae). *Aust. J. Bot.* 56 (1): 59–66.

Kaul, V. and A.K. Koul. 2009. Sex expression and breeding strategy in *Commelina benghalensis*. *J. Biosci.* 34 (6): 977–990.

King, J.R. 1959. The freeze-drying of pine pollen. *Bull. Tor. Bot. Club* 86: 383–386.

Khan, S., P. Kumari, I.A. Wani and S. Verma. 2021. Pollination biology and breeding system in *Vitex negundo* L. (Lamiaceae), an important medicinal plant. *Int. J. Plant Rep. Biol.* 13 (1): 77–82.

Khanduri, P., R. Tandon, V. Bhat and A. Pandey. 2014. Comparative morphology and molecular systematics of Indian Podostemaceae. *Plant Syst. Evol.* 301 (3). https://doi.org/10.1007/s00606-014-1121-x.

Koul, A.K., P. Koul and I.A. Hamal. 1986. Insects in relation to pollination of some umbellifers. *Bull. Bot. Surv. India* 28: 39–42.

Koul, M., P. Thakur and A.K. Bhatnagar. 2018. Pollination limitation: A threat to biodiversity and food security. *J. Food Agric. Environ.* 16 (2): 76–80.

Kunze, H. 1991. Structure and function in Asclepiads pollination. *Plant Syst. Evol.* 176: 227–253.

Iwanami, Y. 1972. Retaining the viability of *Camellia japonica* pollen in various organic solvents. *Plant Cell Physiol.* 13: 1139–1141.

Mangla, Y. and R. Tandon. 2011. Insects facilitates wind pollination in pollen-limited *Crateva adansonii* (Capparaceae). *Aust. J. Bot.* 39. https://doi.org/10.1071/BT 10174.

McCann 1938.

Meeuse, B.J.D. 1961. *The Story of Pollination.* New York: Ronald Press Co.

Nepi, M. and E. Pacini. 1993. Pollination, pollen viability and pistil receptivity in *Cucurbita pepo. Ann. Bot.* 72 (6): 527–536.

Pal, S. and S. Mondal. 2019. Floral biology, breeding system and pollination of *Oxystelma esculentum* (L.f.) Sm. *Int. J. Plant Reprod. Biol.* 11 (2): 176–181.

Pandey, A.K., M.D. Dwivedi and A. Gholami. 2016. Reproductive biology data in plant systematics-An overview. *Int. J. Plant Repro. Biol.* 8 (1): 65–74.

Pratibha, T. and A.K. Bhatnagar. 2013. Pollination constraints in flowering plants-human actions undoing over hundred million years of co-evolution and posing an unprecedented threat to biodiversity. *Int. J. Plant Reprod. Biol.* 5 (1): 29–74.

Roy, S.K., P. Khanduri, A.K. Bhatnagar and A.K. Pandey. 2021. Pollination biology and breeding system analysis of *Ulmus wallichiana* Planchon (Ulmaceae), a rare and threatened tree species of Central and Western Himalaya. *Nord. J. Bot.* https://doi.org/10.1111/njb.02976.

Sarma, K., Tandon, R., K.R. Shivanna and H.Y. Mohan Ram. 2007. Snail pollination in *Volvulopsis nummularium. Curr. Sci.* 93 (6): 826–831.

Sedgley, M. and A.R. Griffin. 1989. *Sexual Reproduction in Tree Crops.* London: Academic Press.

Subramanya, S. and T.R. Radhamani. 1993. Pollination by birds and bats. *Curr. Sci.* 65: 201–209.

Tandon, R. 2001. Pollination biology and breeding system of *Acacia senegal. Bot. J. Linn. Soc.* 135 (3): 251–262.

Tandon, R., K.R. Shivanna and M. Koul. 2020. *Reproductive Ecology of Flowering Plants: Patterns and Processes.* Singapore: Springer.

Thakur, P. and A.K. Bhatnagar. 2013. Pollination contraints in flowering plants-human actions undoing over hundred million years of co-evolution and posing an unprecented treat to biodiversity. *Int. J. Plant Rep. Biol.* 5 (1): 29–74.

Verma, S., P.Kumari and A. Khajuria. 2019. Pollination biology of *Valeriana wallichii*, a threatened medicinal plant of Himalayan region. *Int. J. Plant Rep. Biol.* 11 (2): 182–185.

Vikas, M.G., R. Tandon and H. Ram. 2009. Pollination ecology and breeding system of *Oroxylum indicum* (Bignoniaceae) in the foothills of the Western Himalaya. *J. Trp. Ecol.* 25 (1): 93–96.

Yadav, N., A.K. Pandey and A.K. Bhatnagar. 2016. Cryptic monoecy and floral morph types in *Acer oblongum* (Sapindaceae): An endangered taxon. *Flora* 224: 1–8.

# 9 Fertilization

Fertilization in angiosperms involves interactions of the male gametophyte (the pollen tube) with the female sporophyte (pistil) and the female gametophyte (embryo sac). The female gametophyte controls many steps of the angiosperm sexual reproductive process. During pollen tube growth and fertilization, the female gametophyte guides the pollen tube to the ovule and embryo sac, controls pollen tube growth within the female gametophyte, and mediates fertilization of the egg cell and central cell. After fertilization, female gametophyte-expressed genes participate in inducing embryo and endosperm formation during seed development.

Even after pollination is completed, male gametes are still physically separated. The sperm nuclei are in the pollen grain lying on the top of the stigma, and the egg is buried within the ovary. But when pollen and stigma match, a pollen tube grows down through the stigma and style. Commonly, the pollen tube enters the ovule through the micropyle and discharges its contents in the embryo sac (Figures 9.1, 9.2). One sperm fuses with the egg (syngamy), and other fuses with the secondary nucleus (triple fusion).

In gymnosperms, only one of the two sperms is functional; one unites with the egg, and the other degenerates. The involvement of both sperms in angiosperms – the union of one sperm with the egg (forming a zygote) and of the other with the polar nuclei (forming a primary endosperm nucleus) is called double fertilization.

The interval between pollination and the time of germination of the pollen grain varies in different taxa. Usually, the pollen grains of herbaceous taxa germinate relatively early compared to arborescent forms.

## POLLEN GERMINATION

Pollen grains are transported to the stigma by wind, animals, water or directly by the contact between open anther and stigma. The first interaction between pollen and stigma is attachment or capture of the pollen grains, and this is accomplished by the sticky nature of the pollen surface, the stigmatic exudate or the pellicle covering the stigma papillae.

The first requirement of pollen germination is their hydration. After sticking on the stigma, pollen grains absorb water through a colloidal imbibition and endosmosis. This results in considerable swelling of the pollen grains. The water passes through the exine and is taken up by the intine and the cell. After rehydration, metabolic activity is resumed, which in turn leads to pollen germination.

In the majority of angiosperms, the pollen grains are monosiphonous (Figure 9.3). The presence of polysiphonous pollen grains is reported in Malvaceae,

DOI: 10.1201/9781003260097-9

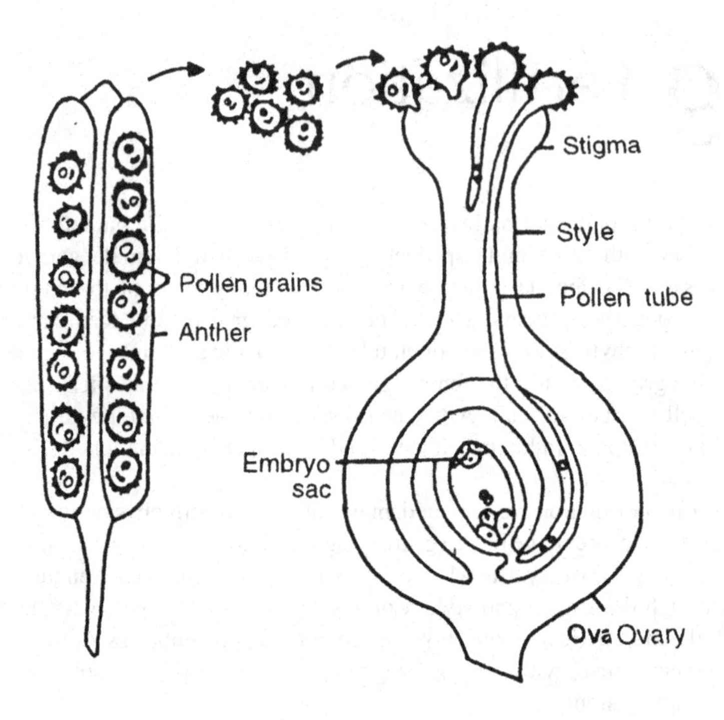


**FIGURE 9.1** Pollen production, its transfer on stigma and fertilization process.

**FIGURE 9.2** Schematic representation of the journey of a pollen tube through an *Arabidopis* pistil. st, stigma; ov, ovules; pg, pollen grain; pt, pollen tube; tt, transmitting tissue (after Lopes et al., 2019).

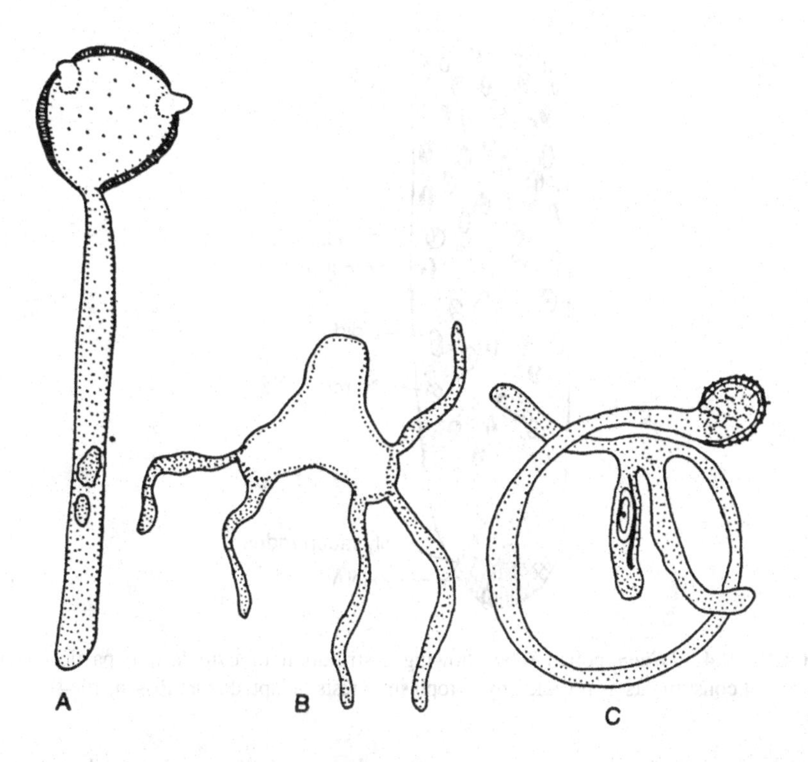


**FIGURE 9.3** A. *Camelia sasanqua.* Germinated pollen grain showing pollen tube with vegetative nucleus and generative cell (after Mathew, 1978). B. *Clarkia elegans.* Pollen with unbranched tube. C. *Lilium regale.* Pollen showing coiled and branched tube (after Johnson, 1959).

Cucurbitaceae, Campanulaceae, etc. (Figure 9.4). Occasional polysiphony is seen in *Argemone mexicana, Lilium auratum, Lupinus,* etc.

Pollen tubes are generally straight, but in *Urticularia stellaris* var. *inflexa*, the tube is not uniform throughout its length, often being curved here and there.

The pollen grains may germinate in the anther locule itself and form pollen tubes. In *Philydrum lanuginosum* (Kapil and Walia, 1965), the microsporangium lies in close proximity to the stigma of the same flower. The pollen grains start germinating inside the microsporangium, and the pollen tubes come out through the dehiscence slit and come to lie on the glandular receptive surface of the stigma.

The time taken by the pollen grains to germinate on the surface of the stigma is variable. In *Oryza sativa*, pollen grains start germinating as soon as they are shed on the stigma. In *Solanum phureja*, some of the pollen grains remain quiescent for 24–72 hours and then germinate.

Pollen tube growth can be observed in the style of a pollinated flower (e.g., sunflower). The style is placed in a couple of drops of aniline blue stain.

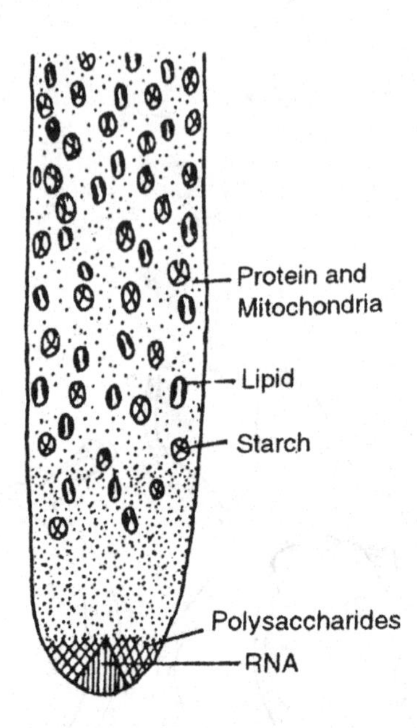


**FIGURE 9.4** *Lilium* pollen tube showing distribution of cytoplasmic particles and chemical constituents as revealed by cytoplasmic tests (adapted from Rosen, 1964).

A cover slip is placed over the style and pressed. When observed under fluorescence microscope, the pollen tube will appear bright yellow. Aniline blue is specific for callose, which fluoresces brightly.

## POLLEN TUBE FORMATION AND GROWTH

Pollen grain germination is marked by the formation of a pollen tube emerging from one (or more) of the germ pores. The growth of the pollen tube is restricted to its tip region of about 4–7 microns. The wall of the emerging pollen tube is pecto-cellulosic in nature and continuous with the intine of the pollen grain.

In the growing pollen tube three distinct zones are recognized:

1. **Apical zone:** The extreme apical region is completely filled with cytoplasm, which contains many Golgi vesicles. These vesicles provide components and plasma membrane required for the growth of the pollen tube tip. The cell wall components needed for tube elongation are excreted by exocytosis at the extreme tube tip. Under a high-power light microscope the extreme tip region of the tube appears hemispherical and transparent. The transparent apical region is called the "cap block".
2. **Subapical zone:** The cytoplasm behind the cap block is rich in the usual cell organelles – especially mitochondria, dictyosomes, Golgi vesicles

and endoplasmic reticulum. The Golgi vesicles are rich in polysaccharides or RNA and are associated with wall formation. Enzymes like phosphatases, amylases, invertases, pectinases and lipases are detected in the cytoplasm (Figure 9.4).

The sub apical zone is metabolically very active. The wall precursors and cytoplasmic proteins are synthesized in this region. Closely behind the tip zone an additional cell wall layer is formed and contains callose. Depending on the length of the pollen tube and the rhythmic growth, callose plug formation takes place (Figure 9.5). The region of callose plug formation is characterized by electron density of the ground plasm, presence of lipid bodies and absence of dictyosomes and plastids.

3. **Vacuolated zone:** The sub apical cytoplasm-rich zone gradually merges into the next zone – a vacuolated region containing the vegetative nucleus and the generative cell or sperm cells. The movement of the vegetative nucleus, generative cell or sperm cells and cytoplasmic streaming is observed in this region. The cytoplasmic streaming is caused by microfilaments. The direction of the streaming is the result of interconnection between microfilaments, microtubules and endoplasmic reticulum.

In the peripheral region of the pollen tube a cytoskeleton composed of longitudinally oriented, cortical microtubules is present. During tube elongation, callose plugs are formed at regular intervals. As a result, a fully grown pollen tube is subdivided into many compartments due to these plugs. The small amount of cytoplasm, which is left behind the plug on the side of the grain, gradually degenerates.

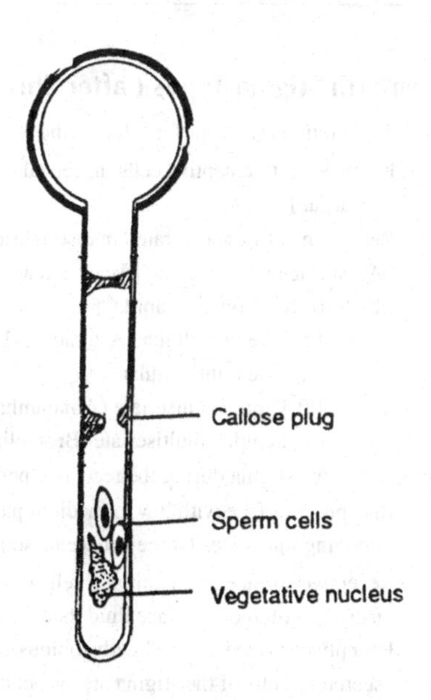


**FIGURE 9.5** *Petunia hybrida* pollen tube showing callose plug formation.

## PATH OF THE POLLEN TUBE

### Stigma

The stigma is the specialized part of the pistil on which the pollen grains are trapped during pollination. After recognition and acceptance, the pollen tube grows between the stigmatic papillae and reaches the stylar part. That stigma may be of wet or dry type. Heslop-Harrison and Shivanna (1977) surveyed the stigma types of about 1000 species belonging to 900 genera and 250 families and proposed a classification of stigma types in angiosperms (Table 9.1). In taxa characterized by the wet stigma, the surface is covered with a secretion termed stigmatic exudate.

## STYLE

The style is a tubular structure that connects the stigma with the ovary. Based on morphological features, styles have been classified as open, closed (solid) or half-closed (semi-solid).

### SOLID STYLE

This type of style is characterized by the presence of a central strand of elongated, specialized cells that constitute the transmitting tissue or conducting tissue (Figure 9.6). Solid style mostly occurs in dicotyledons (e.g., *Datura*, *Gossypium*).

**TABLE 9.1**
**Classification of Angiosperm Stigma Types ( after Shivanna, 2003)**

| | |
|---|---|
| Dry stigma (without apparent fluid secretion on the surface during the receptive period) | |
| Group I | Plumose, with receptive cells dispersed on multiseriate branches (Poaceae) |
| Group II | Receptive cells concentrated in distinct ridges, zones or heads |
| | A. Surface non-papillate (Acanthaceae) |
| | B. Surface distinctly papillate |
| | 1. Papillae unicellular (Asteraceae, Brassicaceae) |
| | 2. Papillae multicellular |
| | (a) Papillae uniseriate (Amaranthaceae) |
| | (b) Papillae multiseriate (Bromeliaceae) |
| Wet stigma (Fluid secretion present on the stigma during the receptive period) | |
| Group III | Receptive surface with low to medium papillae, secretion fluid flooding interstices (some Liliaceae, some Rosaceae) |
| Group IV | Receptive surface non-papillate, cells often necrotic at maturity, usually with more surface fluid than Group III (Apiaceae) |
| Group V | Receptive surface covered with copious exudate in which detached secretory cells of the stigma are suspended. |

The transmitting tissue in a solid style is either loosely arranged or has a more compact structure. In *Gossypium* (Jensen and Fisher, 1969), all cell walls of the transmitting tissue have a thick layered structure. The innermost part consists of pectin and hemicellulose; in the second layer, hemicellulose dominates; and the third layer is again rich in pectic substances and poor in hemicellulose and cellulose. The middle lamella, the fourth layer, is fibrous and the pectic substance is very dominant, while hemicellulose is also present.

The cells of the transmitting tissue contain numerous mitochondria, plastids, rough endoplasmic reticulum (RER), dictyosomes and ribosomes. The intercellular substance in many solid styled plants is composed of carbohydrates, proteins, glycoproteins and some enzymes, like acid phosphatases, peroxidases and esterases.

In *Oryza sativa* the solid style is devoid of any special transmitting tissue. In *Oldenlandia nudicaulis* the pollen tube passes through the intercellular spaces present between the cells of a solid style. In *Ainsliaea aptera* and *Passiflora foetida* the pollen tube makes its way through the central core of richly cytoplasmic transmitting tissue of the style.

In *Petunia*, in between the cells of the transmitting tissue, a secretion product is present comparable with the mucilage of the canal cells in the open style of *Lilium*.

### Hollow Style or Open Type

This type of style is more common in monocotyledons (e.g., *Amaryllis*, *Lilium*, *Gladiolus*). In taxa with hollow styles, there is a canal which runs throughout from stigma to the base of the style (Figure 9.6). In *Lilium*, a secretion appears in the stylar canal. The canal cells show dense cytoplasmic contents and are rich in organelles. The secretion from the canal cells contains proteins, carbohydrates, lipids and shows activity for esterases and acid phosphatases.

In *Gladiolus* the canal is covered by a cuticle which prevents the dehydration of the mucilaginous excretion. The pollen tube penetrates the cuticle and grows through the secretion.

### Semisolid Styles

Half-closed styles occur in the family Cactaceae. The transmitting tissue is confined to only one side of the stylar canal.

## ENTRY OF POLLEN TUBE INTO THE OVULE

The pollen tube may enter the ovule through one of the following three ways:

1. **Porogamy:** When the pollen tube enters into the ovule through the micropyle, it is called porogamy (Figure 9.7). This is the most common mode of pollen tube entry into the ovule. Some of the common examples

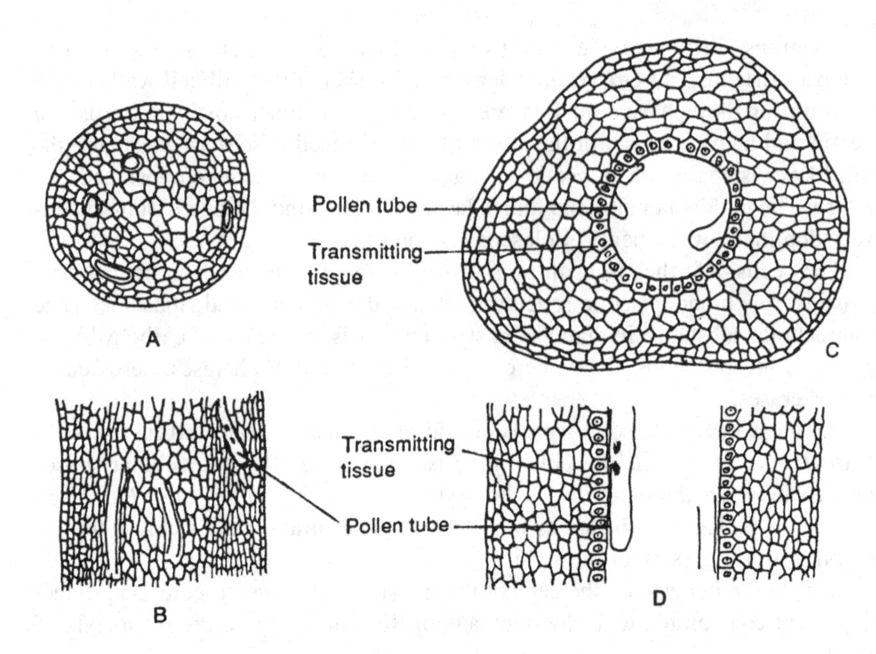


**FIGURE 9.6** Types of styles. Closed (A, B) and open (C, D).

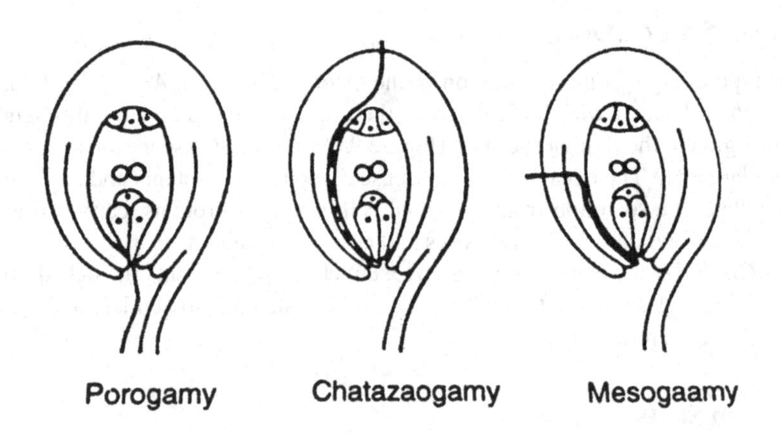


**FIGURE 9.7** Modes of pollen tube entry into the ovule.

of porogamy include *Azadirachta indica*, *Oryza sativa*, *Nyctanthes arbortristis*.

In *Utricularia stellaris* var. *inflexa*, *Torenia violacea* and *Pentaphragma horsfieldii* the embryo sac comes out of the micropyle and as such the pollen tube is not required to enter the micropyle. In such instances, porogamy in the true sense, is not seen. In *Dendrophthoe* the embryo sac comes out of the micropyle and goes up to 2/3rd of the style. The pollen tube meets the embryo sac in the stylar region.

2. **Chalazogamy:** When the pollen tube enters the ovule through the chalazal end, it is known as chalazogamy (Figures 9.7, 9.8). Instances of chalazogamy are few, most of them being from the Amentiferae (e.g., *Casuarina, Juglans*).
3. **Mesogamy:** In this type, the pollen tube enters either through integuments or funiculus (Figure 9.7). Mesogamy is reported in *Cucurbita, Alchemilla, Circaester, Pistacia.*

## ENTRY OF POLLEN TUBE INTO THE EMBRYO SAC

After reaching the ovule, the pollen tube normally reaches the embryo sac by entering the micropyle. The pollen tube may enter the embryo sac by one of the following routes: (i) in between the egg apparatus, i.e., the synergids and egg cell; (ii) between the cells of the egg apparatus and the embryo sac wall; or (iii) directly into one of the synergids.

During fertilization, cytoskeletal components within the female gametophyte direct the sperm cells to the egg cell and the central cell.

Recent ultrastructural studies have shown that synergids play a very important role in the pollen tube entry. The pollen tube enters the embryo sac through filiform apparatus (FA). After growing through the FA for some time, the pollen

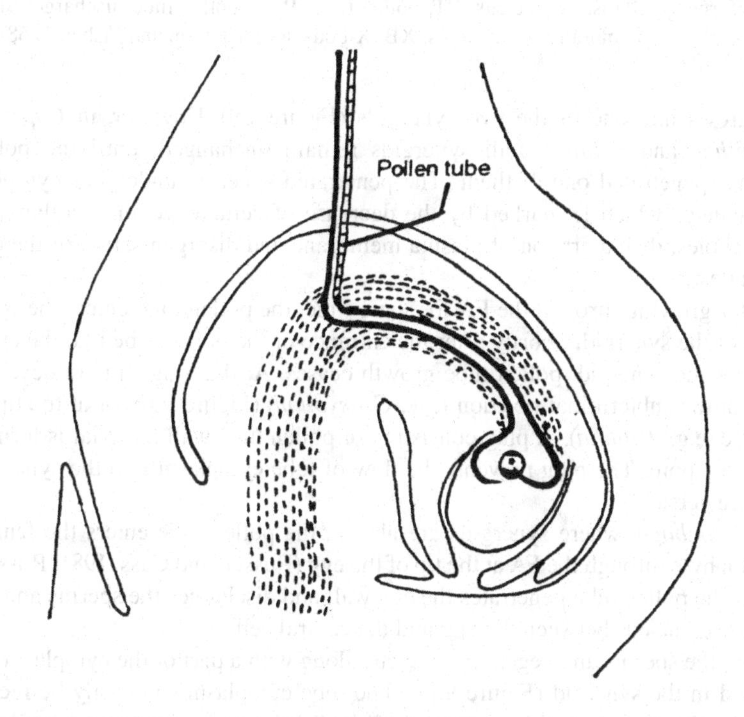


**FIGURE 9.8** *Pistacia vera.* Longisection of ovary showing entry of pollen tube through chalazal vascular strand (after Batygina and Yakovlev, 1985).

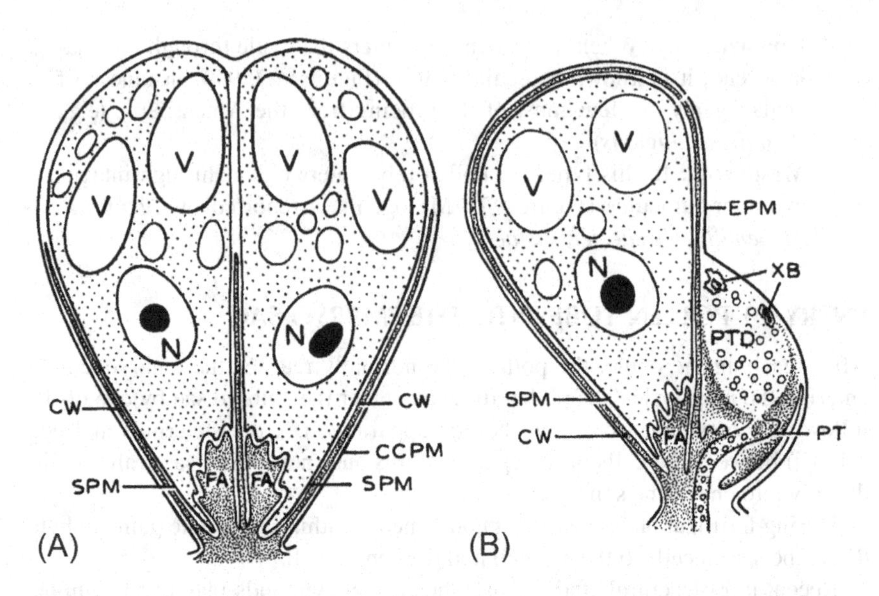


**FIGURE 9.9** Synergids and entry of pollen tube into the embryo sac. A. Synergids before the entry of pollen tube. B. Degenerating synergids after the entry of pollen tube. CCPM, central cell plasma membrane; CW, cell wall; EPM, endosperm plasma membrane; FA, filiform apparatus; N, nucleus; PT, pollen tube; PTD, pollen-tube discharge; SPM, synergid plasma membrane; V, vacuoles; XB, X-bodies(after Jensen and Fisher, 1968).

tube grows into one of the two synergids (Figure 9.9). However, in *Capsella*, *Helianthus* and *Petunia* both synergids remain unchanged, until the pollen tube has penetrated one of them. The penetrated synergid undergoes cytoplasmic changes which is marked by the decrease of cellular volume, collapse of the vacuoles, disintegration of plasma membrane and disorganisation of the cell organelles.

After growing through the FA for some time, the pollen tube enters the cytoplasm of the synergid. Shortly after the entrance of the pollen tube into the cytoplasm of the synergid, pollen tube growth ceases. At this stage a pore develops either at the subterminal position (e.g., *Gossypium*) (Figure 9.10) or at the tip of the tube (e.g., *Petunia*). A plug consisting of pollen tube wall material is formed below the pore. The plug prevents the flow of tube cytoplasm into the synergid, and vice versa.

In *Plumbago*, where synergids are absent, the pollen tube enters the female gametophyte through the FA at the tip of the egg (Russell and Cass, 1981; Russell, 1982). The pollen tube penetrates the egg wall and discharges the sperms and the vegetative nucleus between the egg and the central cell.

Both the sperms and vegetative nucleus along with a part of the cytoplasm are released in the synergid (Figure 9.11). The tube cytoplasm can easily be recognized by the presence of numerous small PAS positive spheres (van Van Went, 1970).

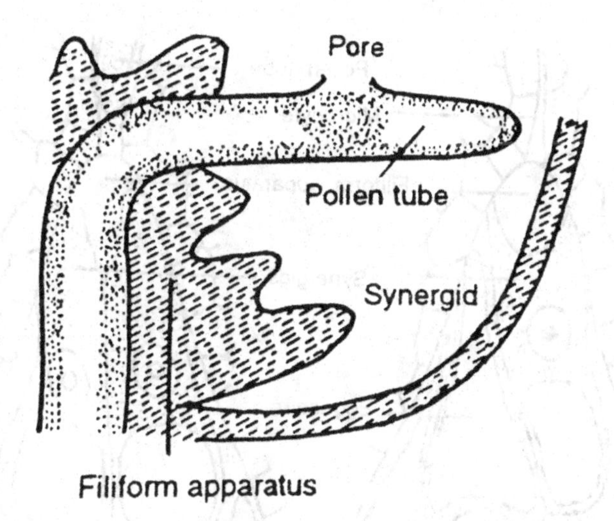


**FIGURE 9.10** A part of penetrated synergid showing subterminal pore (after Jensen and Fisher, 1968).

## FUSION OF GAMETES

The sperms released into the synergid cytoplasm are not nuclei but definite cells. Their delimiting membrane is highly elastic but firm. The two sperms in a pollen tube often change their shapes. Ultrastructural studies have shown the presence of microtubules in the sperm cells. The position of the tubules and the observed shape changes indicate that their chief function is the regulation of cell shape.

Our knowledge about the mode of actual transfer of sperm cells from the penetrated synergid to and into the egg cell and central cell is meagre. The hypothesis proposed by Linsken (1968), Van Went (1970) and Jensen (1972) suggests that, in angiosperms, fertilization involves a real cell-fusion process (Figure 9.11). It is proposed that one of the sperms comes in contact with the plasma membrane of the egg cell, while the other sperm cell fuses with the plasma membrane of the central cell. The membranes at the point of contact dissolve and the sperm nuclei are released, one in the egg and the other in the central cell (Figure 9.12). In this proposed mechanism of fusion there is no requirement for the formation of pores in the female gamete plasma membrane to facilitate the transfer of male material. It also provides a mechanism for preventing the entrance of degenerated synergid or pollen tube cytoplasm into the female gamete.

## FUSION OF NUCLEI (SYNGAMY)

The next step in the process of fertilization is the fusion of the sperm cell nucleus with the egg nucleus (syngamy) and the fusion of the second sperm cell nucleus with the (fused) polar nuclei. This phenomenon is described as double fertilization.

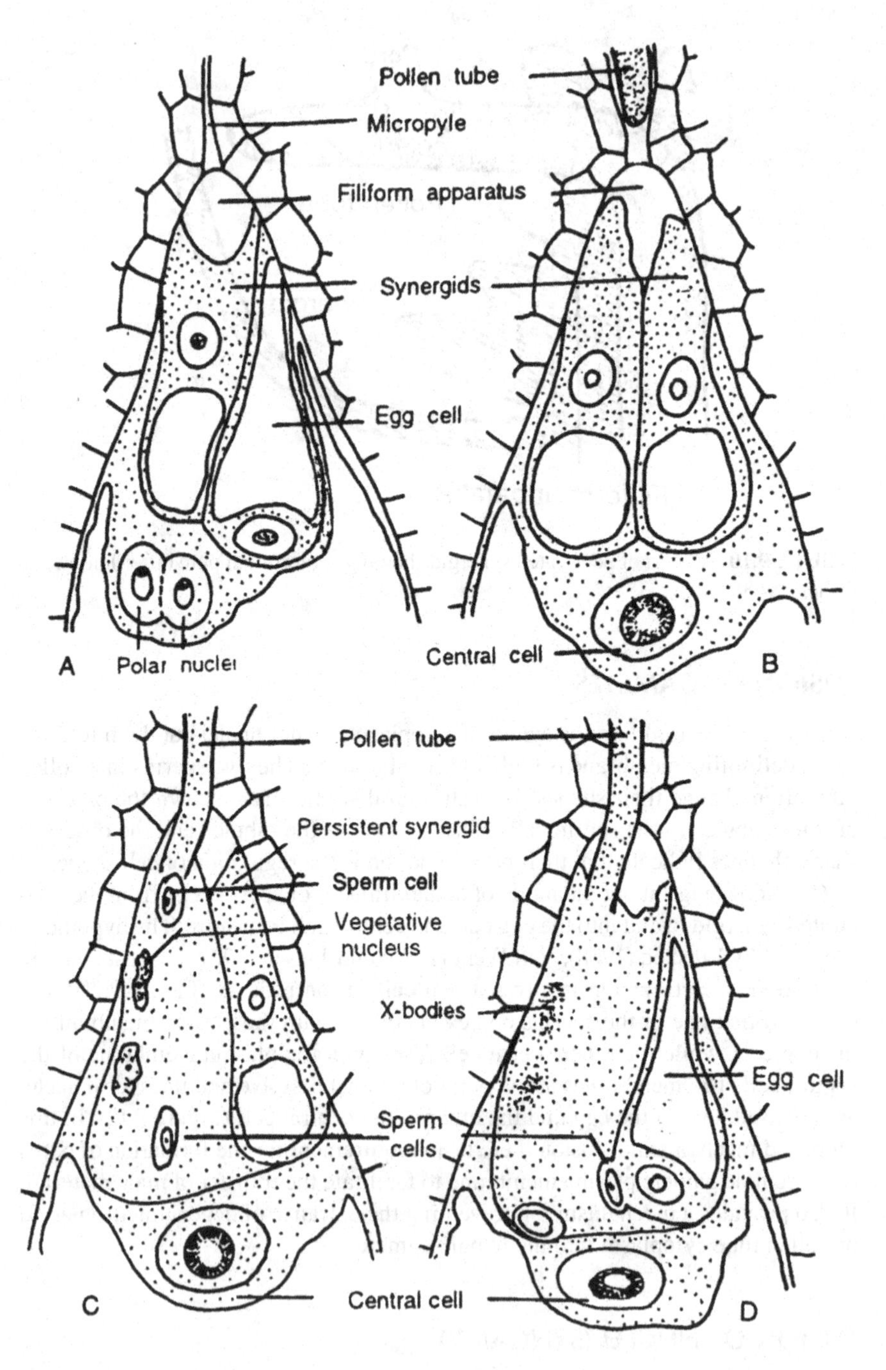


**FIGURE 9.11** Diagram showing pollen tube entry, tube discharge and gametic transfer (after Van Went and Willemse, 1984).

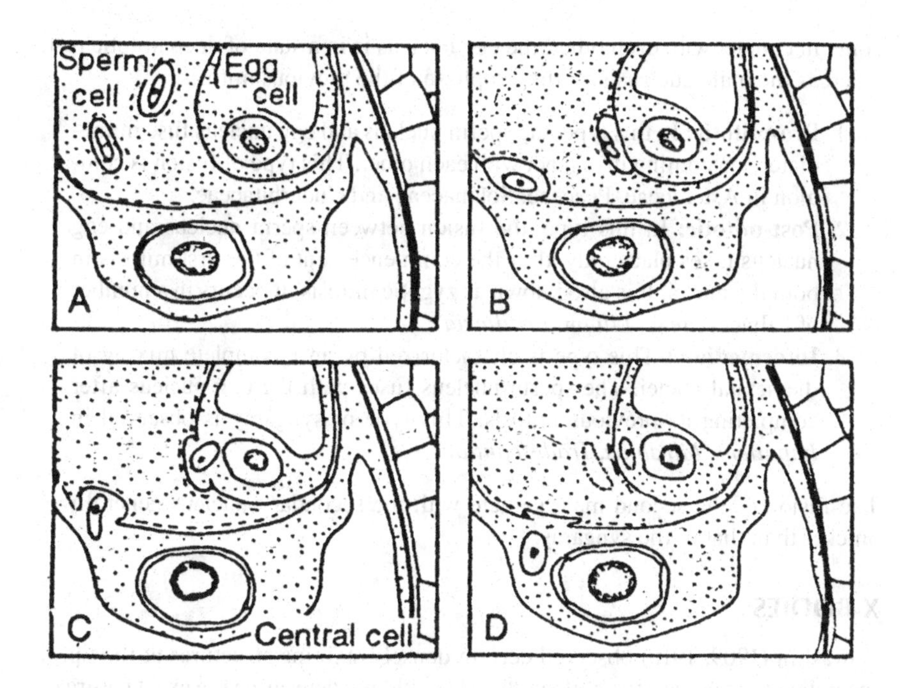


FIGURE 9.12 Diagram showing gametic fusion (after Van Went and Willemse, 1984).

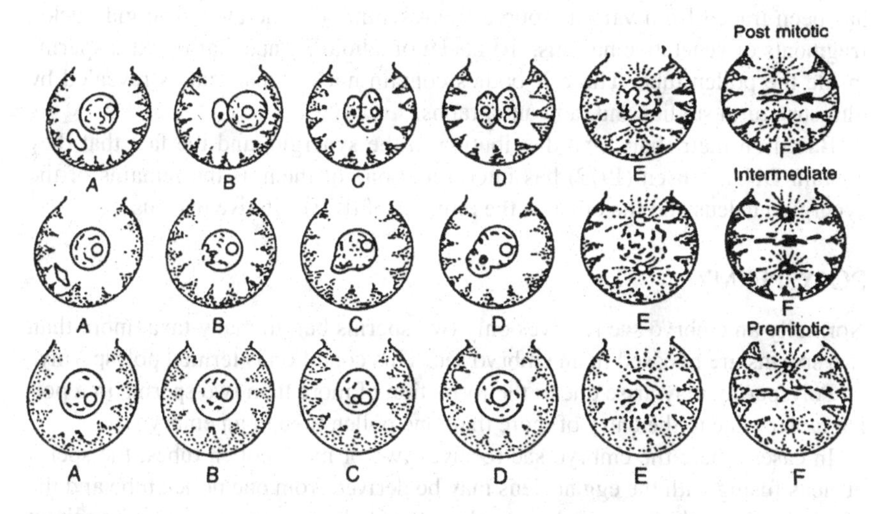


FIGURE 9.13 Mitotic hypothesis of syngamy: pre-mitotic, intermediate and post-mitotic.

Upon fertilization, female gametophyte-expressed genes control the initiation of seed development.

Gerassimova-Navashina (1960) recognized three types of karyogamy that depend upon the differences in the conditions of the sexual nuclei at the time of fertilization (Figure 9.13). According to Navashina, at the time when two nuclei come in contact with each other, the egg nucleus is in a state of deep mitotic rest, whereas the male nucleus is at the telophase of the previous mitosis.

1. **Pre-mitotic:** In this type, the sperm nucleus and egg nucleus fuse immediately on coming in contact with each other. This type of fusion is common in Asteraceae, Poaceae, Solanaceae and Cucurbitaceae.
2. **Post-mitotic:** In this type, the fusion between sperm nucleus and egg nucleus takes place only after the commencement of the first mitosis in both the nuclei. It is also known as zygotic mitosis. It occurs in a number of Liliaceae (e.g., *Lilium, Fritillaria*).
3. **Intermediate:** This type is characterized by an incomplete mixing of the sexual nuclei. The sperm nucleus fuses with the egg nucleus after completing its previous mitosis. This type of syngamy is observed in *Impatiens, Mirabilis, Tradescantia.*

The fusion of the second male gamete with the secondary nucleus generally is quicker than that of the syngamy.

## X-BODIES

Nawashin (1909, 1910) observed certain densely staining structures at the tip of the pollen tube or near the pollen tube. He could not determine the exact nature of these bodies and named them "X-bodies" (Figure 9.9B). The origin of X-bodies has been traced from various sources representing degenerated synergid nuclei, fragments of vegetative nucleus, the nuclei of adjoining nucellar or extra sperms inside the pollen tube. These X-bodies contain nuclear material as revealed by ultrastructural studies and radioautographs.

Based on their shape and distribution in the synergid, and the fact that they contain DNA, Jensen (1972) has interpreted one of them as the remains of the synergid nucleus and the other as the remains of the vegetative nucleus.

## POLYSPERMY

Normally an embryo sac receives only two sperms but, in many taxa, more than two sperms are released in an embryo sac. This condition is termed polyspermy.

Polyspermy may arise due to the formation of more than two sperms in a pollen tube or due to the entry of more than one pollen tube in an embryo sac.

In cases where the embryo sac receives two or more pollen tubes, the sperm nucleus fusing with the egg nucleus may be derived from one pollen tube and the sperm fusing with the secondary nucleus may belong to another. This condition is described as heterofertilization. Occasionally supernumerary sperms may fertilize synergids or antipodal cells.

## PERSISTENT POLLEN TUBES

Normally the pollen tube degenerates after fertilization. In a few cases, the pollen tube persists for a long time. Broad and persistent pollen tubes are observed in a number of taxa belonging to families Cucurbitaceae, Passifloraceae, Nyctaginaceae, Hydrocharitaceae, Orobanchaceae, Malvaceae and Gesneriaceae.

In *Fuchsia boliviana* (Tilquin et al., 1983) the pollen tubes do not collapse after fertilization, but become even more conspicuous and persist for a considerable time, and remain recognizable in the mature seed.

In *Ottelia alismoides* the pollen tube forms a bulla (balloon-like swelling) as it enters the micropyle. D. Singh (1963) observed persistent pollen tubes in 22 species belonging to 15 genera of Cucurbitaceae. The wall of the persistent pollen tube is thick and, as the pollen tube does not show any sign of degeneration, it stains like a healthy organ. According to Singh, persistent pollen tubes are haustorial in nature.

In spinach (Ramanna and Mutsaerts, 1971), branching of pollen tube occurs mainly in the stylar region and near the micropyle. Wilms (1974) observed that in spinach the branching of the pollen tube occurs mainly in the micropyle, and the branches grow between and around the inner and outer integuments.

In *Grevillea banksii*, after fertilization, the pollen tube branches near or at the micropyle. The enucleate branches invade the ovary wall, where further branching occurs, and the branches ramify intercellularly in the ovarian tissue. In *Cucurbita* and some members of the family Onagraceae, branching of the terminal portion of the pollen tube have been reported.

Several researchers have suggested that persistent pollen tubes are haustorial in nature. Johri and Ambegaokar (1984), however emphasized that persistent pollen tubes are merely the remnants of a "dead" structure, which are not at all concerned in absorbing and transmitting the nutrients.

## POLLEN GRAINS IN THE STYLAR CANAL AND OVARY

The pollen grains usually land on the stigma and germinate; the pollen tubes reach the ovary, enter the ovule and initiate fertilization. Johri (1936) reported the occurrence of 5–8 three-celled pollen grains in the stylar canal of *Butomopsis lanceolata.* According to Sahni and Johri (1936) "it can only be regarded as a relic of gymnospermy in a confirmed and unquestionable angiosperm".

## SUMMARY

- The union of one sperm with the egg nucleus and the other with polar nuclei is known as double fertilization.
- In the growing pollen tube, three distinct zones are recognized: apical, subapical and vacuolated.
- Fertilization may be classified as porogamy, chalazogamy and mesogamy.
- When more than two sperms are released in the embryo sac, this condition is called polyspermy.
- After recognition and acceptance, the pollen tube grows between the stigmatic papillae and reaches the stylar part. The stigma may be of wet or dry type.

## SUGGESTED READING

Chauhan, S.V.S., N.H. Rathore and T. Konishita. 1984. *In vitro* pollen germination studies in *Tecoma stans* L. *Jap. J. Palynol.* 30: 1–6.

Cresti, M. and R. Dallai (eds.). 1986. *Biology of Reproduction and Cell Motility in Plants and Animals*. Siena: University of Siena.

Heslop-Harrison, J. 1987. Pollen germination and pollen tube growth. *Int. Rev. Cytol.* 107: 1–78.

Heslop-Harrison, Y. and K.R. Shivanna. 1977. The receptive surface of the angiosperm stigma. *Ann. Bot.* 41: 1233–1258.

Jensen, W.A. 1972. The embryo sac and fertilization in angiosperms. *Harold L Lyon Arbor Leet* 3: 1–32.

Jensen, W.A. and D.B. Fisher. 1968. Cotton embryogenesis: The entrance and discharge of the pollen tube in the embryo sac. *Planta* 78: 158–183.

Jensen, W.A. and D.B. Fisher. 1969. Cotton embryogenesis: The tissues of the stigma and st yle and their relation to the pollen tube. *Planta* 84: 97–121.

Johri, B.M. 1936. The life-history of *Butomopsis lanceolata* Kunth. *Proc. Indian Acad. Sci. B* 4: 139–162.

Johri, B.M. and S.P. Bhatnagar. 1973. Some histochemical and ultrastructural aspects of the female gametophyte and fertilization in angiosperms. *Caryologia* 25: 9–25.

Johri, B.M. and K.B. Ambegaokar. 1984. Embryology: Then and now. In *Embryology of Angiosperms*, ed. B.M. Johri, 1–52. Heidelberg: Springer Berlin.

Kapil, R.N. and K. Walia. 1965. The embryology of *Philydrum lanuginosum* Banks, ex Gaertn. and the systematic position of the Philydraceae. *Beitr. Biol. Pfl.* 41: 381–404.

Kapil, R.N. and A.K. Bhatnagar. 1975. A fresh look at the process of double fertilization in angiosperms. *Phytomorphology* 25: 334–368.

Linskens, H.F. (ed.). 1964. *Pollen Physiology and Fertilization*. Amsterdam: North –Holland.

Lopes, A.L., D. Moreira, M.J. Ferreira, A.M. Pereira and S. Coimbra. 2019. Insights into secrets along the pollen tube pathway in need to be discovered. *Journal of Experimental Botany* 70 (11): 2979–2992. https://doi.org/10.1093/jxb/erz087.

Mukherjee, P.K. 1973. Fertilization in angiosperms. In *Glimpses in Plant Research*, ed. P.K.K. Nair, 120–147. Delhi: Vikas Publishing House.

Rosen, W.G. 1964. Chemotropism and fine structure of pollen tubes. In *Pollen Physiology and Fertilization*, ed. Linskens, H.F., 159–166. Amsterdam: North-Holland.

Rosen, W.G., S.R. Gawlik, W.V. Dashek and K.A. Siegesmund. 1964. Fine structure and cytochemistry of *Lilium* pollen tube. *Am. J. Bot.* 51 (1): 61–71.

Russell, S.D. 1982. Fertilization in *Plumbago zeylanica*: Entry and discharge of the pollen tube in the embryo sac. *Can. J. Bot.* 60: 2219–2230.

Russell, S.D. and D.D. Cass. 1981. Ultrastructure of fertilization in *Plumbago zeylanica*. *Acta Soc. Bot. Pol.* 50: 185–190.

Sahni, B. and B.M. Johri. 1936. Pollen grains in the stylar canal and in the ovary of an angiosperm. *Curr. Sci.* 4: 587–589.

Singh, D. 1963. Studies on the persistent pollen tubes of the Cucurbitaceae. *J. Indian Bot. Soc.* 42: 208–213.

Tilquin, J.P., K. de Brouwer, A. Mathieu and M. Calier. 1983. Haustorial pollen tubes in *Fuchsia boliviana*. *Ann. Bot.* 52 (3): 425–428.

Van Went, J.L. 1970. The ultrastructure of the fertilized embryo sac of *Petunia*. *Acta Bot. Neerl.* 19: 468–480.

Van Went, J.L., and M.T.M. Willemse. 1984. Fertilization. In *Embryology of Angiosperms*, ed. B.M. Johri, 273–318. Berlin: Springer-Verlag.

Wilms, R. 1974. Branching of pollen tubes in spinach. In *Fertilization in Higher Plants*, ed. Linskens, R.F., 155–160. Amsterdam: North-Rolland.

Yadegari, R. and G.N. Drews. 2004. Female gametophyte development. *The Plant Cell* 16 Supplement: S133–S141.

# 10 Self-Incompatibility

After pollination, pollen grains germinate to produce pollen tubes which grow through the tissues of the pistil and discharge the sperms in the vicinity of the egg for fertilization, an event leading to development of seeds and fruits. In many taxa, however, pollen grains may fail to bring about fertilization of ovules in the same plant resulting in the failure of embryo and seed development. This phenomenon is called self-incompatibility (SI).

Self-incompatibility is defined as the inability of a hermaphrodite plant producing functional gametes to effect fertilization upon self-pollination.

Self-incompatibility is the mechanism adopted by a plant to prevent self-fertilization and thus promote outcrossing. Self-incompatibility is mainly a pre-fertilization barrier manifested through inhibition of pollen germination on the stigma or pollen tube growth in the style. The presence of physical barriers between the male and female plants to avoid self-fertilization such as dichogamy, monoecy, dioecy, or floral heteromorphy, has evolved in both gymnosperms and angiosperms.

Incompatibility occurs between species (interspecific incompatibility), as well as within the species (intraspecific incompatibility or self-incompatibility). Interspecific incompatibility prevents fertilization between gametes of distantly related species, whereas intraspecific incompatibility prevents fertilization between gametes of the same or other individuals of the same species (Figure 10.1). There are two types of self-incompatibility: homomorphic and heteromorphic.

## HOMOMORPHIC INCOMPATIBILITY

In this category, all the individuals of a species produce only one type of flower (e.g., *Petunia, Nicotiana*). This type of incompatibility is controlled by multiple alleles called S alleles. A pollen tube possessing a particular S allele is inhibited in the style carrying the same S allele. The homomorphic SI system is controlled by a single Mendelian locus (the S-locus), which is comprised of tightly linked genes determining self-recognition specificities and many accessory genes which are also necessary for the proper function of self-incompatibility. A species showing homomorphic incompatibility has numerous mating types (mating type refers to a group of individuals showing similar breeding behaviour).

## HETEROMORPHIC INCOMPATIBILITY

In this category, different individuals of a species produce either two or three types of flowers differing in the length of stamens and style (heterostyly). In distylic flowers there are two lengths of styles, and in tristylic species three distinct style lengths exist, and the stamens are of different lengths, which do not correspond to their stigmas (Figure 10.2).

DOI: 10.1201/9781003260097-10

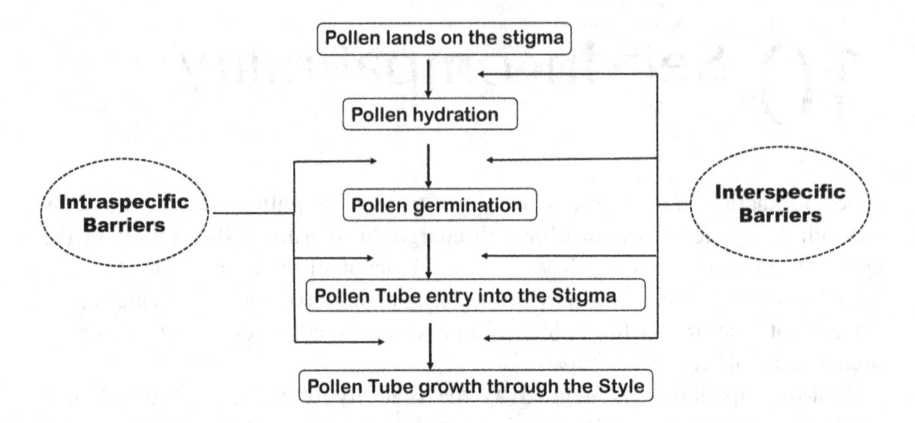


**FIGURE 10.1** Post-pollination events that occur during pollen–pistil interaction (after Shivanna, 1982).

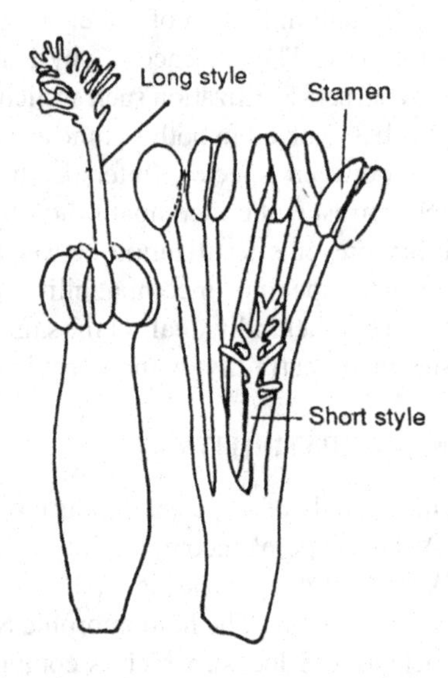


**FIGURE 10.2** *Waltheria indica* showing heterostyly. A. Long-styled flower; B. Short-styled flower (after Bahadur and Srikanth, 1983).

In heteromorphic incompatible species, the flowers of different incompatibility groups are different in morphology. For example, in *Primula*, there are two types of flowers, pin and thrum. Pin flowers have long styles and short stamens, while thrum flowers have short styles and long stamens (Figure 7.5). The only compatible mating is between pin and thrum flowers and incompatibility is governed by a single gene, s, Ss producing thrum and ss producing pin flowers. The mating between pin and thrum plants would produce Ss and ss progeny in equal frequencies. The pollen grains produced by pin flowers would all be s in genotype and also in incompatibility reaction. On the other hand, pollen produced by thrum flowers would be of two types genotypically, S and s, while all of them would be S phenotypically.

In *Linum grandiflorum* the stigma of the pin morph is of the dry type while that of the thrum morph is of the wet type. Ghosh and Shivanna (1982) suggested that, in the *Linum* intramorph, incompatibility is a combination of the structural mismatch between the pollen and the stigma, as well as active inhibition of pollen tubes; the former is a passive phenomenon, the latter is the result of positive recognition of the pollen.

## CLASSIFICATION OF SELF-INCOMPATIBILITY

Self-incompatibility may be categorized into two groups (gametophytic self-incompatibility and sporophytic self-incompatibility), depending upon the origin of factors which determine the mating types on the pollen side.

### 1. Gametophytic Self-Incompatibility (GSI)

This system was first described by East and Mangelsdorf (1925) in *Nicotiana sanderae*. The incompatibility reaction of the pollen is determined by its own genotype and not by the genotype of the plant on which it is produced. Gametophytic incompatibility is determined by a single gene, S. GSI is found in members of families Fabaceae, Poaceae, Rosaceae, Papaveraceae, Liliaceae and Solanaceae.

### 2. Sporophytic Self-Incompatibility (SSI)

In the sporophytic system, the incompatibility reaction is controlled by the genotype of the sporophytic tissue of the plant from which the pollen is derived. It was first reported by Hughes and Babcock (1950) in *Crepis foetida* and by Gerstel (1950) in *Parthenium argentatum* (Asteraceae). *Carthamus tinctorius* (Asteraceae) shows sporophytic self-incompatibility (Figure 10.3). Other families showing SSI include Brassicaeae and Convolvulaceae.

Gametophytic (GSI) is common in species with 2-celled pollen, whereas sporophytic (SSI) is associated with species shedding pollen at the 3-celled stage with many other differences (Table 10.1).

## GENETICS OF SELF-INCOMPATIBILITY

The first satisfactory genetic interpretation of homomorphic SI was given by East and Mangeldorf (1925) for *Nicotiana*. According to this hypothesis, incompatibility reactions are controlled by a single gene, called the S gene, which has several alleles. The concept holds that any pollen carrying an "S" allele identical with one present in the pistil would be rejected.

A common cause of incompatibility is the failure of pollen tubes to traverse through styles. In incompatible matings, the pollen tube grows so slowly that it may never reach the ovule, or if it does, the ovules will already have withered.

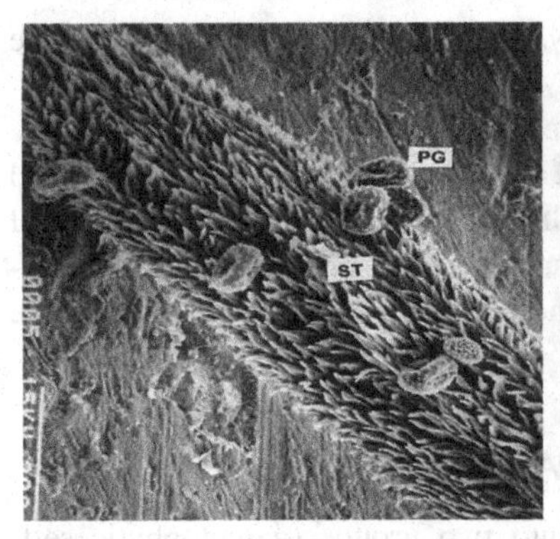


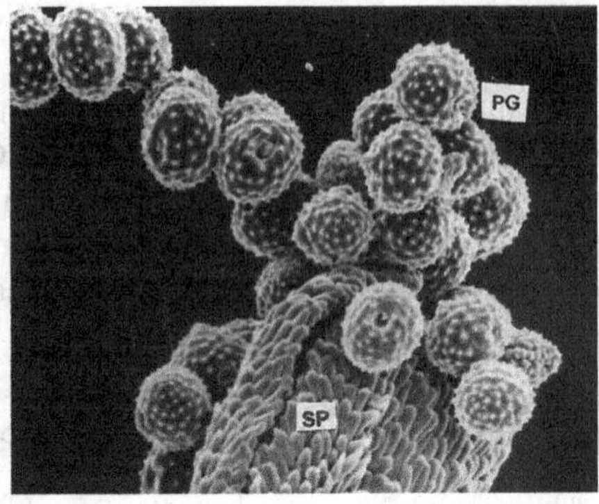


**FIGURE 10.3** *Carthamus tinctorius.* SEM micrographs. A, B. Pollen grains on stigma surface (PG=pollen grain; SP=stigmatic papillae).

**TABLE 10.1**
**Differences Between Gametophytic and Sporophytic Self-Incompatibility**

| Features | Gametophytic | Sporophytic |
|---|---|---|
| Pollen | 2-celled | 3-celled |
| Inhibition zone | Style | Stigma surface |
| Type of stigma | Wet/Dry | Dry |
| Callose plug in stigma papilla | Absent | Present |

Reproductive success in obligate SI plants requires maintenance of a sufficient amount of allelic diversity and heterogeneity among the conspecifics.

In compatible mating of the same species, the pollen tube grows at a normal rate, and fertilization is complete after the entry of the pollen tube into the ovule. The rate of pollen tube growth is governed by a series of alleles ($S_1$, $S_2$, $S_3$, etc.) for incompatibility. If the allele present in the pollen tube is similar with the allele present in the stylar tissue, the growth of the pollen tube is very slow. If the allele in the pollen tube is different from the alleles in the stylar tissue, the pollen tube grows at the normal rate.

If a plant with the genotype $S_1S_2$ is pollinated with its own pollen, or with pollen from another plant with the $S_1S_2$ genotype, the pollen tubes rarely penetrate the style far enough to reach the ovule (Figure 10.4A). If a plant having $S_1S_2$ genotype is pollinated with pollen from a plant with genotype $S_1S_3$, usually only the pollen with the $S_3$ allele penetrates the style and fertilizes the ovule (Figure 10.4B). On the other hand, if $S_1S_2$ genotype is pollinated with the pollen from an $S_3S_4$ plant, either the $S_3$ pollen or the S4 pollen may enter the style and effect fertilization (Figure 10.4C).

Self-incompatibility in *Papaver rhoeas* is homomorphic and is controlled by a single multi-allelic gene, (S–), which is gametophytically controlled.

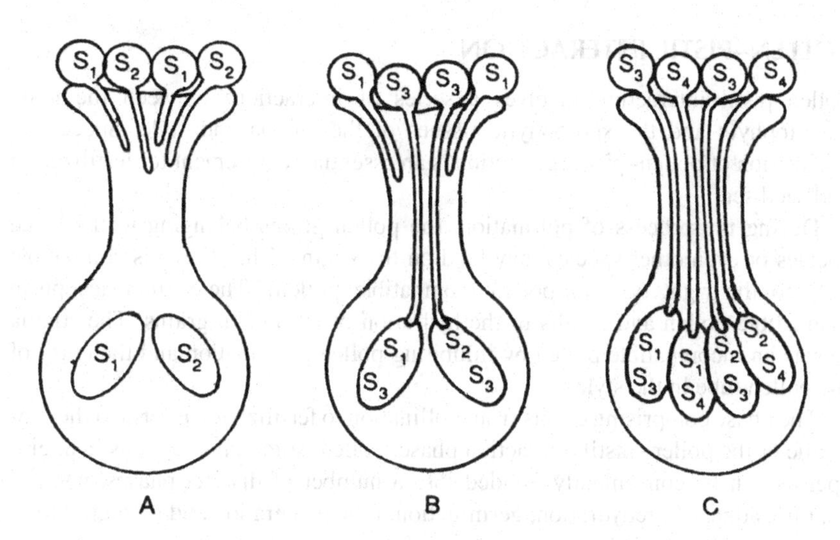


**FIGURE 10.4** Pollen tube growth in compatible and incompatible pollinators.

Self-fertilization is prevented when pollen carrying an S-allele which is identical to that carried by the stigma on which it lands (incompatible) is discriminated from pollen carrying an S-allele which is not carried by the stigma (compatible).

In sporophytic systems, the S-gene is activated before completion of meiosis, with the result that products of both the alleles are distributed in all four microspores and thus in all the pollen grains produced by the plant. In SSI, S-allele specific products are synthesized in the pollen grain as well as tapetum and incorporated in the pollen exine due to the breakdown of the tapetum.

In gametophytic systems, S-gene activation is delayed until completion of meiosis, with the result that two of the microspores of each tetrad receive the products of one of the S-alleles, and the other two the products of the other S-allele; thus, each pollen will have the products of only one of the S-alleles. In the majority of gametophytic systems SI is controlled by a single locus with multiple alleles.

In polymorphic loci, complex genetic dominance relationships between alleles are often observed. In *Brassica* and *Arabidopis*, multiple small RNA genes linked to the Self-incompatibility (SI) locus have been reported. Monoallelic gene expression of the male determinant of SI, *SP11/SCR*, from a dominant *S*-allele is under epigenetic control by such small RNA genes (Fujii and Takayama, 2018).

Self-incompatibility has arisen multiple times in the evolution of flowering plants. SI controlled by many genes is the primitive one, and it is controlled by a single gene in the derived condition.

## POLLEN–PISTIL INTERACTION

Pollen–pistil interaction involves a series of interactions between the male gametophyte and the sporophytic tissues of the stigma and style. Successful completion of pollen–pistil interaction is an essential requirement for fertilization and seed set.

During the process of pollination, the pollen grains belonging to the same species or even other species may land on the stigma. The stigma is compatible only to the right type of pollen (compatible pollen). The stigma accepts a compatible pollen and results in the hydration of the pollen grains. The stigma rejects an incompatible pollen by inhibiting pollen germination and the entry of the pollen tube in the style.

The phase comprising events from pollination to fertilization is termed the progamic or the pollen–pistil interaction phase. Pollen–stigma interactions in angiosperms can be conveniently divided into a number of distinct phases, namely, capture, adhesion, rehydration, germination, tube penetration and tube growth.

### Pollen Recognition

Recognition is a stage which is essential for pollen acceptance and germination. After pollination, incompatibility reactions occur either in the stigma, style or ovule. The pollen grains do not germinate, pollen tube growth is blocked on the stigma, in the style, or in the ovary. The pollen tube may even stop near the embryo sac, but fusion of sperm cell does not take place.

Both wall layers of the pollen grain, the exine and intine, contain a large number of mobile proteins. In the exine, the proteins are located in the sculptured region of the exine in the chambers between the baculate (tectate grains) and surface depressions (non-tectate grains). These wall proteins are readily released into the medium upon moistening. The intine proteins are present in the form of tubules or leaflets and are generally concentrated near the germ pore.

After pollination, pollen wall proteins are released on to the stigma and come in contact with the stigma-surface proteins. If compatible, pollen grains germinate, and the tube enters the stigma, growing through the style. If incompatible, the pistil will initiate a rejection reaction.

In species with a wet stigma and solid style, the recognition of pollen takes place on the stigma. In taxa with a wet stigma and hollow style, the recognition is confined to the style. Many wet types have their stigma cells covered by a hydrophilic exudate rich in proteins, glycoproteins and carbohydrates, e.g., *Lilium*. In *Petunia* the glistening exudate is actually hydrophobic. Herrero and Dickinson (1979) have shown that in *Petunia* the lipophilic zone at the surface is formed by a holocrine secretion from the papillae, which becomes necrotic.

In *Holostemma ada-kodien* (Devipriya and Radhmany, 2021) the stigma is the wet type. The viscous and sticky exudates remain on the surface of stigma for two days after anthesis. The receptivity gradually decreases from the third day and stigmatic lobes become completely dry indicating complete loss of receptivity.

In dry stigmas the pellicle in contact with the pollen coating creates the condition for the recognition reaction. In such cases the rejection reaction results either in failure of pollen to germinate, or the occlusion of short pollen tubes with callose, or the failure to penetrate stigma and style. Details about compatible and incompatible pollinations are given in Table 10.2.

In *Gladiolus* and *Crocus* (both having dry type of stigma), hydration of compatible pollen is dependent on adhesion or some degree of recognition, since foreign pollen may not be permitted to adhere or hydrate.

## Rejection Reaction

Incompatible self-pollinations, such as those controlled by the gametophytically active S-gene in *Lycopersicon peruvianum*, show a rejection response to pollen tubes in a particular region of the style. Tube growth ceases, often with swelling, bursting and callose deposition at the tip. Pandey (1977) has suggested that in *Nicotiana* species, the S-gene has two independent functions governing pollen tube rejection at different levels within the pistil. These are: (i) sporophytic primary specificity controlling rejection of foreign pollen at or close to the stigma, and (ii) gametophytic secondary specificity controlling rejection of incompatible intraspecific pollen further down the style. These two functions of the S-gene may represent distinct points of genetic control, even though they may be expressed through the same molecular mechanisms relating to tip growth of the pollen tubes.

**TABLE 10.2**
**Events After Pollen Lands on Stigma and Hydration of Pollen Grains Takes Place**

| | Compatible Pollination | | Incompatible Pollination |
|---|---|---|---|
| 1. | Pollen grains germinate | 1. | Pollen grains do not germinate |
| 2. | Pollen tubes enter through stigma and grow in the style | 2. | Pollen tubes abort on stigmatic surface |
| 3. | Pollen tubes grow through transmitting tissue/stylar canal | 3. | Pollen tubes abort in the upper part of the style |
| 4. | Pollen tubes enter ovary | 4. | Pollen tubes fail to enter ovary |
| 5. | Pollen tubes enter micropyle | 5. | Pollen tubes fail to enter micropyle |
| 6. | Pollen tubes enter embryo sacs | 6. | Pollen tubes do not enter embryo sacs |
| 7. | Pollen tubes discharge sperms in the embryo sac | 7. | Pollen tubes fail to discharge sperms |
| 8. | Fertilization takes place | 8. | Fertilization does not occur |

### Stigma Surface Inhibition

The surface of the stigmatic papillar cells is covered by a proteinaceous pellicle and, by means of protease digestion experiments, proteins of this layer have been shown to be responsible for binding the pollen (Stead et al., 1980). Thus, the pellicle layer which the pollen grain encounters on alighting on the stigma can be regarded as providing a very special microenvironment for first adhesion and then subsequent recognition reactions (Roberts and Dickinson, 1981).

The pollen-coat, which is tapetally derived and believed to carry the sporophytic S-gene determinants, is the first component of the pollen grain to make contact with the stigma. The self-incompatible pollen grain also encounters difficulties in obtaining water from the stigma during the rehydration phase and fails to rehydrate in a manner similar to cross-pollination. However, the self-pollen grain may take up water from the atmosphere, and significantly high humidity can be effective in overcoming the self-incompatibility response.

Despite the inhibition experienced during adhesion and rehydration, self-pollen grains do frequently succeed in germinating, but the tubes produced are generally malformed. In addition, although these incompatible tubes may grow to a considerable length and encircle the papilla, they rarely penetrate the stigmatic cuticle. The very few tubes which do effect entry into the papillar cell have been shown to stimulate the deposition of callose in the stigma, a reaction which apparently prevents further tube growth (Dickinson and Lewis, 1973).

The stigmatic exudate in *Petunia* is composed mainly of lipids and small amounts of sugars and amino acids originated in the glandular cells and extruded by exocytosis (Dumas et al., 1978). Below the oily layer of the stigma, an extremely thin film of water is present, which is in direct contact with the epidermis. Due to this water film, a moisture gradient is created between the water and the upper surface of the exudate (Konar and Linskens, 1966).

Like pollen wall proteins, stigma surface proteins are also involved in pollen–pistil interaction. In *Gladiolus* (Knox et al., 1976) concanavalin A (con A) binds specifically to the pellicle, which prevents entry to pollen tubes. Besides stigma surface proteins, non-proteinaceous factors, such as phenolic compounds and carbohydrates, may also play a role in pollen hydration and pollen germination. Phenolic compounds have been found to promote or selectively inhibit pollen germination.

In Asteraceae and Brassicaceae the pollen wall proteins are released onto the pellicle where recognition reaction occurs. The compatible pollen tube penetrates the cuticle and grows down the papilla (Figure 10.5A) The incompatible pollen tube, although it penetrates the cuticle, is inhibited from further growth. The most distinct response of the stigma to an incomplete pollen is the development of the callose plug between the plasma membrane and pectocellulosic layer of the stigmatic papillae, just below the point of contact with the pollen (Figure 10.5B). A callose plug also appears at the tip of the pollen tube.

In gametophytic self-incompatibility systems (e.g., grasses), where the rejection reaction occurs on the stigma rather than in style, there is no callose deposition in

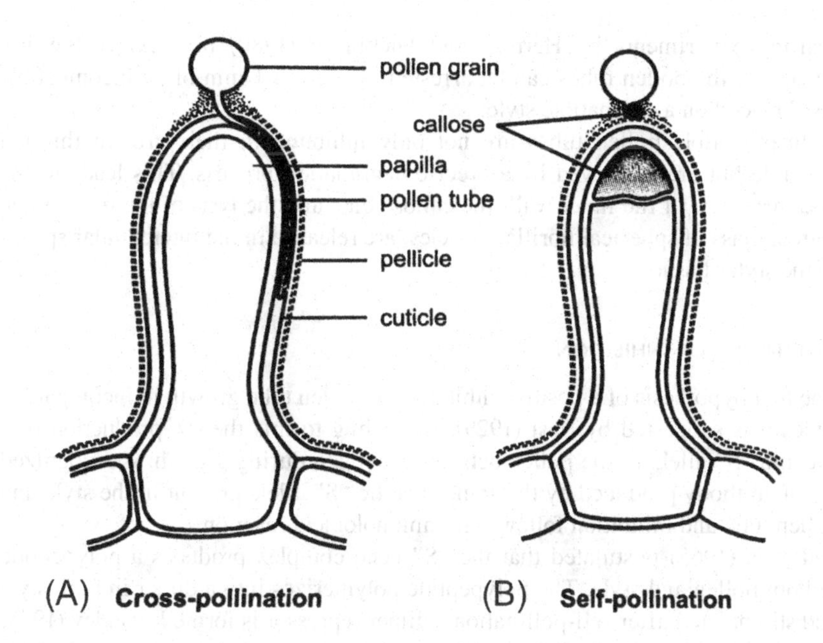


**FIGURE 10.5** Pollen-stigma interaction (after Shivanna, 1982).

the stigmatic papillae following incompatible pollination. The callose deposition occurs in the pollen tube.

## Stylar Inhibition

In *Petunia* and *Lycopersicon* the central part of the style has a transmitting tissue which is composed of cells with broad intercellular spaces. These are filled with very sticky fluid. Pollen tubes grow in this fluid during their journey toward the ovules, and there is evidence that they use substances coming from the intercellular spaces. According to Cresti et al. (1976), the stylar intercellular fluid plays a vital role in the incompatibility reaction.

After the recognition, the style shows different gene activity; quantitative as well as qualitative differences have been found in both RNA and protein synthesis between self- and cross-pollinated styles. During different stages of pollen tube growth, different protein patterns have been observed. This indicates that rejection or acceptance of pollen tubes by a style is not a straightforward reaction but a result of a sequence of activations and inactivations of a number of genes (van der Donk, 1974).

Compatible and incompatible pollen tubes are emitted at the same time, and upon entry into the stylar transmitting tissue, both types of tubes accelerate; but the compatible achieve a higher terminal velocity than do the incompatible ones, which eventually slow down and stop (Herrero and Dickinson, 1980).

The occurrence of a zone where (in the case of a solid style) the recognition between pollen genotype and pistil genotype occurs was demonstrated through grafting experiments by Herrero and Dickinson (1980) in *Petunia.* Seventy percent of the pollen tubes can be arrested by the top 1 mm of an incompatible pistil placed on a compatible style.

Incompatible pollen tubes are not only inhibited in their growth through the style but are destroyed by a precise degradation process. This leads to the disappearance of the inner wall; the callose one, and the lysis of the tube, while a great mass of spherical fibrillar vesicles, are released in the intercellular spaces of the stylar tissue.

### Mechanism of Inhibition

The first hypothesis of a positive inhibition of pollen tube growth in incompatible styles was suggested by East (1929). According to this theory, production of a specific "S" allele in the pollen acts as an antigen during growth, is recognized by an antibody produced by the same specific "S" allele present in the style and pollen tube and inhibition follows an immunological reaction.

Lewis (1965) postulated that the "S" gene complex produces a polypeptide in both pollen and style. The polypeptide polymerizes into a dimer in both style and stigma, and after self-pollination a dimer repressor is formed. Pandey (1975) hypothesized that the S allele proteins in the pollen and style have an identical specificity protein and a tissue-specific adaptive protein. The mechanism by which the repressor inhibits the growth of the pollen tube is not clear.

Nettancourt (1984, 2001) studied ultrastructural details of pollen tubes inhibited in self-incompatible and unilateral incompatible reactions. They concluded that interspecific unilateral incompatibility results from the interaction of S elements in the pollen grains.

Shivanna (1979, 1982) proposed that self-incompatibility inhibition is the result of more than one mechanism, each controlled by a different unlinked or closely linked genes.

## METHODS OF OVERCOMING INCOMPATIBILITY

Self-incompatibility is a hindrance in crop improvement programmes in which production and maintenance of pure lines are involved. There are several methods employed for overcoming both intraspecific and interspecific incompatibility. Some of the techniques are described below:

### Mentor Pollen (Mixed Pollination)

If the incompatible pollen is mixed with compatible, the pollen type of the former would be encouraged to grow. This phenomenon is known as the mentor effect. Pandey (1977) reported success in overcoming both intra- and interspecific incompatibility by using mentor pollen (recognition pollen) in *Nicotiana* species. Knox et al. (1972) separated the wall proteins from compatible pollen of Asteraceae and added to incompatible pollen at the time of self-pollination. This method proved to be highly successful. In *Cosmos* the mentor pollen as well as their diffusates were found to be effective in overcoming self-incompatibility (Howlett et al., 1975). Mentor pollen has been successfully used in *Brassica*, *Lilium*, *Nicotiana*, *Petunia* and other taxa.

The mechanisms implicated in overcoming incompatibility by the mentor pollen method may be summarized as follows:

- The pollen-wall proteins of killed mentor pollen contribute the proteins necessary for mobilizing the incompatibility component in incompatible pollen at least in sporophytic species.
- Mentor pollen provide P-factor which interacts with S-factor from stigma to render it accessible to incompatible pollen.
- Mentor pollen provide a pollen growth-promoting substance which allows incompatible pollen to sustain pollen tube growth.
- After tube penetration in the style, mentor pollen provide substances needed for sustained growth of ovules, ovary and other fruit tissues.

## Bud Pollination

The incompatibility barrier can be overcome by pollinating immature flower buds. This method has been tried successfully on gametophytic and sporophytic species, such as *Brassica*, *Raphanus*, *Petunia* and *Nicotiana*. Success can be achieved if the stigma is first smeared with exudates from an open flower.

## Stub Pollination

In cases where incompatibility reaction is confined to the stigma, or the length of the style of the female parent is more than the maximum length attained by the pollen tube of the male parent, stub pollination has been helpful in overcoming incompatibility. In *Ipomoea trichocarpa* the stigmatic surface is the primary site of incompatibility. If the stigmatic lobe is removed, and the cut surface is pollinated, then the pollen tube grows uninhibited into the ovule. Hecht (1964) grafted the compatible style along with stigma in incompatible stocks of *Oenothera organensis*. The incompatible pollen germinated in the scion, passed through the junction and fertilized the ovule.

## Intra-ovarian and *In Vitro* Pollination

The technique of intra-ovarian pollination involves injecting pollen grain (suspended in a suitable medium) directly into the ovary, achieving pollen germination, pollen tube entry into ovule and fertilization. This method is well suited in flowers with large ovary (e.g., *Papaver*, *Argemone*).

In this technique, the ovary surface is sterilized with ethanol and two punctures are made in the wall, one for injecting the pollen suspension and other to permit the escape of air present in the ovarian cavity (Kanta et al., 1962). A pollen suspension is prepared in distilled water and injected into the ovary. Both the holes are plugged with petroleum jelly. Pollen grains germinate inside the ovary and bring about fertilization. Viable seeds following intra-ovarian pollination have been obtained in *Papaver somniferum*, *P. rhoeas*, *Argemone mexicana* and *A. ochroleuca*. This technique has also been applied to achieve interspecific hybridization between *Argemone mexicana* and *A. ochroleuca* (Kanta and Maheshwari, 1963).

The technique of intra-ovarian pollination is not suitable for those species which do not have enough space in the ovary to inject pollen suspension. This has been overcome by culturing the whole pistils or ovules in order to achieve test-tube fertilization. Rangaswamy and Shivanna (1967) used the test-tube fertilization technique to overcome self-incompatibility in *Petunia axillaris*. The fertilized ovules showed normal development of the embryo and endosperm, and mature viable seeds were obtained in three weeks after fertilization. The test-tube fertilization technique has been successfully used to overcome self-incompatibility in *Petunia hybrida* and *Nicotiana tabacum*.

## Irradiation

Sexual incompatibility can be overcome by exposing the pistils to different doses of gamma- or x-rays. This has been achieved in *Dendrobium*, *Petunia*, *Lilium*, *Rubus*, *Ribes*, *Lycopersicon* and *Nicotiana*. In the *Lilium* style (Hopper and Peloquin, 1968; Pandey, 1970), a high exposure to irradiation ranging from 6,000 to 24,000 rads was required to overcome incompatibility. In *Lycopersicon peruvianum* (De Nettancourt and Ecohard, 1968), chronic exposure to low doses of gamma rays during the entire flowering season significantly increased fruit and seed setting. The treatment delays floral abscission, which otherwise takes place after an incompatible pollination. The irradiation damages the physiological mechanism of self-incompatibility in the style, thus allowing the pollen tube to pass through the style.

## High Temperature Treatment

Incompatibility reactions are affected by high temperature treatment. Heat treatment has been tried to reduce the self-incompatibility reaction in a number of plants like *Malus*, *Pyrus*, *Prunus*, *Oenothera*, *Trifolium*, *Lilium* etc. Hooper et al. (1967) observed that in *Lilium longiflorum* self-incompatibility could be suppressed by pre-pollination treatment of the style at 50°C for six minutes. Ascher (1973) proposed that heat treatment enables the incompatible tubes to make use of the special stylar materials which permit the growth of compatible tubes. In *Secale cereale* a temperature treatment at 30°C is sufficient to overcome self-incompatibility. Molecular studies have indicated that sensitivity to temperature is due to a dominant gene marked as T-gene.

### Application of Growth Substances

One of the factors responsible for failure of fertilization in incompatible forms is the premature shedding of the flower. If the abscission of the flower can be checked by some method, it would enable the slow growing pollen tube to reach the ovary in time. Use of growth hormones like IAA, NAA, 2,4-D in overcoming incompatibility has been made in *Lilium*, *Petunia*, *Tagetes*, *Brassica*, *Ipomoea* and *Oenothera*. It has been reported that these growth hormones inhibit the floral inhibition and allow slow-growing pollen to reach the ovary.

## ARTIFICIAL HYBRIDIZATION

In artificial hybridization, only the desired pollen grains (pollen grains from superior varieties) are used for pollination. In such cases the stigma is bagged to prevent the entry of unwanted pollen. During artificial hybridization, anthers are removed from the bisexual flowers before their dehiscence, with the help of a forceps. This process is called emasculation. Emasculated flowers are covered with a suitably sized bag made up of butter paper. This is done to protect the stigma from contamination with the unwanted pollen. This process is called bagging. When the stigma of the bagged flower becomes receptive, pollen grains from superior varieties are dusted on the stigma and flowers are bagged again. Flowers are left for seed setting and fruit maturation.

If the flowers are unisexual, there is no need of emasculation. In such cases female flower buds are bagged before the opening of flowers. When the stigma becomes receptive, artificial pollination is done using the desired pollen and the flowers are re-bagged.

The process of crossing different species using desired pollen is to combine the desired characteristics to produce commercially 'superior' varieties.

## SUMMARY

- Self-incompatibility is when pollen grains fail to bring about fertilization of ovules in the same plant resulting in the failure of embryo and seed development.
- Self-incompatibility is of two types: heteromorphic and homomorphic incompatibility.
- Self-incompatibility is categorized into two groups: gametophytic and sporophytic self-incompatibility.
- Self-incompatibility is genetically controlled.
- Pollen–stigma interactions in angiosperms can be conveniently divided into a number of distinct phases: namely, capture, adhesion, rehydration, germination, tube penetration and tube growth.
- Incompatibility may be overcome by different methods such as the use of mentor pollen, intra-ovarian pollination and test-tube fertilization, bud pollination, pistil grafting, irradiation, high temperature treatment and application of growth substances.

## SUGGESTED READING

Ascher, P.D. 1975. Special stylar property required for compatible pollen-tube growth in Lilium longiflorum. *Thunb. Bot Gaz* 136: 317–321.

Ascher, P.D. 1976. Self-incompatibility systems in floriculture crops. *Acta Hort.* 63: 205–215.

Bahadur, B. and R. Srikanth. 1983. Pollination biology and the species problem in *Waltheria indica* complex. *Phytomorphology* 33 (1–4): 96–107.

Cresti, M. and J.L. van Went. 1976. Callose deposition and plug formation in Petunia pollen tubes in situ. *Planta* 133: 35–40.

de Nettancourt, D. 1984. Incompatibility. In *Cellular Interactions. Encyclopedia of Plant Physiology*, vol. 17, pp. 624–639. eds. Linskens, H.F. and Heslop-Harrison, J. Berlin: Springer-Verlag.

de Nettancourt, D. 2001. *Incompatibility and Incongruity in Wild and Cultivated Plants.* Heidelberg: Springer Berlin.

Devipriya, D. and P.M. Radhamany. 2021. Reproductive features and in vitro pollinia germination in *Holostemma ada-kodien* Schult, a RET species. *Int. J. Plant Rep. Biol.* 13 (2): 103–108.

Dickinson, H.G. and D. Lewis. 1973. The formation of the tryphine coating the pollen grains of *Raphanus*, and its properties relating to the self-incompatibility system. *Proc. R. Soc. London, Ser. B* 184: 148–165.

Dickinson, H.G., J. Moriarty and J. Lawson. 1982. Pollen-pistil interaction in *Lilium longiflorum*. The role of the pistil in controlling pollen tube growth following cross and self-pollination. *Proc. R. Soc.* 215: 46–62.

Dickinson, H. and I. Roberts. 2014. A molecular basis for the self-incompatibility system operating in *Brassica* sp. *Acta Societatis Botanicorum Poloniae* 50: 227–234. https://doi.org/10.5586/asbp.1981.037.

Dumas, C., M. Rougier, P. Zandonella, M. Cresti and E. Pacini. 1978. The secretory stigma of Lycopersicum esculentum Mill.: ontogenesis and glandular activity. *Protoplasma* 96: 173–187.

East, E.M. 1929. Self-sterility. *Bibl. Genet.* 5: 331–370.

East, E.M. and A.J. Mangelsdorf. 1925. A new interpretation of the hereditary behaviour of self-sterile plants. *PNAS* 11: 166–171.

Fujii, S. and S. Takayama. 2018. Multilayered dominance hierarchy in plant self-incompatibility. *Plant Reproduction* 31: 15–19.

Gerstel, D.U. 1950. Self-incompatibility studies in guayule; inheritance. *Genetics* 35 (4): 482–506. https://doi.org/10.1093/genetics/35.4.482.

Ghosh, S. and K.R. Shivanna. 1980. Pollen-pistil interaction in *Linum grandiflorum. Planta* 149: 257–261.

Ghosh, S. and K.R. Shivanna. 1982. Studies on pollen pistil interactions in *Linum grandiflorum. Phytomorphology* 32: 385–395.

Hecht, A. 1964. Partial inactivation of an incompatibility substance in the stigmas and styles of *Oenothera*. In *Pollen Physiology and Fertilization*, ed. H.F. Linskens, 237–243. Amsterdam: North-Holland Publishing Company.

Herrero, M. and H.G. Dickinson. 1979. Pollen pistil incompatibility in *Petunia hybrida*: changes in the pistil following compatible and incompatible intraspecific crosses. *J. Cell Sci.* 36(1): 1–18.

Herrero, M. and H.G. Dickinson. 1980. Pollen tube growth following compatible and incompatible intraspecific pollinations in *Petunia hybrida. Planta* 148: 217–221. https://doi.org/10.1007/BF00380030.

Hopper, J.E., P.D. Ascher and S.J. Peloquin. 1967. Inactivation of self-incompatibility following temperature pretreatments of styles in *Lilium longiflorum*. *Euphytica* 16: 215–220.

Howlett, B.J., R.B. Knox, J.D. Paxton and J. Heslop-Harrison. 1975. Pollen wall proteins: Physiochemical characterization and role in self-incompatibility in *Cosmos biphnnatus*. *Proc. R. Soc. London B* 188: 167–182.

Hughes, M.B. and E.B. Babcock. 1950. Self-incompatibility in *Crepis foetida* (L.) subsp. *rhoeadifolia* (bieb.) Schinz et Keller. *Genetics* 35: 570–588.

Kanta, K. and P. Maheshwari. 1963. Test-tube fertilization in some angiosperms. *Phytomorphology* 13: 230–237.

Kanta, K., N.S. Rangaswamy and P. Maheshwari. 1962. Test-tube fertilization in flowering plants. *Nature (London)* 194: 1214–1217.

Knox, R.B., A. Clarke, S. Harrison, P. Smith and J.J. Marchalonis. 1976. Cell recognition in plants: Determinants of the stigma surface and their pollen interactions. *Proc. Natl. Acad. Sci. USA* 73 (8): 2788–2792.

Knox, R.B., R. Willing and A.E. Ashford. 1972. Role of pollen-wall proteins as recognition substances in interspecific incompatibility in poplars. *Nature* 237: 381–383.

Konar, R.N. and H.F. Linskens. 1966. The morphology and anatomy of the stigma of *Petunia hybrida*. *Planta* 71: 356–371. https://doi.org/10.1007/BF00396321.

Karfmann, H., H. Kirch, T. Wemmer, A. Peil, F. Lottspeich, H. Uhrig, F. Salamini and R. Thompson. 1992. Sporophytic and gametophytic self-incompatibility. In *Sexual Plant Reproduction*, eds. M. Cresti and A. Tiezzi. Berlin: Springer-Verlag. https://doi.org/10.1007/978-3-642-77677-9_11

Lewis, D. 1965. A protein dimer hypothesis on incompatibility. Proc. 11th Int. Congr. Genet. The Hague, 1963. In *Genetics Today*, Vol. 3, ed. Geerts, S.J., 656–663 London: Pergamon Press.

Nettancourt, D de. 1972. Self-incompatibility in basic and applied researches with higher plants. *Genet. Agrar* 26: 163–216.

Nettancourt, D de, M. Devreux, U. Laneri, M. Cresti, E. Pacini and G. Sarfatti. 1974. Genetical and ultrastructural aspects of self- and cross-incompatibility in interspecific hybrids between self–compatible Lycopersicum esculentum and self-incompatible L. peruvianum. *Theoret. Appl. Genet.* 44: 278–288.

Nettancourt, D. and R. Ecochard. 1968. Effects of chronic irradiation upon a self-incompatible clone of *Lycopersicum peruvianum* Mill. *Theor. Appl. Genet.* 38: 289–299.

Pandey, K.K. 1968. Compatibility relationships in flowering plants: Role of the S-gene complex. *Am. Nat.* 102: 475–489.

Pandey, K. K. 1970. Elements of the S–gene complex. VI. Mutations of the self–incompatibility gene, pseudo–compatibility and origin of new incompatibility alleles. *Genetica* 41: 477–516.

Pandey, K.K. 1970. Time and site of the S gene action, breeding systems and relationships in incompatibility. *Euphytica* 19: 364–372.

Pandey, K.K. 1975. Sexual transfer of specific genes without gametic fusion. *Nature* 256: 311–312.

Pandey, K.K. 1977. Mentor pollen: Possible role of wall-held pollen growth promoting substances in overcoming intra- and interspecific incompatibility. *Genetica* 47: 219–229.

Rangaswamy, N. and K.R. Shivanna. 1967. Induction of gamete compatibility and seed formation in axenic cultures of a diploid self-incompatible species of *Petunia*. *Nature* 216: 937–939. https://doi.org/10.1038/216937a0.

Roberts, I. N., A. D. Stead, D. J. Ockendon and H.G. Dickinson. 1980. Pollen stigma interactions in Brassica oleracea. *Theor. Appl. Genet.* 58: 241–246.

Sastri, D.C. and K.R. Shivanna. 1979. Role of pollen-wall proteins in intraspecific incompatibility in *Saccharum bengalense. Phytomorphology* 29: 324–330.

Shivanna, K. R. 1979. Recognition and rejection phenomena during pollen-pistil interaction. Proc. Indian Academy Sci.-Section B. Part 2, *Plant Sciences* 88(2): 115–141.

Shivanna, K.R. 1982. Pollen-pistil interaction and control of fertilization. In *Experimental Embryology of Vascular Plants*, ed. B.M. Johri. New Delhi: Narosa Publishing House.

Shivanna, K.R. 1995. Pollen physiology and pollen-pistil interaction. In *Botany in India-History and Progress*. Vol. II, ed. B.M. Johri, pp. 147–163. New Delhi: Oxford and IBH Publishing Co.

Shivanna, K.R. 2003. *Pollen Biology and Biotechnology*. New York: CRC Press.

Shivanna, K.R., Y. Heslop-Harrison and J. Heslop-Harrison. 1978. The pollen-stigma interaction: Bud pollination in the Cruciferae. *Acta Bot. Neerl.* 27 (2): 107–119.

Stead, A.D., I.N. Roberts and H.G. Dickinson. 1980. Pollen-stigma interactions in *Brassica oleracea*: The role of stigmatic proteins in pollen grain adhesion. *J. Cell Sci.* 42: 417–423.

Van der Donk, J. 1974. Synthesis of RNA and protein as a function of time and type of pollen tube - style interaction in petunia-hybridal. *Mol. Gen. Genetics* 134: 93.

Van der Donk, JAWM. 1975. Recognition and gene expression during incompatibility reaction in Petunia hybrida L. *Mol. Gen. Genet.* 141: 305–317.

Zinki, G.M., B.I. Zweibei, D.G. Grier and D. Preuss. 1999. Pollen stigma adhesion in *Arabidopsis*: A species-specific interaction mediated by hydrophobic molecules in the pollen exine. *Development* 126: 5431–5440.

Zuberi, M.L. and Lewis, D. 1988. Gametophytic-sporophytic incompatibility in Cruciferae, *Brassica campestris. Heredity* 61: 367–377.

# 11 Endosperm

In gymnosperms, the haploid female gametophyte is called endosperm, but in angiosperms, the endosperm is produced by the repeated divisions of the primary endosperm nucleus. Nuclei or cells derived from the fusion product of a sperm and one or more polar nuclei constitute the endosperm. In most angiosperms, the endosperm is triploid. It is diploid in *Butomopsis*, *Oenothera* and pentaploid in *Penaea* and *Plumbago*.

The chief function of the endosperm is to provide nourishment to the developing embryo and later during germination. In the majority of angiosperms, growth of the embryo is dependent on endosperm. In many taxa, the endosperm is utilized during the development of the seed. Such seeds are called exalbuminous or non-endospermic (e.g., pea, beans). In endospermic seeds, the endosperm serves as an organ of storage and persists in the mature seed (e.g., cereals, castor bean, coconut). Endosperm is found in almost all the angiosperms except in families Podostemaceae, Orchidaceae and Trapaceae.

## DEVELOPMENT OF ENDOSPERM

The division of the primary endosperm nucleus is followed by repeated divisions of the daughter nuclei. On the basis of whether a wall is formed or not, after the division of the primary endosperm nucleus, the endosperm is classified as nuclear, cellular or helobial types (Figure 11.1). Some of the families showing different patterns of endosperm development are mentioned in Table 11.1.

## NUCLEAR ENDOSPERM

In this type, the primary endosperm nucleus and the daughter nuclei undergo free nuclear divisions (Figures 11.2A-C). When many nuclei have been formed, they arrange themselves peripherally in a layer of cytoplasm (Figure 11.2D). The endosperm remains nuclear throughout the course of development as in *Floerkea*, *Limnanthes*, *Oxyspora* or it may become cellular (Figures 11.2E, F). The degree of cellularization, however, varies in different taxa.

In many taxa, wall formation is limited only to upper and middle regions of the embryo sac so that the chalazal region remains free nuclear. This region often elongates and functions as a haustorium, e.g., *Grevillea robusta*. In *Citrullus fistulosus* the haustorium becomes cellular with multinucleate cells.

In *Prosopis africana* the primary endosperm nucleus divides to form 6–8 nuclei. At the two-celled proembryo stage, the number of endosperm nuclei increases, and at the preglobular stage, the number becomes 20–25. At the globular embryo stage, cellularization starts in the endosperm from the micropylar

DOI: 10.1201/9781003260097-11

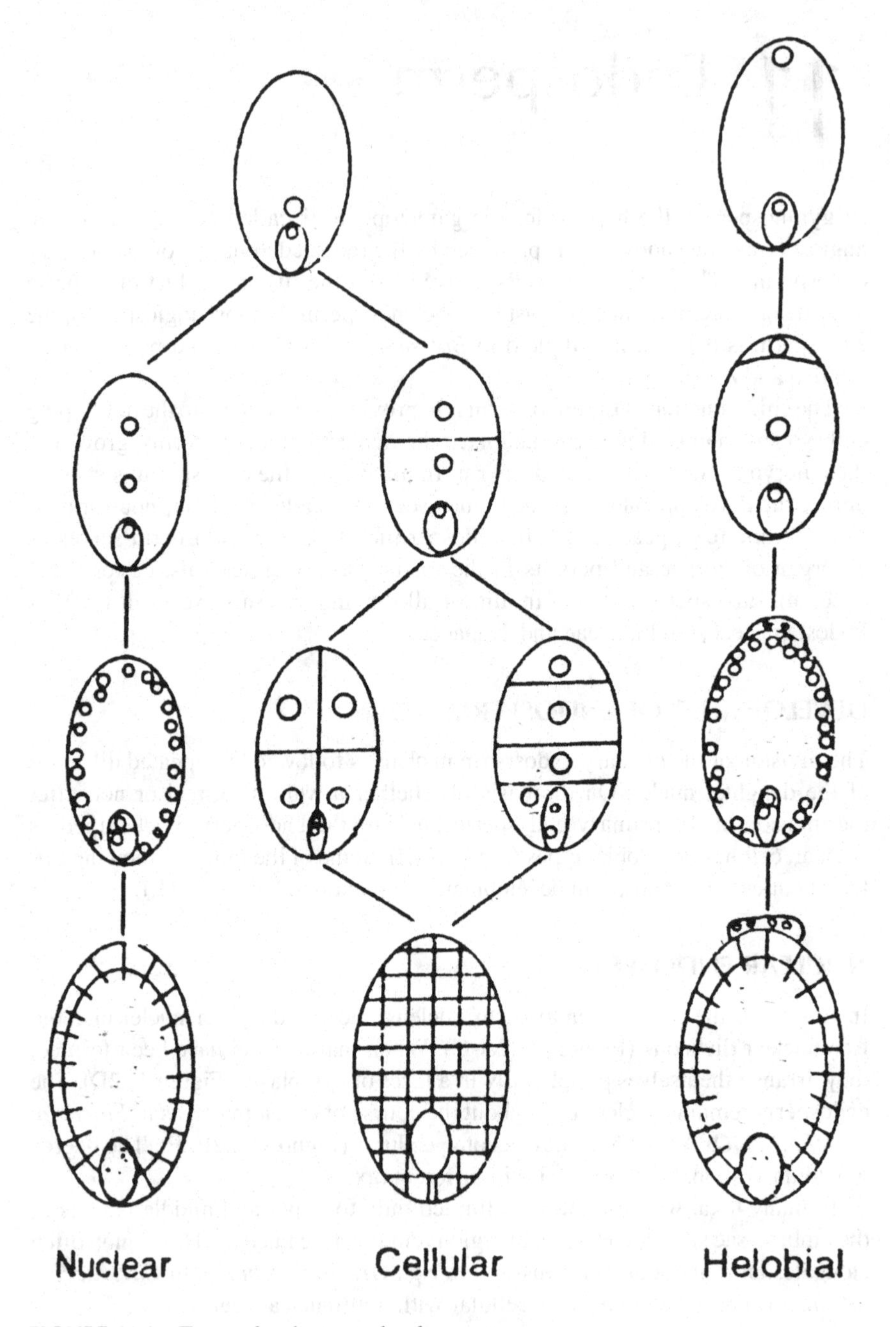


**FIGURE 11.1** Types of endosperm development pattern.

**TABLE 11.1**
**Families Showing Patterns of Endosperm Development**

| Features | Families |
|---|---|
| Nuclear | Apiaceae, Cucurbitaceae, Fabaceae |
| Cellular | Acanthaceae, Bignoniaceae, Scrophulariaceae |
| Helobial | Butomaceae, Hydrocharitaceae, Najadaceae |
| Both nuclear and cellular | Asteraceae, Apocynaceae (subfamily Asclepiadoideae, formerly Asclepiadaceae), Rubiaceae |
| Both cellular and helobial | Olacaceae, Santalaceae, Thismiaceae |
| Both nuclear and helobial | Agavaceae, Alismataceae, Liliaceae |
| Nuclear, cellular and helobial | Boraginaceae, Solanaceae |
| Endosperm absent | Podostemaceae, Orchidaceae, Trapaceae |

end towards chalaza. At the advanced globular stage, even at the chalazal end, a small portion of free nuclear endosperm is seen which functions as endosperm haustorium (Figure 11.3).

In *Ipomoea* the primary endosperm nucleus divides prior to that of the zygotic nucleus. Its first division gives rise to two free nuclei which divide further several times in a free nuclear fashion, and before cellularization commences, the embryo sac is filled with a large number of free nuclei. Cellularization of the endosperm begins from the micropylar end and proceeds gradually towards the chalazal end. After the completion of cell wall formation, massive tissue of endosperm is organized.

In *Scleria foliosa*, at the two-nucleate stage, a tubular extension is formed at the micropylar end of the embryo sac. It remains enucleate with scanty cytoplasm and functions as a haustorium (Figure 11.4).

In *Lomatia polymorpha*, besides the main chalazal haustorium, numerous single-celled, finger-like outgrowths develop all over the surface of the endosperm. This increases the absorptive surface of the endosperm.

The concentration of polysaccharides, total proteins, histones, DNA and RNA increases in the endosperm of *Linum usitatissimum* and *Ranunculus* during embryogeny, and by the time cotyledons are well-differentiated, the endosperm cells accumulate protein bodies. In *Alysum maritimum* and *Iberis amara* the micropylar portion of the endosperm stains intensely for proteins, RNA and DNA but lacks starch or protein bodies during early embryogenesis.

The endosperm of coconut (*Cocos nucifera*) is very interesting. The primary endosperm nucleus undergoes a number of free nuclear divisions, and when fruits are about 50 mm long, the embryo sac is filled with liquid endosperm, which contains many free nuclei. It is referred to as "liquid syncytium". The liquid endosperm contains growth hormones, protein granules, oil droplets and many nuclei. The ploidy of the nuclei in the syncytium varies from 2n to 10n. Mitotic activity in the nuclei of syncytium is rapid and followed by wall formation. The cellular endosperm thus formed is called coconut meat.

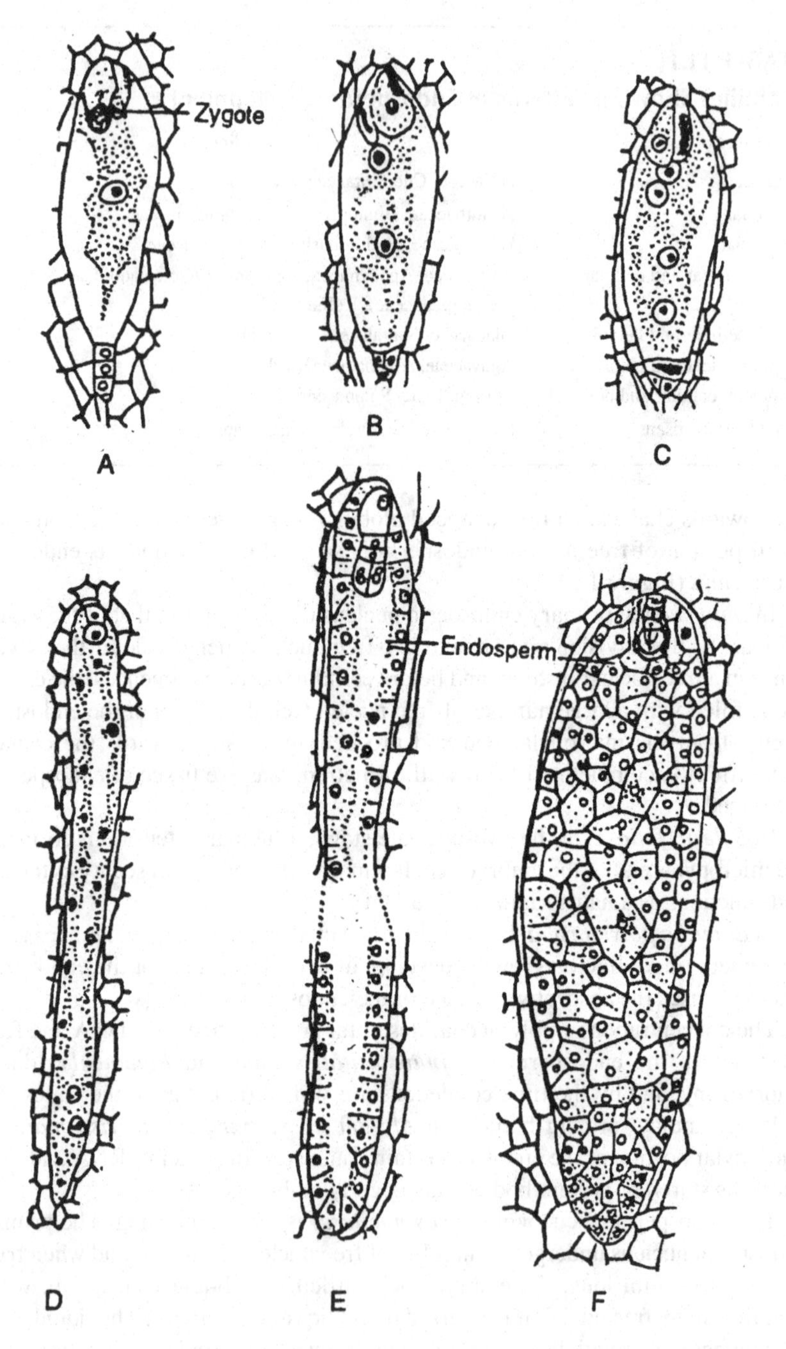


**FIGURE 11.2** Development of endosperm in *Cyperus* species (after Tejvathi, 1987).

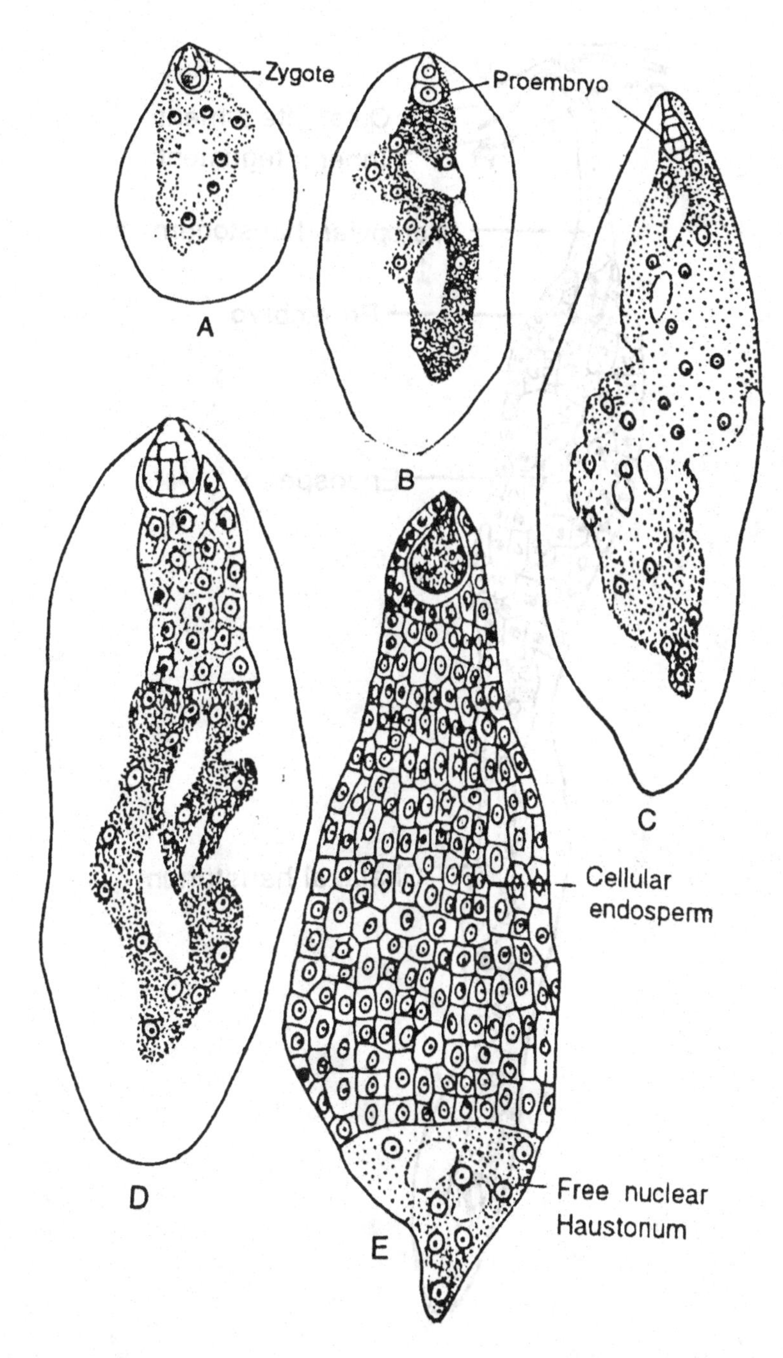


**FIGURE 11.3** Endosperm development in *Prosopis africana* (after Rao and Mukherjee, 1995).

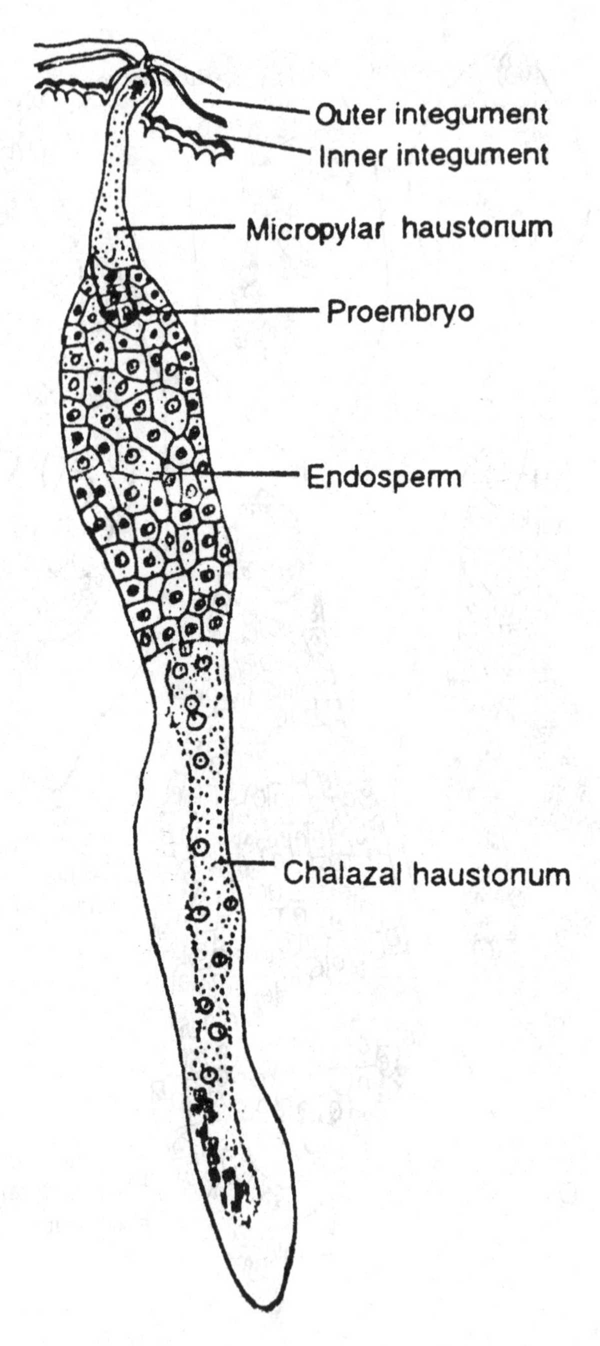


**FIGURE 11.4** *Scleria foliosa*. Endosperm with micropylar and chalazal haustoria (after Nijlingappa and Devaki, 1979).

In *Areca catechu* the endosperm development is similar to that found in coconut, but the embryo sac cavity is small and it becomes completely filled with endosperm. It later becomes extremely hard.

### Cytoplasmic Nodules

Cytoplasmic extensions of definite shape growing into the embryo sac during endosperm development are called cytoplasmic nodules or vesicles. They are of two types, nucleate and enucleate. Nucleate nodules are reported in *Impatiens roylei*, *Isomeris arborea*, *Musa errans* and *Carica papaya*. Enucleate nodules occur in *Oldenlandia corymbosam*, *Cyclanthera explodens*, *Sechium edule*, *Blastania fimbistipula*. In *Volulopsis nummularia*, *Ipomoea angulata* and *Mina lobata* enucleate cytoplasmic bulges are observed at the globular embryo stage (Figure 11.5). These bulges in later stages of seed development merge with the general cytoplasm of the free nuclear endosperm. These cytoplasmic nodules function to fill up the central vacuole.

## CELLULAR ENDOSPERM

The division of primary endosperm nucleus and subsequent divisions are followed by the wall formation. In *Circidiphyllum japonicum* a large micropylar and a small chalazal chamber are produced following the transverse division of the primary endosperm nucleus (Figure 11.6). Both the chambers undergo transverse divisions, and the endosperm gives a uniseriate appearance; vertical and oblique walls develop afterwards.

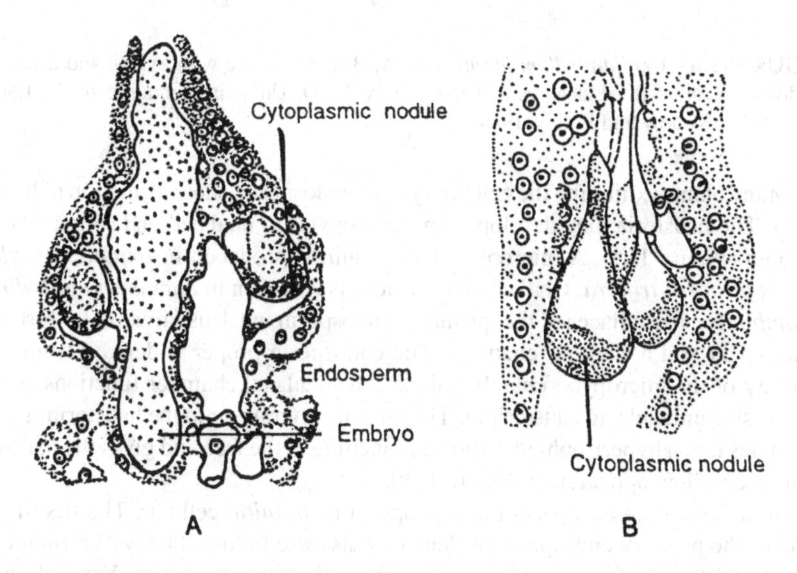


**FIGURE 11.5** A. *Mina lobata.* Longitudinal section of a part of embryo sac showing enucleate cytoplasmic nodules (after Kaur and R.P. Singh, 1987). B. *Jatropha curcas.* A part of free nuclear endosperm showing cytoplasmic nodules (after R.P. Singh, 1970).

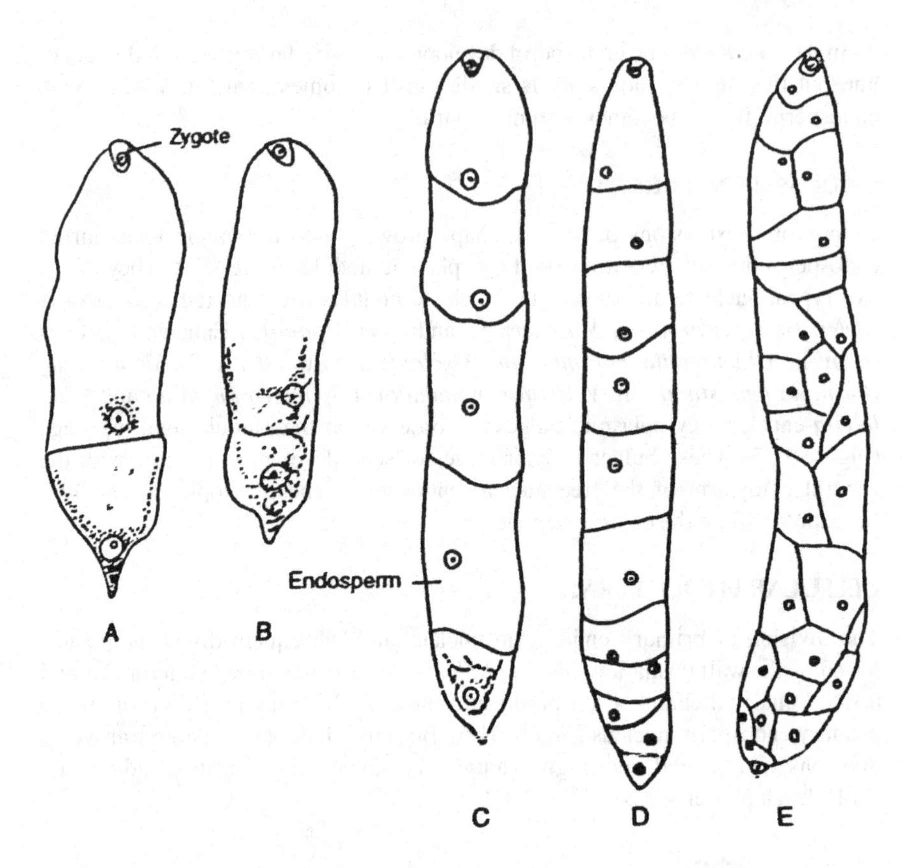


**FIGURE 11.6** *Cercidiphyllum japonicum.* A, B. Embryo sac with zygote and 2-celled endosperm. Note three degenerated antipodals. C, D. Uniseriate endosperm. E. Later stage (after Swamy and Bailey, 1949).

Many plants exhibiting the cellular type of endosperm, show formation of haustoria. The haustoria may develop at the micropylar or chalazal end (Figure 11.7) or both (Figure 11.8). A micropylar haustorium is produced in *Impatiens roylei* and *Hydrocera triflora*. Chalazal haustorium is common in Santalales. In *Iodina rhombifolia* (Santalaceae) the primary endosperm nucleus divides to form a micropylar and a chalazal chamber. The endosperm proper is derived from the activity of the micropylar chamber alone. The chalazal chamber functions as an aggressive uninucleate haustorium. The nucleus migrates into the haustorium and becomes much hypertrophied. Profuse branching at the free end gives the haustorium a coralloid appearance (Figure 11.9).

In *Microcarpaea muscosa* the endosperm is *ab initio* cellular. The first division of the primary endosperm nucleus is transverse followed by wall formation. This results in the formation of two primary endosperm chambers. Vertical divisions occur in both these chambers to form a four-celled endosperm of two tiers, each with two cells. The two juxtaposed cells derived from the chalazal chamber

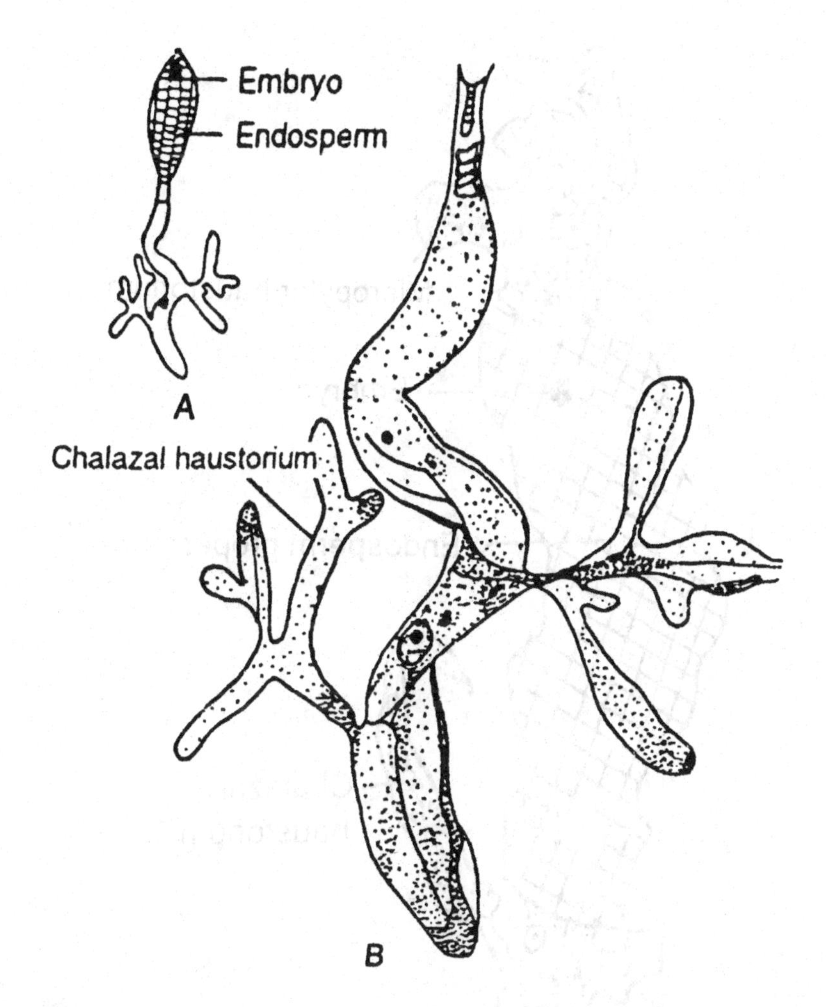


**FIGURE 11.7** *Cajophora sylvestris.* A. Endosperm showing chalazal haustorium. B. Magnified view of the chalazal haustorium shown in A (after Garcia, 1962).

divide further at right angles to the first and thus give rise to four cells which finally develop into the chalazal haustorium. The two cells at the micropylar end divide transversely resulting in two tiers of two cells each. The upper tier of cells develops into the micropylar haustorium and the lower one serves as initials of the endosperm proper (Figure 11.10).

In Scrophulariaceae and Orobanchaceae both micropylar and chalazal haustoria are formed. In *Thunbergia alata* the micropylar chamber forms a branched, coenocytic haustorium, whereas the chalazal chamber gives rise to the endosperm proper.

In *Linaria bipartita* both micropylar and chalazal haustoria show hypertrophied nuclei which undergo several cycles of DNA synthesis and increase of nucleolar and cytoplasmic RNA and proteins. The haustorial cells, however, lack

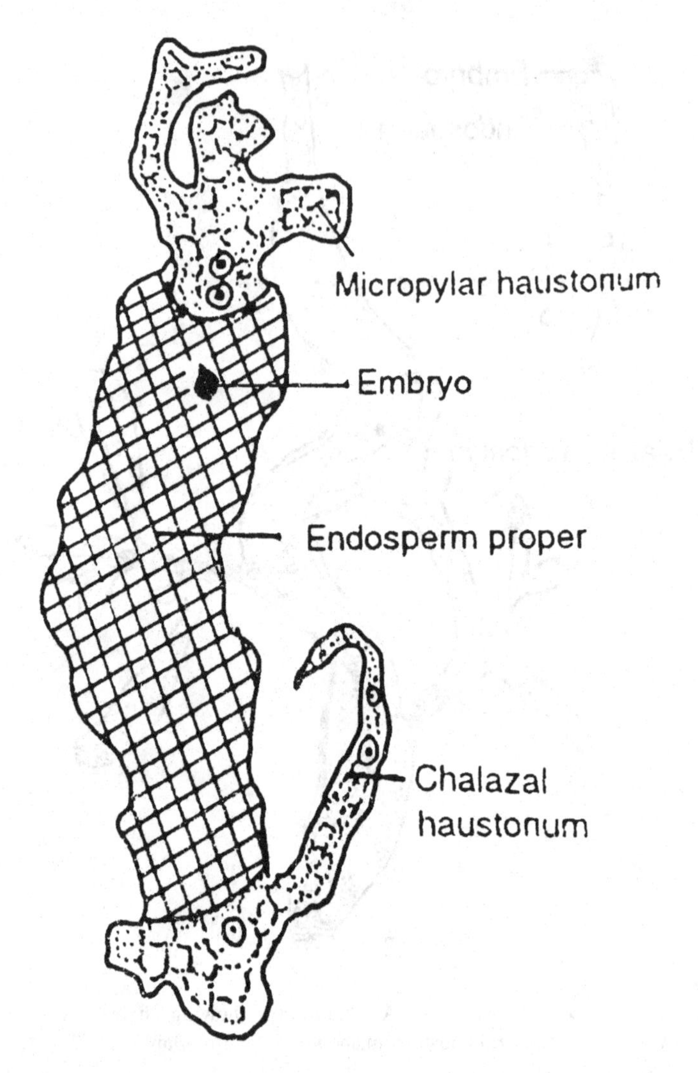


**FIGURE 11.8** *Cyrilla racemiflora*. Endosperm with one-celled, binucleate, branched, micropylar haustorium, central endosperm proper and one-celled 4-nucleate chalazal haustorium (after Vijayaraghavan and Dhar, 1978).

starch grains. On the other hand, other cells of the endosperm accumulate large quantities of starch and protein bodies.

## HELOBIAL ENDOSPERM

This type of endosperm is mostly found in monocotyledons (e.g., *Haemanthus multiflorum*, *Zephyranthes citrina*, *Juncus bufonius*). The primary endosperm nucleus migrates towards the chalazal end of the embryo sac where it divides

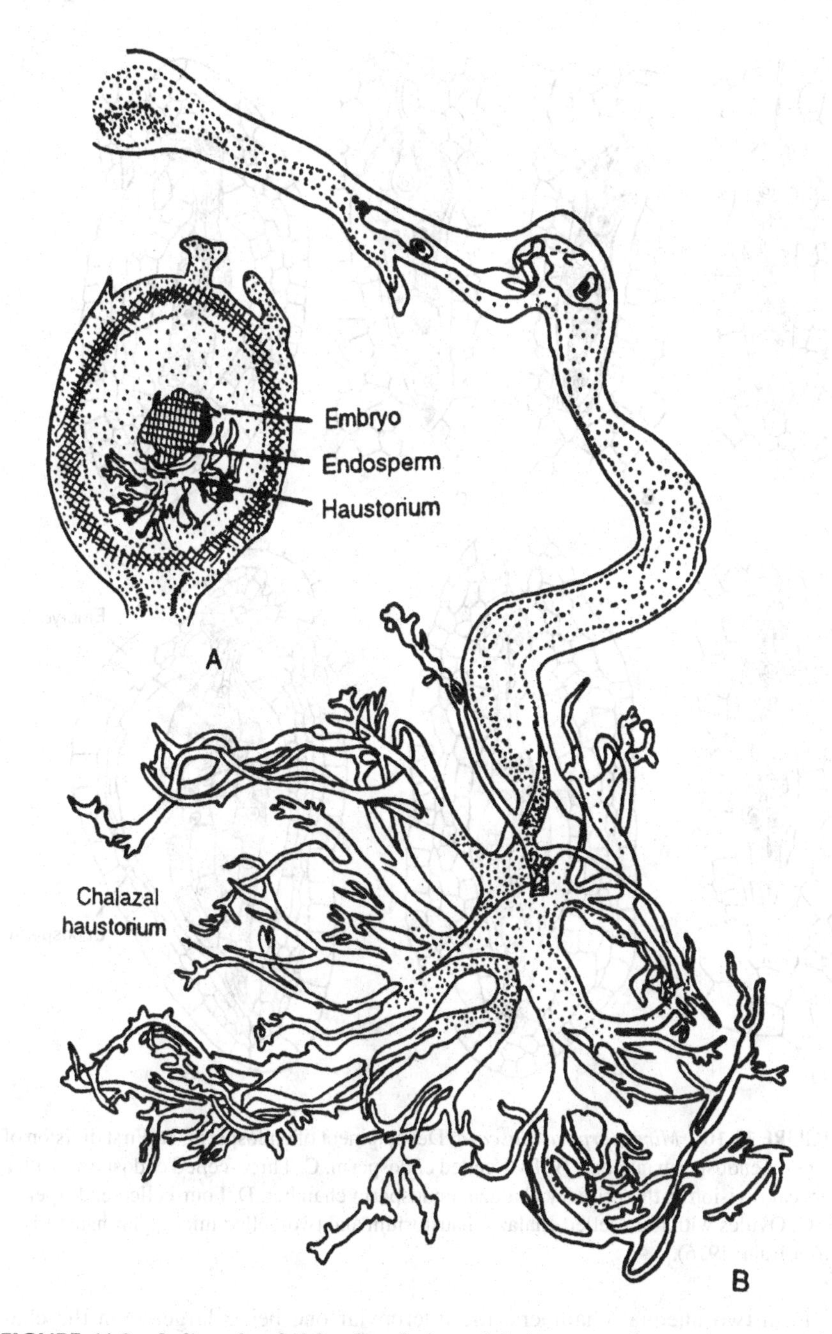


**FIGURE 11.9** *Iodina rhombifolia*. L.s. fruit at globular stage of proembryo; note the aggressive nature of the chalazal endosperm haustorium. B. Haustorium with a much-branched lower end (after Bhatnagar and Sabharwal, 1969).

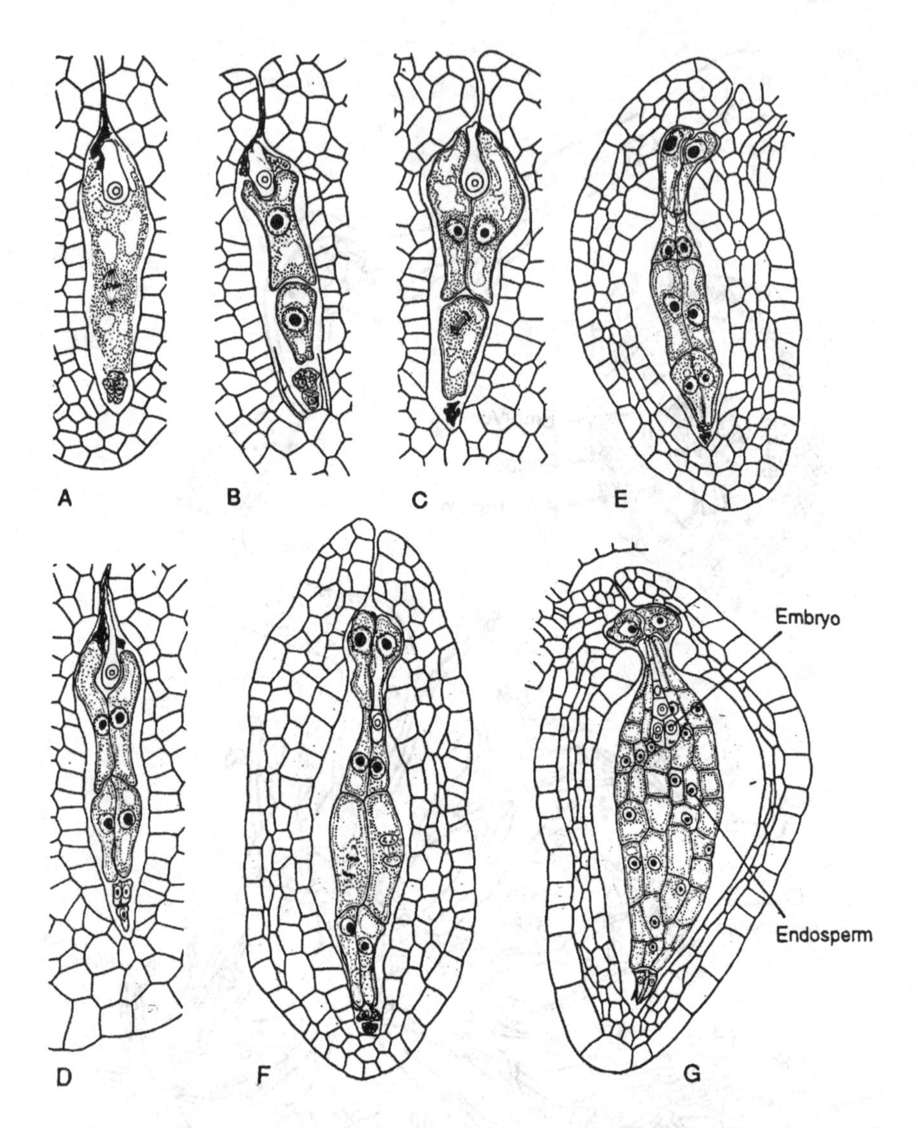


**FIGURE 11.10** *Microcarpaea muscosa.* Development of endosperm. A. First division of primary endosperm nucleus. B. Two-celled endosperm. C. Three-celled endosperm with a vertical division in the primary chalazal endosperm chamber. D. Four-celled endosperm. E–G. Ovules with four-celled chalazal haustorium and two-celled micropylar haustorium (after Raju, 1976).

to form two unequal chambers, the micropylar one being larger than the chalazal one. In the micropylar chamber, several free nuclear divisions take place (Figure 11.11). Cell wall formation, if any, begins at a much later stage. The chalazal chamber behaves differently and its nucleus may or may not divide. It usually remains coenocytic.

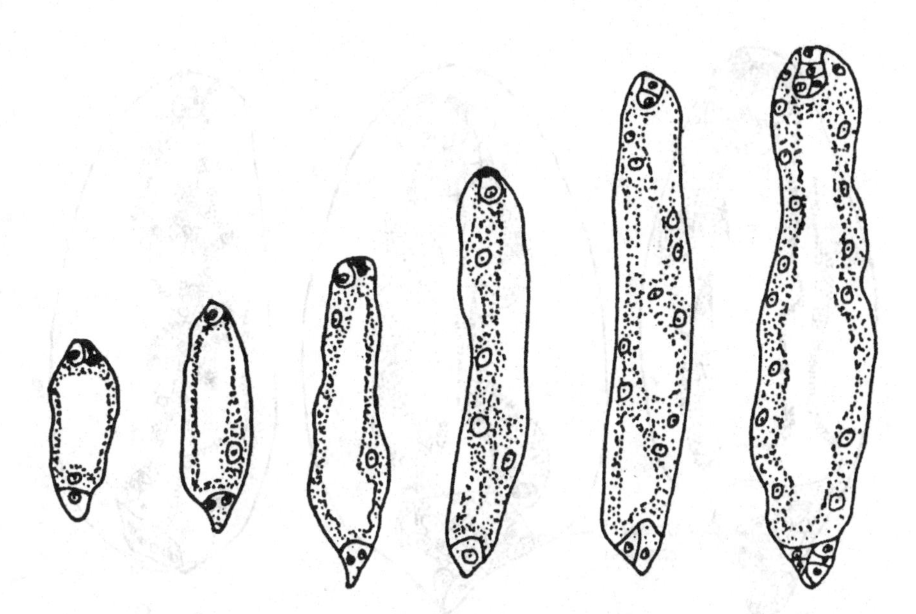

**FIGURE 11.11** *Philydrum lanuginosum.* Development of helobial type endosperm (after Hamann, 1966).

In *Sparganium* (Sparganiaceae) the primary endosperm nucleus lies near the antipodals, and its division results in a large micropylar and a small chalazal chamber (Figure 11.12). The divisions in both the micropylar chamber and the chalazal chamber are synchronous and free nuclear (Figure 11.12). Wall formation in the micropylar chamber does not always precede that in the chalazal chamber (Asplund, 1973). The endosperm cells become multinucleate.

The development in the chalazal chamber follows two different courses: (i) A coenocyte with as many as 59 nuclei may be formed due to free nuclear divisions. Later the nuclei disintegrate. (ii) Wall formation takes place after the second nuclear division; further divisions are followed by wall formation, resulting in a hemispherical tissue containing 300–400 cells.

In *Ottelia alismoides* the primary endosperm nucleus moves towards the chalazal side and divides forming a large micropylar chamber and a small chalazal chamber (Indra and Krishnamurthy, 1986). As a result of free nuclear divisions, micropylar endosperm becomes multinucleate. Even at the cotyledonary embryo stage, mitotic divisions are synchronous throughout this chamber. Cellularization of the micropylar chamber begins in the late cotyledonary stage. The micropylar endosperm is completely cellular at the mature embryo stage. The micropylar endosperm in the nuclear as well as in the cellular states contains very little cytoplasmic proteins, DNA and RNA and no carbohydrates. The chalazal endosperm, which remains unicellular till the end, is densely cytoplasmic and shows a positive staining reaction for cytoplasmic proteins, DNA and RNA but is completely devoid of insoluble carbohydrates.

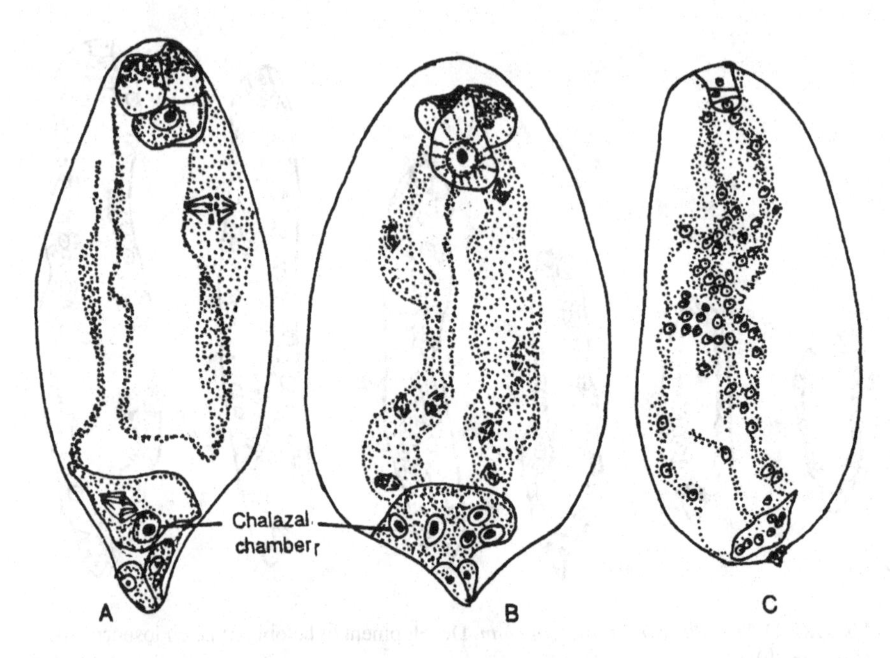


**FIGURE 11.12** Development of endosperm in *Sparganium*. A. Embryo sac with zygote and helobial type of endosperm. B. Synchronous divisions in micropylar chamber and chalazal chamber. C. Three-celled proembryo, multinucleate micropylar chamber, coenocytic chalazal chamber (after Asplund, 1973).

## WALL FORMATION IN ENDOSPERM

In *Triticum aestivum* (Mares et al., 1975), the cellularization of free-nucleate syncytium is accompanied by freely growing walls which, with the accompanying cytoplasm and organelles, extend centripetally from the periphery of the central cell. The central cell is gradually cleaved into smaller, highly vacuolate compartments. According to Mares et al. (1977), the first anticlinal walls arise as projections of the wall of the central cell and form open cylinders or alveoli. The alveoli are not complete cells as cell wall material is absent in this thin layer of cytoplasm. The first periclinal walls develop within the alveoli as a result of normal phragmoplast following nuclear division, forming a peripheral layer of cells and the new inner layer of alveoli. Subsequent periclinal divisions result in the formation of cellular endosperm (Figure 11.13).

Fineran et al. (1982) observed that during initial cellularization of wheat endosperm a typical phragmoplast develops between the free endosperm nuclei that occupy the peripheral cytoplasm adjoining the central cell wall. The cell walls soon fuse with the wall of the central cell. The anticlinal wall develops on all sides of the nuclei and soon fuses with each other in such a way that an open cylinder with a free inner end is formed around each nucleus (Figure 11.14). These open cylinders undergo cell division and form periclinal walls through normal

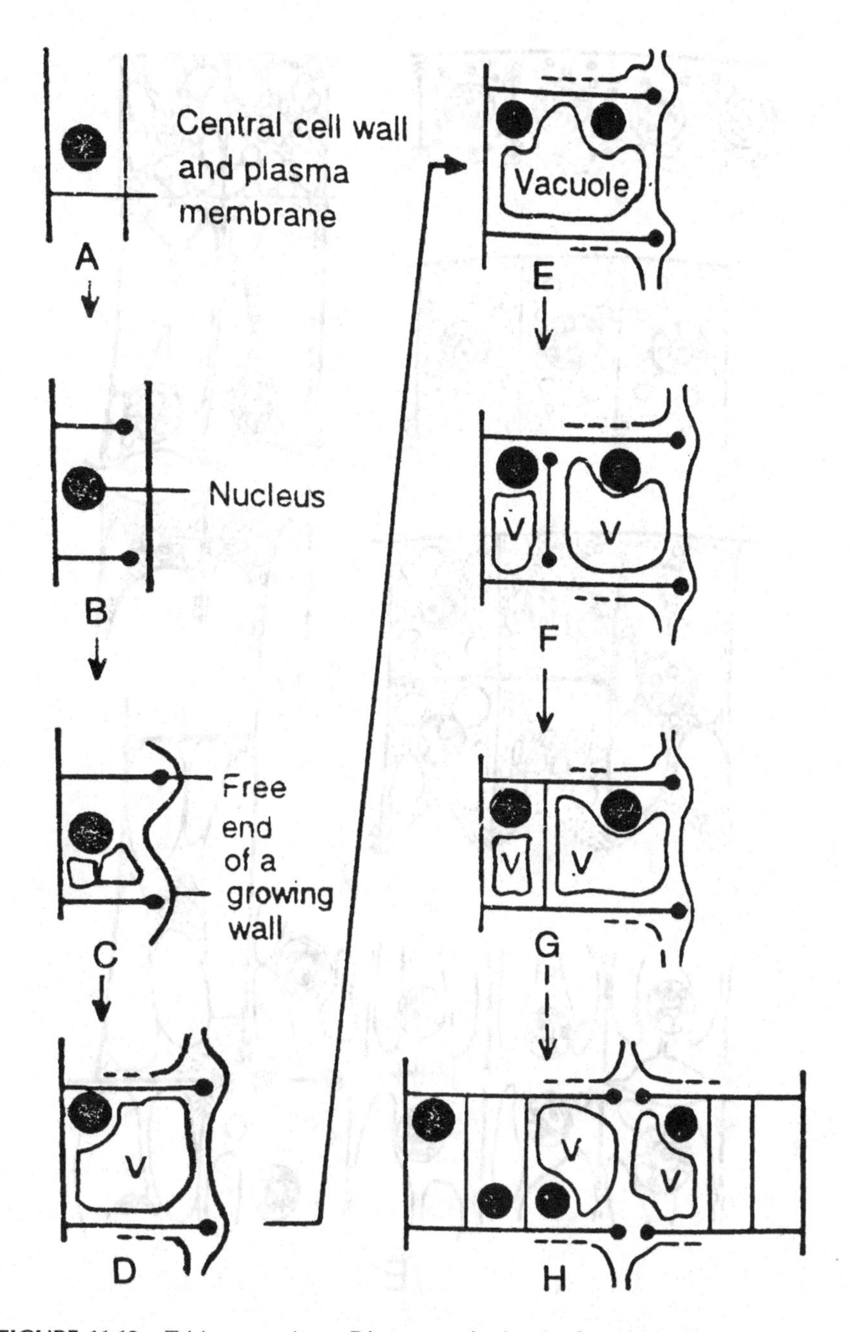


**FIGURE 11.13** *Triticum aestivum.* Diagrammatic sketch of events leading to cell wall formation in nuclear endosperm (adapted from Mares et al., 1977).

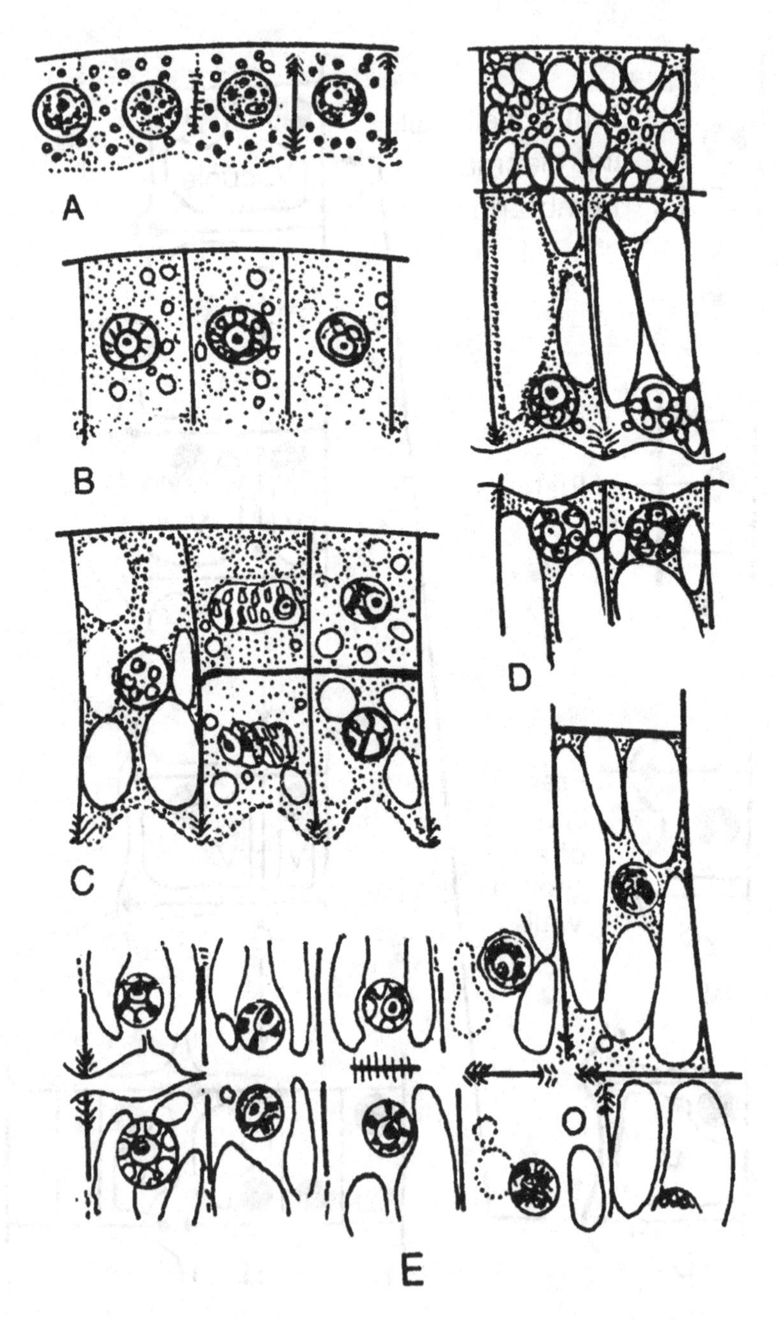


**FIGURE 11.14** *Triticum aestivum*. Cellularization of endosperm (after Fineran et al., 1982).

cytokinesis. The inner compartments usually continue to extend centripetally until the free ends of the anticlinal walls from two sides of the central cell fuse together. The final closing phase is characterized by a further row of periclinal walls formed along the line where the compartments fuse. Further subdivisions in the large endosperm cells result in the formation of smaller cells (Figure 11.14).

In *Papaver somniferum* large vacuoles appear prior to wall formation, which push the nuclei laterally and position them equidistantly. A large number of fine thread-like, rather feeble, PAS-positive walls are observed extending perpendicularly, from the wall of the central cell inwards, on the sides of nuclei. These newly formed walls grow freely towards the central vacuole. Serial sections revealed that three-dimensionally, the walls are disc-like and extend along their circumference. As a result, the adjacent walls fuse and partition the endosperm into a layer of "open compartments" or "open cells" at the periphery of the central vacuole. The open compartments are uninucleate and are highly vacuolate. When the freely growing walls are fairly long, the nuclei of the open compartments undergo mitotic divisions in a periclinal plane accompanied by phragmoplast and cell plate formation. The newly formed periclinal walls grow and fuse with the freely growing perpendicular side walls, resulting in the formation of an outer layer of closed cells and inner large open compartments. By similar periclinal divisions of the "open cell" and continued centripetal extension of freely growing walls, the central vacuole is eventually obliterated, and the endosperm becomes completely cellular.

## RUMINATE ENDOSPERM

Ruminate endosperm may be defined as the endosperm which exhibits any degree of irregularity and unevenness in its surface contour within the mature seed.

The rumination is caused by the irregularities in the inner surface of the seed coat which develop in two ways: (i) by an unequal radial elongation of the cells of any one layer or the outer layer of the seed coat, e.g., *Passiflora calcarata* (Figure 11.15); or (ii) by a definite ingrowth or infoldings of the seed coat. This is brought about either by excessive elongation or enlargement of cells of the seed coat, e.g., *Annona squamosa* (Figure 11.16A), *Atrabotrys blumei* (Annonaceae), *Apana siliquosa* (Aristolochiaceae) or by localized meristematic activity of the cells of the seed coat (e.g., *Myristrica fragrance*, *Diospyros chloroxylon* and ruminate seeds of Arecaceae (Palmae)).

In *Elytraria acaulis*, *Andrographis paniculata* (Figure 11.16B) and *Andrographis echinoides*, seed coat does not form infoldings but becomes consumed and pushed in an irregular manner by the unequal peripheral activity of the enlarging endosperm.

Periasamy (1962) reported ruminate endosperm in 30 families of dicotyledons, one of monocotyledons and one of the gymnosperms. He classified seven types of rumination based on morphological features. These typically present in the following taxa: *Passiflora*, *Verbascum*, *Annona*, *Myristica*, *Coccoloba*, *Spigelia* and *Elytraria*.

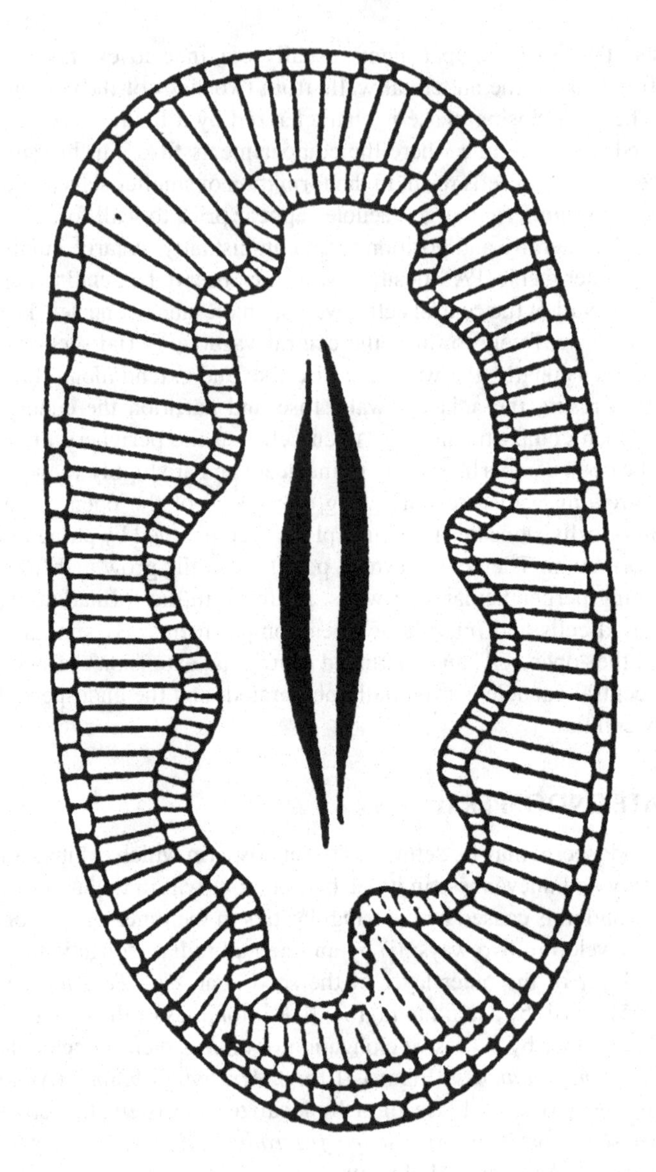

**FIGURE 11.15** *Passiflora calcarata*. Transverse section of mature seed showing rumination of endosperm (after Raju, 1956).

## CYTOLOGY OF ENDOSPERM

Depending upon the number of nuclei that fuse with the male gamete, the ploidy of the endosperm cells varies. Usually, the endosperm is triploid, but its ploidy level varies in taxa where the number of polar nuclei is more than two.

Shape of the endosperm nuclei varies from dumb-bell shape to spindle, star, lobed and with whip-like extensions. In *Alysum maritimum* the lobbing of

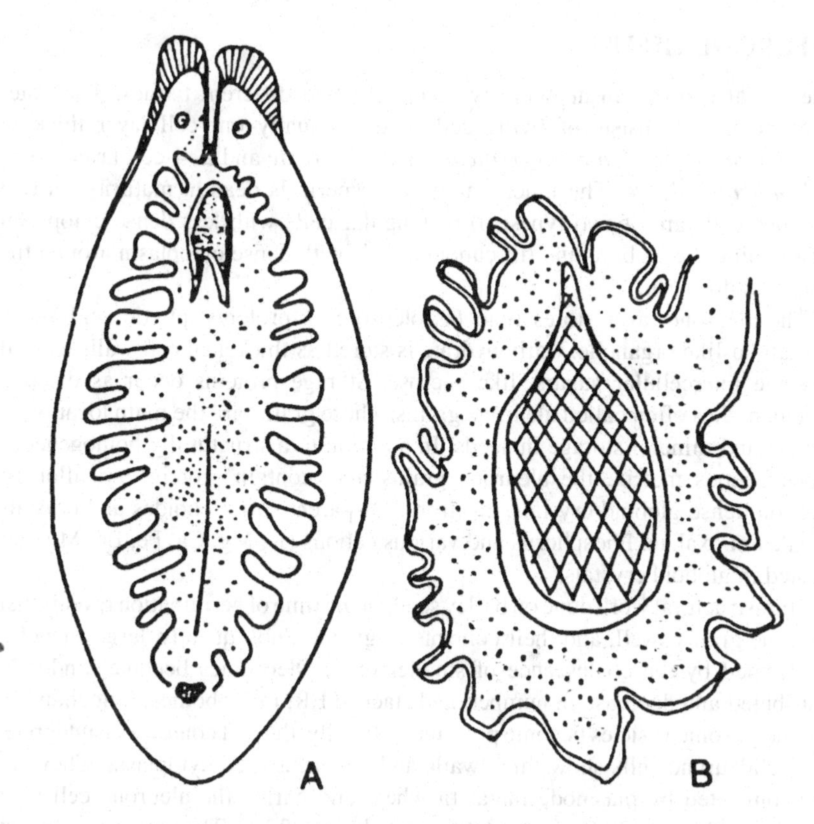


**FIGURE 11.16** L.S. mature seed showing ruminate endosperm. A. *Annona squamosa* (after Corner, 1949). B. *Andrographis serpyllifolia* (after Mohan Ram, 1960).

endosperm nuclei is so pronounced that it forms pseudopodia-like structures. Such nuclei are associated with high metabolic activity.

The endosperm tissue exhibits various mitotic irregularities, such as chromosome bridges, lagging chromosomes and fragments. Size and shape of endosperm nuclei also vary in the central and peripheral regions of the embryo sac. Endosperm nuclei show a high degree of polyploidy in the portion of endosperm which accumulates at the chalazal end in *Alysum maritimum* (Prabhakar, 1979). The size of endosperm nuclei may increase due to polyploidy or polyteny. In *Gagea* nuclear enlargement is accompanied by an increase in the number of chromosomes.

In the majority of angiosperms, the endosperm is predominantly triploid. The ploidy level, however, varies at maturity. The endosperm in *Datura fruticosa* is haploid ($3x = 36$), but during the course of development it becomes mixploid due to endopolyploidy and aneuploidy. Occurrence of endopolyploidy in the endosperm nucleus is a gene-controlled mechanism designed to increase macromolecule synthesis of nutrients, such as carbohydrates and proteins, needed for the growth of the endosperm as well as the embryo.

## ALEURONE TISSUE

The cereal endosperm at maturity comprises two different tissues. The outer aleurone layer consists of living cells and is usually one-cell layer thick in *Triticum aestivum*, *Zea mays*, *Triticale*, *Secale cereale* and 2–4-cell layers thick in *Hordeum vulgare*. The inner starchy endosperm is dead at maturity. In rice aleurone cells are of two types: (i) rectangular cells with less dense cytoplasm surrounding the embryo and (ii) cuboidal cells with dense cytoplasm around the starchy endosperm.

The aleurone layer stores lipid (as oleosomes), proteins, phytic acid and a crystalloid-like organelle. Carbohydrate is stored as thickened cell walls as well as some intercellular sugars, like sucrose. Storage proteins occur as discrete protein bodies often called aleurone grains. These grains are the main location of protein and mineral storage in seeds. In addition to a structurally homogeneous proteinaceous matrix, the aleurone grains may contain protein crystalloides, electron dense globoid crystals, electron transparent soft globoides and crystals of calcium oxalate. Phosphorus and various cations (K, Mg, Ca, Fe, Ba, Mn) are located in globoid crystals.

Ultrastructural work done on barley seeds at the time of germination reveals that aleurone grains swell, and their contents fragment. Subsequently, large vacuoles are formed by the coalescence of smaller ones. Oleosomes become randomly distributed and decrease in number, and stacked ER, microbodies, mitochondria and dictyosome vesicles become prominent. Finally, the cell contents disintegrate.

The aleurone cells show thick walls and nonvacuolated cytoplasm. They are interconnected by plasmodesmata. In wheat and barley the aleurone cell wall contains arabinose (20%), xylose (40%) and glucose (30%). The pentose is present as arabinoxylans, and the glucose as a mixed link 1,3-1,4 beta glucan.

## ENDOSPERM AND EMBRYO RELATIONSHIP

The endosperm has been assigned the function of nourishing the embryo, but recent ultrastructural and histochemical studies have made the nutritive role of the endosperm in early stages of embryogeny questionable. The endosperm during early ontogeny, is metabolically active and does not accumulate any food reserves (Newcomb, 1973). At this stage, endosperm does not provide direct nutrition for the development of the embryo. In *Helianthus annuus*, instead of endosperm, persistent synergids contribute greatly to the growth and development of the embryo. The developing endosperm during early embryogenesis needs sufficient nutrients for its own growth. Similar observations have been made in *Phaseolus*, *Iberis* and *Alysum*.

Newcomb (1973) proposed that the endosperm represents a large pool of reserve materials which is certainly used during later stages of embryo development. Thus, the classic role of storage and supply of nutrition to the embryo appears to be true only during later stages of embryogenesis.

Light and electron microscopic studies have suggested that the young embryo depends on nutrients derived from the suspensor, and when the suspensor degenerates, then the embryo switches over to endosperm tissue for its nutrition (Erdelska, 1980). In *Linum usitatissimum* the switchover to endosperm nutrition begins at the torpedo embryo stage, and consumption of endosperm is indicated by the appearance of a conspicuous zone of destroyed endosperm cells around the cotyledons. Besides providing nutrition, the endosperm may exhibit a morphogenetic role in regulating the growth of the embryo.

## FUNCTION OF ENDOSPERM

Endosperm provides nutrition to the developing embryo. The embryo sac has very little nutritive material at the time of fertilization. As the development of endosperm proceeds, it stores sufficient food material to ensure an adequate supply for the developing embryo. During its growth, the embryo consumes the surrounding cells of the endosperm. In legumes, the developing embryo consumes almost all the cells of the endosperm, leaving behind only one or two layers. Endosperm is a rich repository of carbohydrates, proteins, fats, enzymes, vitamins and growth regulators. Besides being a nutritive tissue, endosperm also controls the precise mode of embryo development. Endosperm acts as a combustible source of energy at the time of seed germination.

## SUMMARY

- The endosperm is formed by the repeated division in the primary endosperm nucleus.
- In most angiosperms, the endosperm is triploid. Endosperm is absent in Podostemaceae, Orchidaceae and Trapaceae.
- On the basis of whether a wall is formed or not, after the division of the primary endosperm nucleus, the endosperm is classified as nuclear, cellular or helobial types.
- The endosperm which exhibits any degree of irregularity and unevenness in its surface contour within the mature seed is called ruminate endosperm.
- Aleurone tissue is formed in cereal endosperm.
- The chief function of the endosperm is to provide nourishment to the embryo during development and later during germination.

## SUGGESTED READING

Asplund, I. 1973. Embryological studies in the genus *Sparganium*. *Svensk. Bot. Tidskr* 67: 177–200.

Bhandari, N.N., M. Bhargava and P. Chitralekha. 1986. Cellularization of free nuclear endosperm of *Papaver somniferum* L. *Phytomorphology* 36: 357–366.

Bhatnagar, S.P. and G. Sabharwal. 1966. Morphology and embryology of Iodina rhombilfolia Hook. & Am. *Beitr. Biol. Pflanzen* 45: 465–479.

Bhatnagar, S.P. and G. Sabharwal. 1968. Morphology and embryology of *Iodina rhombifolia* Hook. & Arn. *Beitr. Biol. Pflanzen* 45: 465–479.

Corner, E.J.H. 1949. The annonaceous seed and its four integuments. *New Phytol.* 48: 332–364.

Fineran, B.A., D.J.C. Wild and M. Ingerfeld. 1982. Initial wall formation in the endosperm of wheat, *Triticum aestivum*: A re-evaluation. *Can. J. Bot.* 60: 1776–1795.

Garcia, V. 1962. Embryological studies in the Lossaceae with special reference to the endosperm haustoria. In *Symp. Plant Embryology*, 157–161. New Delhi: Council Sci. Ind. Res.

Grau, J. and H. Hopf. 1985. Das endosperm der Compositae. *Bot. Jahrb. Syst.* 107: 251–268.

Hamann, U. 1966. Nochmals zur embryologie von *Philydrum lanuginosum*. *Betr. Biol. Pflazen* 42: 151–159.

Johri, B.M. and S.P. Bhatnagar. 1969. Endosperm in santalales. *Rev. Cytol. Biol. Veg.* 32: 355–369.

Kaur, H. and R.P. Singh. 1987. Development and structure of seed and fruit in some Convolvulaceae. *Phytomorphology* 37: 145–154.

Mares, D.J., K. Norstog and B.A. Stone. 1975. Early stages in development of wheat endosperm. *Aust. J. Bot.* 23: 311–326.

Mohan Ram, H.Y. 1960. The development of the seed in *Andrographis serpyllifolia*. *Am. J. Bot.* 47: 215–219.

Nijlingappa, B.H.M. and N. Devaki. 1979. Endosperm of *Scleria foliosa* Hochestetter ex A. Richard. *Curr. Sci.* 48 (10): 451–452.

Olsen, O.-A. (ed.). 2007. *Endosperm. Plant Cell Monograph (8)*. Berlin, Heidelberg: Springer-Verlag.

Periasamy, K. 1962. The ruminate endosperm. Development and types of rumination. In *Plant Embryology: A Symposium*, 62–74. New Delhi: CSIR.

Prabhakar, K. 1979. Histochemical and ultrastructural studies in some Cruciferae: Zygote to seedling. Ph. D. Thesis, Univ., Delhi, Delhi.

Raju, D. 1976. Gametogenesis, endosperm development and embryogeny in *Microcarpaea mucosa* (Scrophulariaceae, Gratioleae). *Plant Syst. Evol.* 126: 209–220.

Raju, M.V.S. 1956. Embryology of the Passifloraceae I. Gametogenesis and seed development of *Passiflora calcarata*. *J. Indian Bot. Soc.* 35: 126–137.

Singh, D. 1964. Cytoplasmic nodules in the endosperm of angiosperms. *Bull. Torrey Bot. Club* 91: 86–94.

Singh, R.P. 1970. Structure and development of seeds in Euphorbiaceae. *Jatropha* species. *Beitr. Biol. Pflanzen* 47: 79–90.

Swamy, B.G. and L. Bailey. 1949. The morphology and relationships of *Cercidiphyllum*. *J. Arnold Arbor.* 30: 197–210.

Vijayaraghavan, M.R. and K. Prabhakar. 1984. The endosperm. In *Embryology of Angiosperms*, ed. B.M. Johri, 319–376. Berlin: Springer-Verlag.

Vijayaraghavan, M.R. and U. Dhar. 1978. Embryology of *Cyrilla* and *Cliftonia* (Cyrillaceae). *Bot. Not.* 131: 127–138.

# 12 Embryo

The process of syngamy results in the formation of a zygote. Depending upon the species, the zygote gives rise to either a dicotyledonous or monocotyledonous embryo. The portion of the embryonal axis above the level of cotyledons is called epicotyl and the portion below the level of the cotyledons is known as hypocotyl. The epicotyl terminates into the plumule (embryonic shoot), and the hypocotyl at its lower end bears radicle (embryonic root). The embryos of monocotyledons bear only one cotyledon.

## ZYGOTE

After its formation, the zygote may or may not undergo a period of dormancy. The shortest resting periods occur in the Asteraceae and Poaceae. In *Crepis capillaris*, the first division of the zygote takes place 5–10 hours after pollination. In *Hordeum vulgare* the first division in the zygotic nucleus occurs about 22–24 hours after pollination.

The zygote is a unicellular system that, after repeated divisions, gives rise to a well-differentiated embryo. In angiosperms, the zygote is situated towards the micropylar side of the embryo sac. The basal portion of the zygote is attached to the embryo sac wall, whereas the apical part projects into the central cell.

In *Gossypium* (Jensen, 1968) the volume of the zygote decreases to half that of the egg in 8–10 hours after fertilization. The cell organelles, like endoplasmic reticulum (ER), ribosomes, plastids and mitochondria, cluster around the zygote nuclei. Shrinkage in volume of the zygote has also been observed in *Nicotiana tabacum* and *Hibiscus* species. The zygote enlarges during the premitotic phase in *Datura stramonium* and *Cypripedium insigne*.

## EMBRYOGENY

In most of the angiosperms, the zygote divides transversely, forming a small apical cell (*ca*) and a large basal cell (*cb*). In Loranthaceae and Piperaceae, the zygote invariably divides longitudinally. The next division may be transverse in both the terminal cell (*ca*) and basal cell (*cb*), and a linear tetrad is produced. The daughter cells are designated l and l'. The derivatives of *cb* are designated as *m* and *ci*. The term proembryo is applied from the two-celled condition till the initiation of organs in the embryonal mass.

In the formation of a T-shaped type of tetrad, a vertical division occurs in *ca*, and the two juxtaposed hemispherical cells form the tier *q*, which gives rise to a quadrant by vertical divisions in l and l' or by vertical division in each of the two juxtaposed cells perpendicular to the previous one (Figure 12.1).

DOI: 10.1201/9781003260097-12

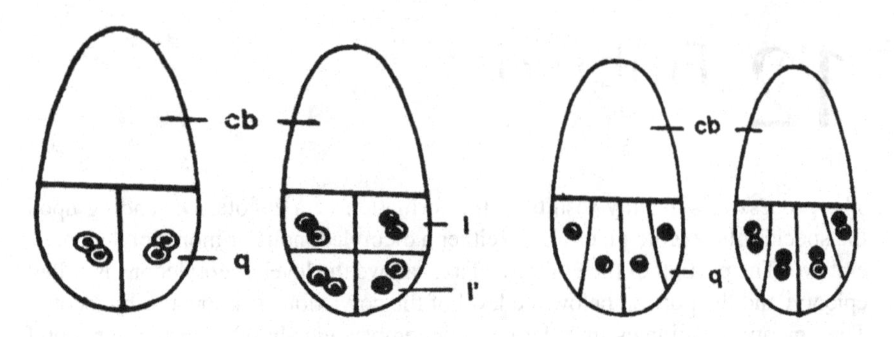


FIGURE 12.1 Quadrants produce octants either in two tiers (A) or in one tier (B) (after Natesh and Rau,1984).

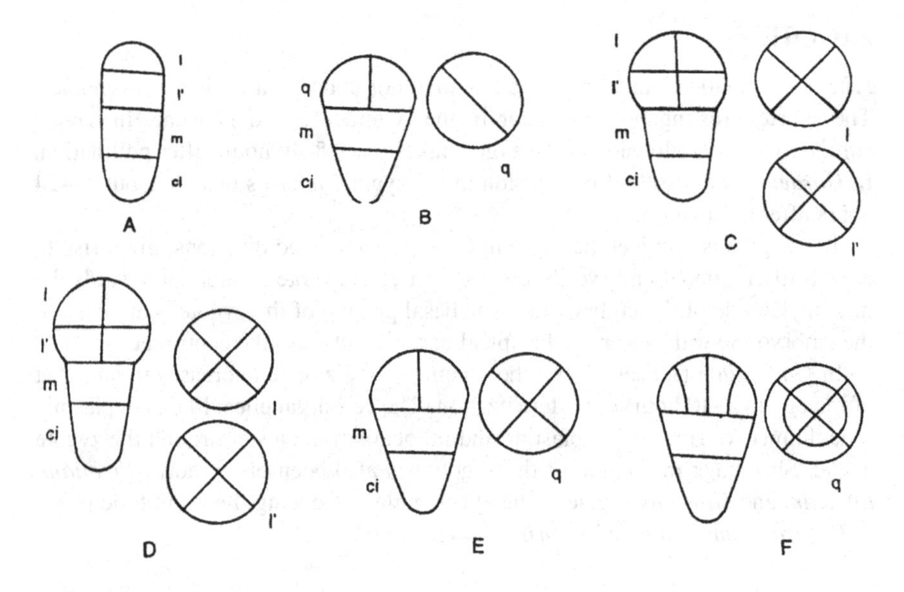


FIGURE 12.2 Formation of octant (after Swamy, 1962).

Subsequent to the formation of a quadrant, the proembryo develops into an octant. In angiosperms, two types of octant configurations occur: (i) the component cells are arranged in two superposed tiers of four cells each; and (ii) all the eight-cells occur in a single tier (Figure 12.2). The development of proembryo up to an octant stage and the formation of two types of octant are similar in dicotyledons and monocotyledons (Swamy, 1962). In the early stages of development, the embryos of dicotyledons and monocotyledons follow a similar sequence of cell division, and both become cylindrical or club-shaped bodies. The difference in the development becomes evident when the formation of the cotyledon begins. In the absence of a second cotyledon, the monocotyledon embryo does not become two-lobed at the distal end.

## EMBRYOGENY IN DICOTYLEDONS

Schnarf (1929) and Johansen (1945) classified embryogeny into six categories (Figure 12.3). Based on the plane of division of the apical cell (*ca*) in the two-celled proembryo, and the contribution of basal cell (*cb*) and apical cell (*ca*) in the development of the embryo, Maheshwari (1950) recognized different types of embryogeny as follows:

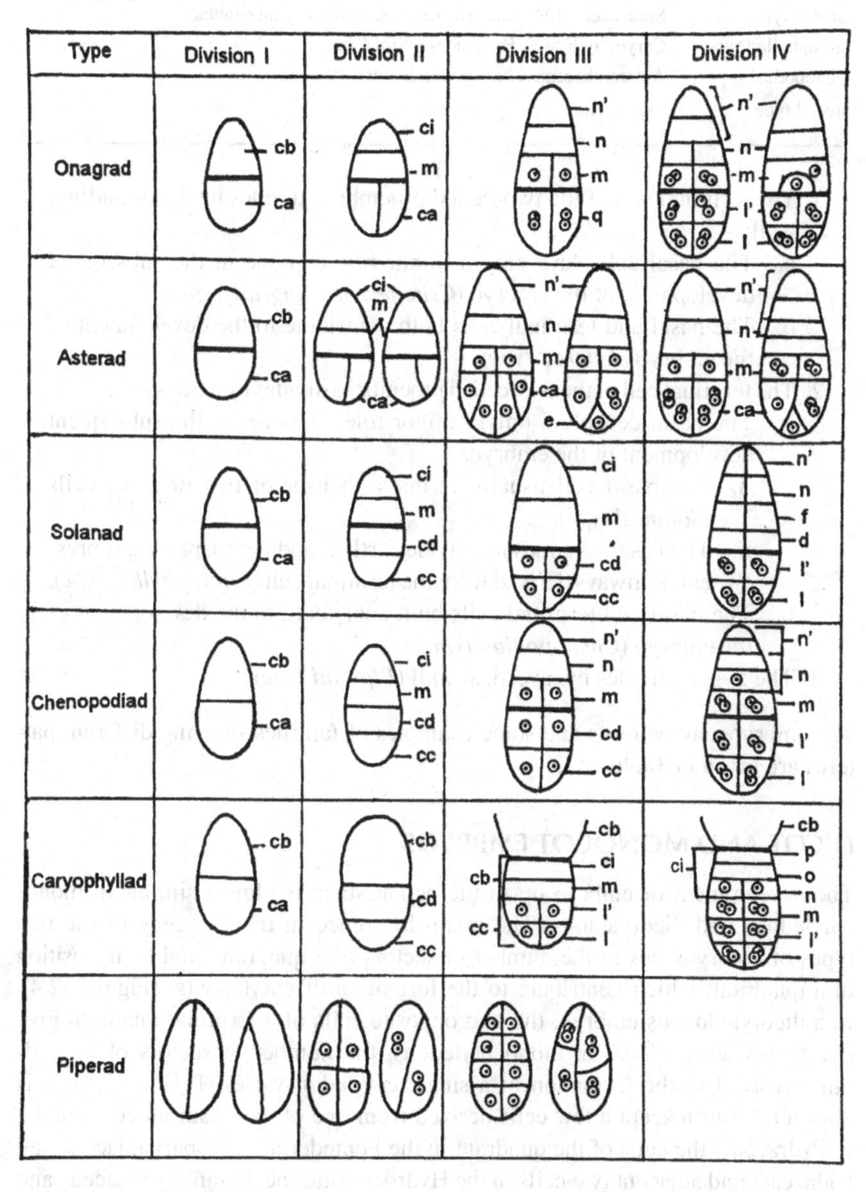


FIGURE 12.3 Schematic representation of main types of embryogeny.

**TABLE 12.1**
**Some Examples of Families Showing Different Types of Embryogeny**

| Embryogeny | Families |
|---|---|
| Onagrad type | Annonaceae, Brassicaceae, Magnoliaceae, Onagraceae, Scrophulariaceae |
| Asterad type | Asteraceae, Blasaminaceae, Violaceae |
| Solanad type | Solanaceae, Linaceae, Theaceae, Aizoaceae, Campanulaceae |
| Caryophyllad type | Caryophyllaceae, Portulaceae, Molluginaceae |
| Chenopodiad type | Amaranthaceae, Chneopodiaceae, Basellaceae |
| Piparad type | Piperaceae |

1. The terminal cell of the two-celled proembryo divides by a longitudinal wall:
   a. The basal cell plays only a minor role or none in the subsequent development of the embryo (*Crucifer*, or *Onagrad*, *type*).
   b. The basal and terminal cells both contribute to the development of the embryo (*Asterad type*).
2. The terminal cell of the two-celled proembryo divides by a transverse wall.
   a. The basal cell plays only a minor role, or none, in the subsequent development of the embryo.
      i. The basal cell usually forms suspensor of two or more cells (*Solanad type*).
      ii. The basal cell does not divide further, and the suspensor, if present, is always derived from the terminal cell (*Caryophyllad type*).
   b. The basal and terminal cells both contribute to the development of the embryo (*Chenopodiad type*).
3. The zygote divides by a vertical wall (*Piperad type*).

The embryogeny patterns and some examples of families showing different patterns are given in Table 12.1.

## DICOT AND MONOCOT EMBRYOS

The development of embryo up to the octant stage is almost similar in monocotyledons and dicotyledons. The main difference in the ontogeny of the two types of embryos lies in the number of sectors of a quadrant, and their position in a quadrant, which contribute to the formation of cotyledon(s) (Figure 12.4). In a dicotyledonous embryo, the two opposite cells of a terminal quadrant give rise to two cotyledons. In monocotyledons, the number of sectors of a quadrant involved in the formation of a single cotyledon varies (Figure 12.5). It is almost all four (except a few cells derived from one of the quadrant cells) in the Philydraceae; the cells of the quadrant in the Pontederiaceae, Sparganiaceae and Iridaceae; and adjacent two cells in the Hydrocharitaceae, Potamogetonaceae and Amaryllidaceae that constitute the cotyledonary locus (Lakshmanan, 1972).

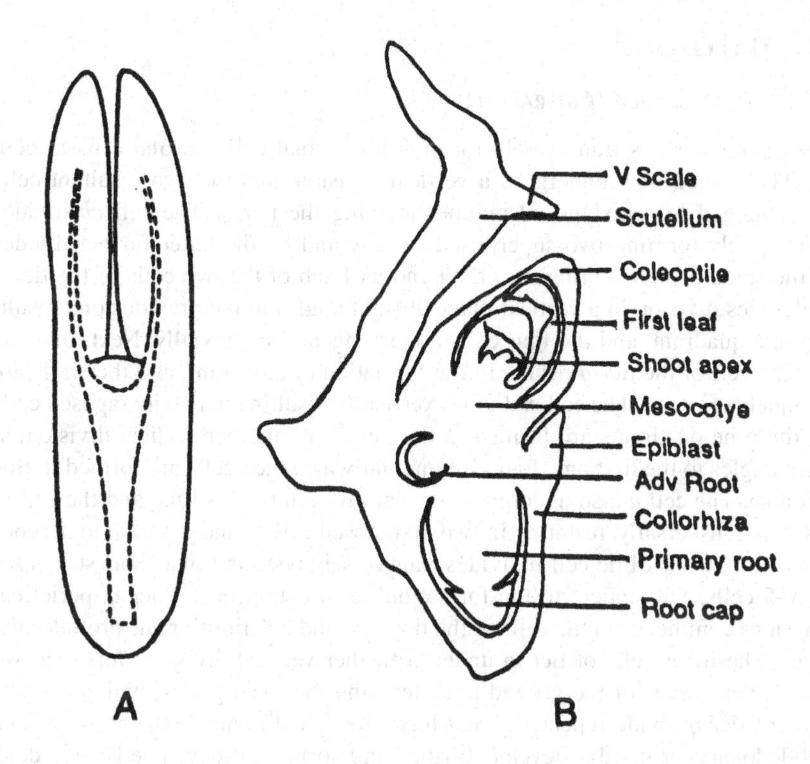


**FIGURE 12.4** Dicot and monocot embryos. A. Mature embryo of *Bartsia alpina* (after Terokhin and Nikiticheva, 1982). B. Longitudinal section of mature embryo of oat (after Shah and Sreekumari, 1980).

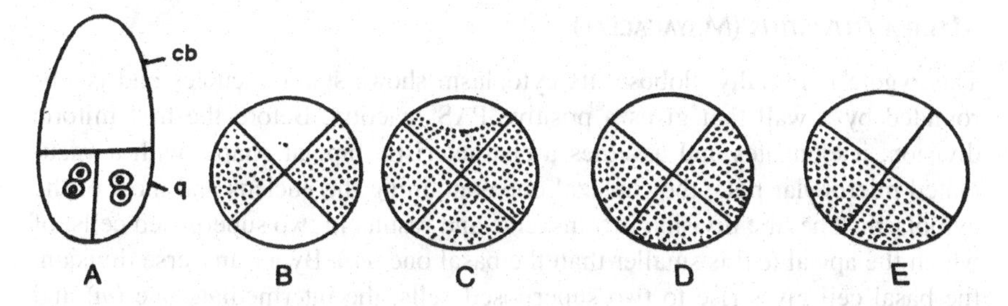


**FIGURE 12.5** A–E. Derivation of cotyledons in monocotyledons and dicotyledons. A. Quadrant proembryo. B–E. Transections of terminal tier of A; stippled portion represents progenitors of cotyledon. B. Development of dicotyledons. C–E. Development in various monocotyledonous taxa (after Lakshmanan, 1972).

## Some Examples of Dicot and Monocot Embryogeny

To illustrate embryogeny in angiosperms, examples of *Guizotia abyssinica* (Chopra and Singh, 1974), *Mayaca fluviatilis* (Venturelli and Bouman, 1986), *Commelina subulata* (Chikkanaiah, 1963) and *Triticum aestivum* (Deshpanade and Raju, 1979) are described.

## DICOTYLEDONS

### *Guizotia abyssinica* (Asteraceae)

The zygote divides transversely forming a terminal cell, *ca*, and a basal cell, *cb*. The cell *ca* now undergoes a vertical division, and the two resultant cells are arranged in a juxtaposed manner forming the tier *q*. The cell *cb* divides transversely forming two superposed cells *m* and *ci*; the latter further divides in the same plane resulting in cells *n* and *n'*. Each of the two cells of the tier *q* undergoes division in a vertical plane at right angles to the previous one resulting in a quadrant, and the four cells are arranged circumaxially. Next, division in each cell of the tier *q* results in the formation of an octant, and the walls are obliquely oriented. The cell *m* divides vertically resulting in two juxtaposed cells by the time quadrants are formed in the tier *q*. By another vertical division, at right angles to the first one, four circumaxially arranged cells are formed in this tier also. The cell *n* also undergoes vertical division by this time and the cell *n'* divides transversally, resulting in two juxtaposed cells *o* and *p*. Cell *o* undergoes vertical division while cell *p* divides transversely resulting in a short suspensor of 3–5 cells. Subsequent to the formation of an octant in the tier *q*, periclinal divisions commence in the cells of the tier *m*, *q* and *n* delimiting the protodermal layer. The inner cells of tier *m* undergo further vertical divisions forming two layers, the initials of the ground meristem and the axial procambial zone. The cells of tier *q* divide repeatedly and form the apical dome of the embryo. The cotyledonary primordia develop further and form well-developed cotyledons which are fleshy and grooved (Figure 12.6). The development of embryo is of the Asterad type.

### *Mayaca fluviatilis* (Mayacaceae)

The zygote is initially globose. Its cytoplasm shows small vacuoles and is surrounded by a wall that gives a positive PAS reaction. Before the first mitotic division, it elongates and assumes a characteristic bipolar shape, with a vacuolated micropylar pole, the chalazal one containing the nucleus and most of the cytoplasm. The first division is transverse and results in two superposed cells, of which the apical (*ca*) is smaller than the basal one (*cb*). By a transverse division, the basal cell gives rise to two superposed cells, the intermediate one (*m*) and the basally situated one (*ci*). This division is followed by a vertical one in the apical cell to form two cells in juxtaposition (*q*), so that an invertedly T-shaped, four-celled proembryo is formed. A vertical division in *m* results in two cells that subsequently undergo further divisions and contribute to the initials of the future root meristem. The *ci* divides only once again by a transverse or vertical wall and forms a two-celled suspensor. The derivatives of the apical cell divide transversely so that four cells are formed, arranged in two tiers of two cells each. This stage is followed by one with numerous cell divisions in diverse planes, so that the embryo acquires a globular shape. Epidermis initials are separated

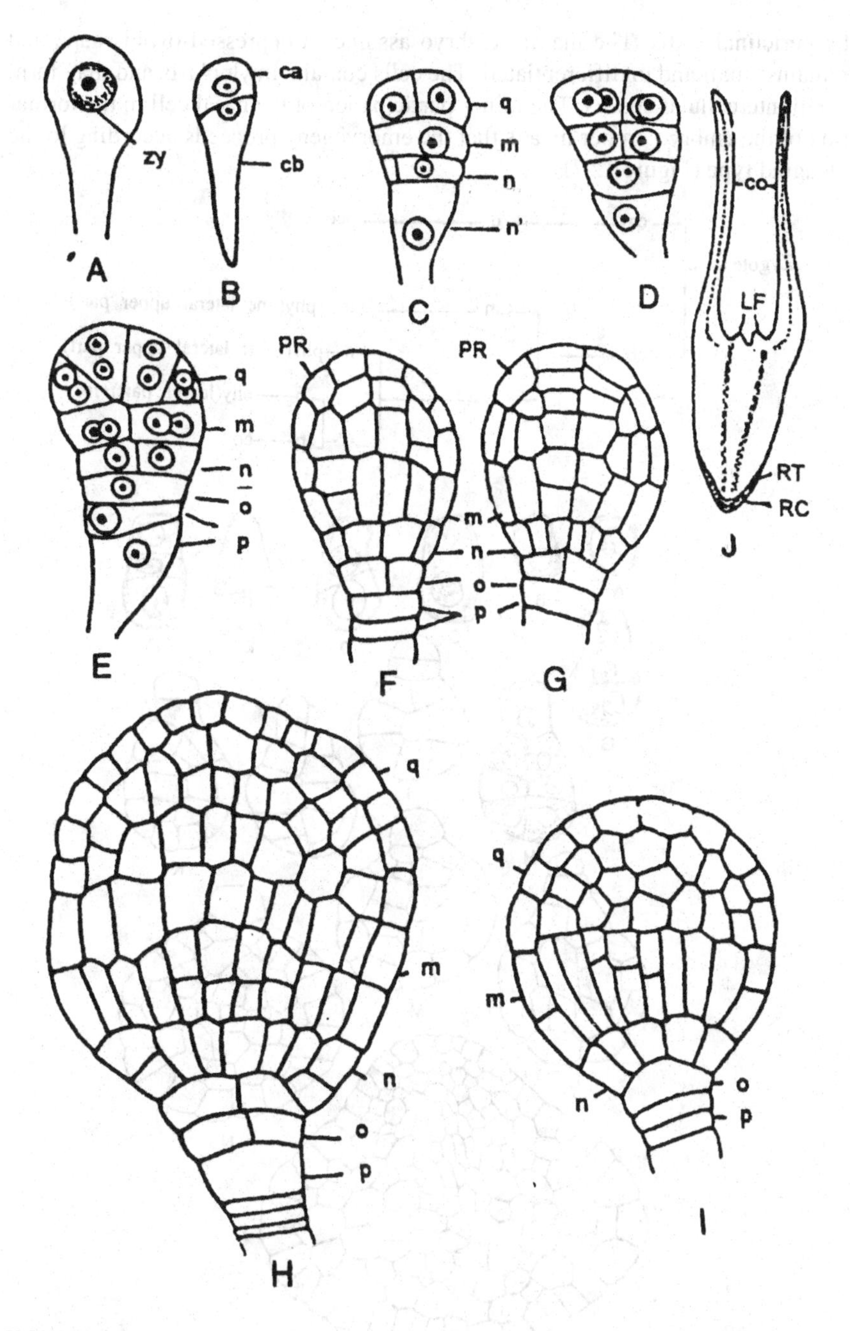


**FIGURE 12.6** Embryogeny in *Guizotia abyssinica* (after Chopra and R.P. Singh, 1982).

by periclinal walls. The mature embryo assumes a depressed-ovoid shape and remains small and undifferentiated. The cells contain amyloplasts and may form small intercellular spaces. The minor participation of the basal cell in the formation of the embryo proper means that the embryogeny proceeds according to the Onagrad type (Figure 12.7).

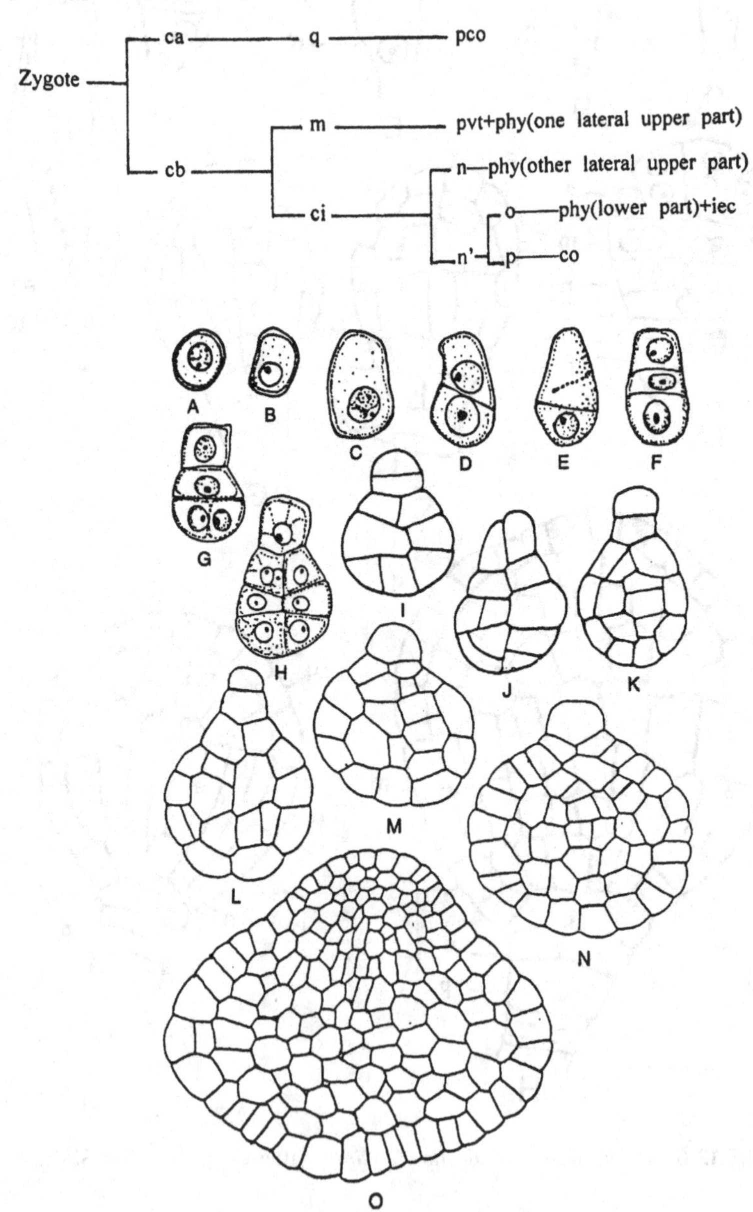


**FIGURE 12.7** Embryogeny in *Mayaca fluviatilis* (after Venturelli and Bouman, 1986).

## MONOCOTYLEDONS

### *COMMELINA SUBULATA* (COMMELINACEAE)

The zygote divides usually by an oblique wall to form the cells *ca* and *cb*. Both the cells divide vertically forming a four-celled proembryo. The two derivatives of *cb*, *m* and *ci* are unequal, due to a slightly oblique-vertical wall, and are juxtaposed. The dermatogen initials are cut off from *q* at this stage. In the meantime, *ci* divides transversely producing *n* and *n'* while *m* undergoes a vertical division. Dermatogen initials are cut off from all the segments except *n*'. Later the segment *n'* also divides transversely forming *o* and *p*.

Subsequent divisions are irregular resulting in a globular embryo. Due to rapid divisions in the basal region, the embryo becomes sub-spherical. At this stage, a lateral depression appears separating the derivatives of *q* from those of *m*. The terminal cotyledon is formed from the segment *q*. The stem tip is derived from the lateral part of *m* just above the cleft. The cotyledonary sheath and a lateral upper part of the hypocotyl are contributed by *n*. The basal part of the hypocotyl and the root tip develop from *o*. Lastly, *p* gives rise to the root cap. Procambial strands differentiate in the hypocotyl, and the traces to the cotyledon and lateral roots are also formed. The leaf surrounding the stem apex is formed at a later stage (Figure 12.8). Thus, the embryogeny of the *Commelina* follows the Asterad type and may be represented as follows (co, root cap; iec, initials of central cylinder of root; pco, cotyledonary region; phy, hypocotylendonary region; pvt, shoot apex).

### *TRITICUM AESTIVUM* (POACEAE)

The first transverse division in the zygote results in the formation of a bicellular proembryo consisting of a terminal cell (*ca*) and a basal cell (*cb*) (Figure 12.9A). The next division is vertical in *ca* but transverse in *cb* (Figure 12.9B) so that at the end of second cell generation, the proembryo consists of four cells, disposed in three tiers, i.e., two juxtaposed cells being derived from *ca* and two superposed cells, *m* and *ci* derived from *cb*. During the next cell generation, *ca* and *m* divide vertically, whereas *ci* undergoes a transverse division (Figure 12.9C). The second vertical division in the two cells of *ca* is at right angles to the first, producing a quadrant (*q*). This is followed by a horizontal division in each of the cells of *ca*. The octant, thusly produced, consists of two superposed quadrants, the superior being designated as l and the inferior as l' (Figure 12.9D). The cells of tier *m* also divide longitudinally, followed by a transverse one (Figure 12.9E). The tier *ci*, by horizontal segmentation, produces two superposed tiers, n and n' (Figure 12.9C). The latter, by horizontal division, results in the formation of tiers *o* and *p* (Figure 12.9F). Further divisions in l and l' occur in different planes (Figure 12.9G). The tier *l'* multiplies faster. Incessant divisions in different planes in *m* give rise to a massive structure. The divisions in *o* and *p* are rather slow, and their derivatives are responsible for formation of the suspensor alone.

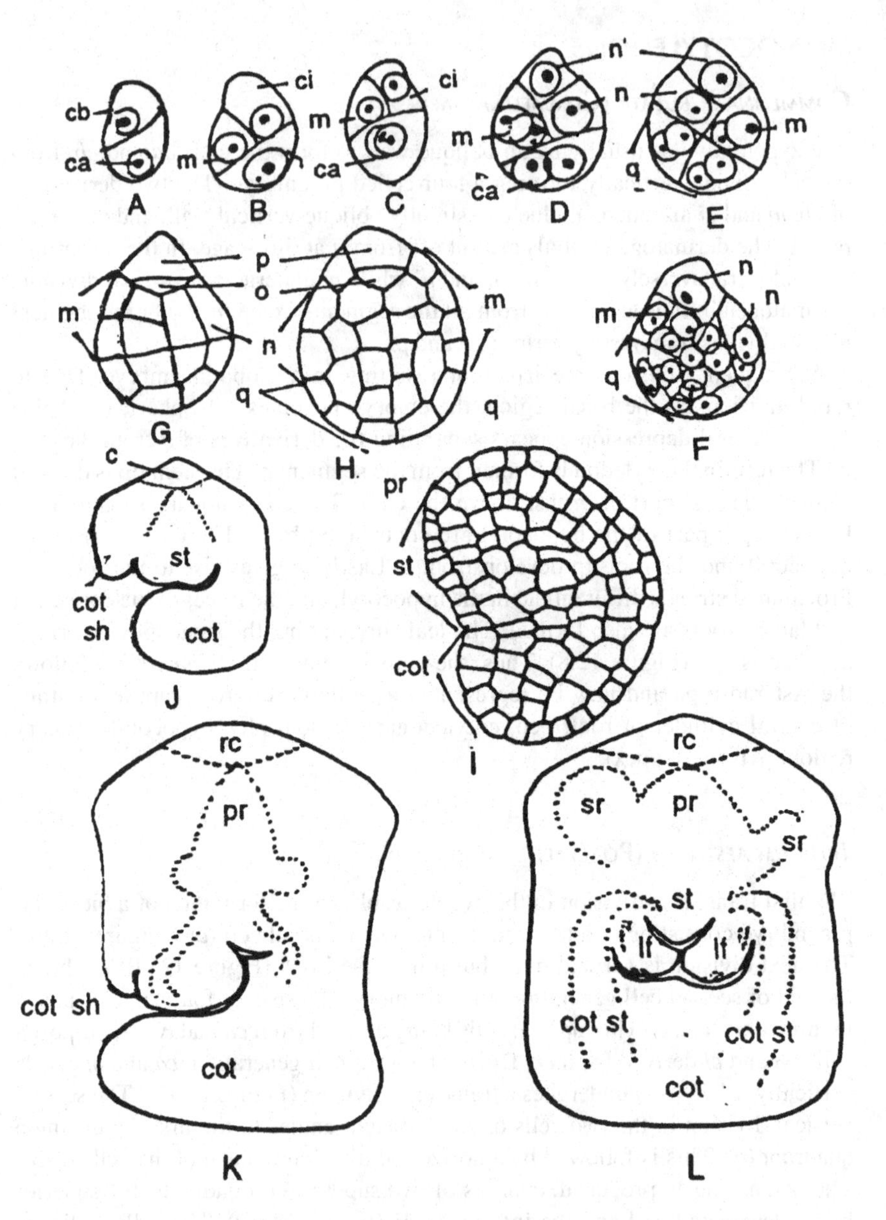


FIGURE 12.8 Embryogeny in *Commelina subulata* (after Chikkanaiah, 1963).

The globular embryo now exhibits laterality. A depression produced at the common limit of *l* and *l'* (Figure 12.9H) marks the separation of cotyledon from other parts of the embryo. The cotyledon, generally referred to as scutellum, is entirely derived from the derivatives of *l*. The upper lip of the coleoptile (cl') differentiates from the epidermal cells of l situated towards the depression

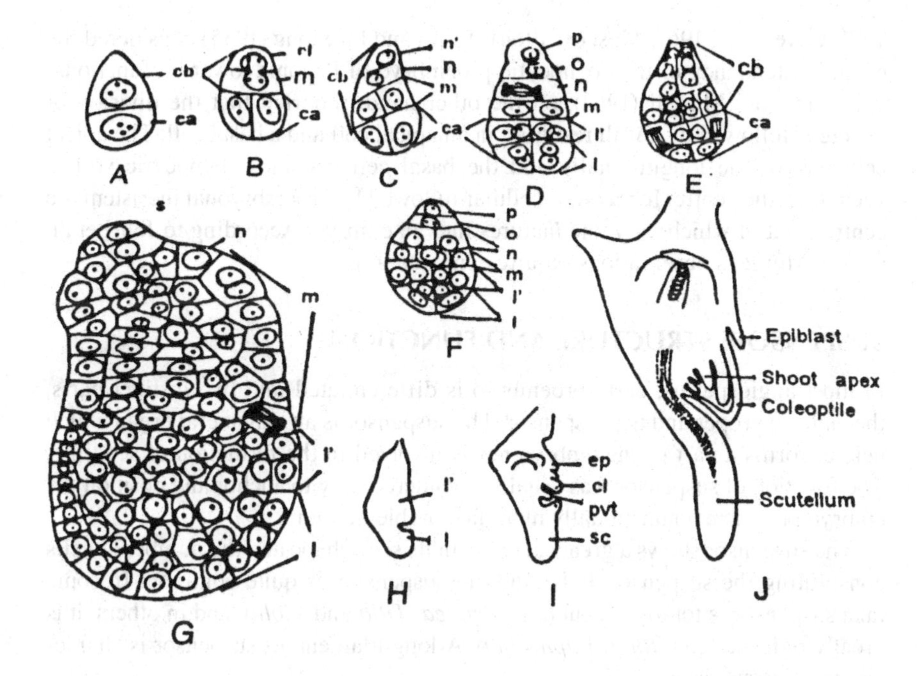


**FIGURE 12.9** Development and structure of embryo in *Triticum* (after Deshpande and Raju, 1979).

(Figure 12.9I). From the epidermal cells of l' develop rudimentary projections which are the initials of stem tip (pvt), first leaf and the lower lip of coleoptile (cl). The stem tip is rather slow in growth, whereas the upper and the lower lips of the coleoptile grow rapidly. The two lips grow until they meet and enclose the stem apex and the rudiment of first leaf (Figure 12.9J). The superior derivatives of *m* differentiate into the hypocotyledonary region (*phy*) and initials of the central cylinder of the root (*iec*). The epiblast situated below the coleoptile originates from the superior derivatives of *m* (Figure 12.9J). The central cells of the inferior derivatives of *m* give rise to root cortex (iec), whereas the lateral elements contribute to the formation of coleorrhiza (*cr*). The cells situated below root tip are derived from tier *n* and constitute the calyptrogen from which originates the root cap (Figure 11.9J). The development of embryo conforms to the Asterad type.

## Embryogeny in *Paeonia*

*Paeonia* shows a unique pattern of embryogeny. The nucleus of the zygote divides repeatedly in a free nuclear manner to form a coenocyte, and the nuclei become distributed around a large central vacuole. At a later stage the nuclei become delimited by walls, and a cellular mass is formed. Several peripheral cells develop meristematic centres and produce embryonal primordia. Only one of these primordia matures into a dicotyledonous embryo (Yakovlev and Yoffe, 1957,

1961; Cave et al., 1961; Moscov, 1964). Xi-jin and Fu-xiong (1985) considered the evolution tendency of the coenocytic proembryo of Paeonia toward a functional specialization. Murgai (1962), on the other hand, reported that the division of zygote is followed by a wall resulting in an apical cell and a basal cell. The apical cell may divide longitudinally and the basal cell becomes coenocytic with a vacuole in the centre. It becomes cellular followed by proembryonal meristematic centres, out of which only one matures into an embryo. According to Johri et al. (1992), Murgai's observations require confirmation.

## SUSPENSOR: STRUCTURE AND FUNCTION

In most angiosperms, early proembryo is differentiated into two distinct parts, the embryo proper and the suspensor. The suspensor is an ephemeral organ which neither forms a part of the embryo nor is involved in the formation of seedling. The function of suspensor was usually considered only to anchor and position the embryo proper in a nutritionally more favourable medium.

The suspensor shows a great variation in its size, shape and the number of cells constituting the suspensor. Polyploidy in suspension is quite common. In some taxa suspensor is totally absent (e.g., *Penaea*, *Tilia* and *Viola*), and in others, it is greatly reduced (e.g., *Ruta*, *Euphorbia*). A long filamentous suspensor is characteristic of Brassicaceae.

### SUSPENSOR HAUSTORIA

The formation of suspensor haustoria is widely known in angiosperms (e.g., Cuscutaceae, Leguminosae, Tropaeolaceae). In the families Orchidaceae, Trapaceae and Podostemaceae, where the endosperm is absent, the embryo possesses extensively developed suspensor haustoria. In *Dicraea stylosa* (Podostemaceae) the basal cell of the proembryo enlarges considerably and generally has two hypertrophied nuclei. It gives out a number of haustorial branches (Figure 12.10). These branches grow in between two integuments and the haustoria

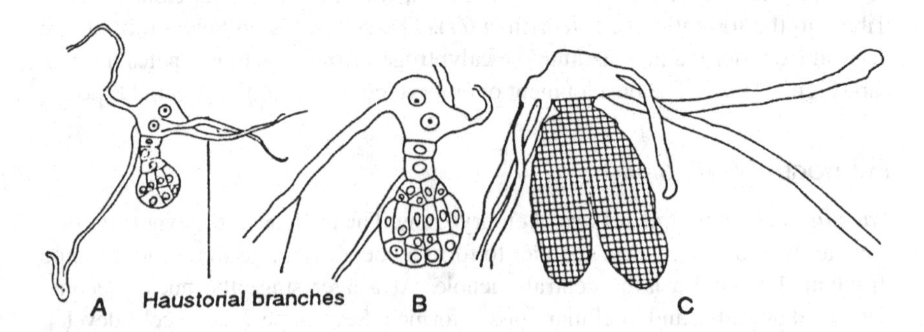


**FIGURE 12.10** *Dicraea elongata*. A–C. Different stages of embryogeny showing long, hypha-like haustorial branches (after Mukkada, 1962).

are much longer than the embryo itself (Mukkada, 1962). In *Rhynchostylis retusa* (Orchidaceae) the cells of the suspensor cap elongate and surround the upper portion of the mature embryo (Figure 12.11A). The suspensor cells collapse shortly before the dispersal of seeds (Sood and Sham, 1987)

In *Ipomoea hispida*, the peripheral cells of the suspensor become finger-like and are clearly marked out because of dense cytoplasm and prominent nuclei. During further development, these cells are embedded into the cellular endosperm (Figure 12.11B) and appear to draw nutrients from it (Kaur and Singh, 1987).

In Rubiaceae the suspensor is filamentous. During development some cells in the micropylar region give out lateral protrusions which penetrate the endosperm and show swelling at their tip.

## Ultrastructure

Ultrastructural studies have revealed that suspensor cells are more vacuolated than embryonal cells. Endoplasmic reticulum is more prominently developed in suspensor cells. Numerous plasmodesmatal connections have been observed in

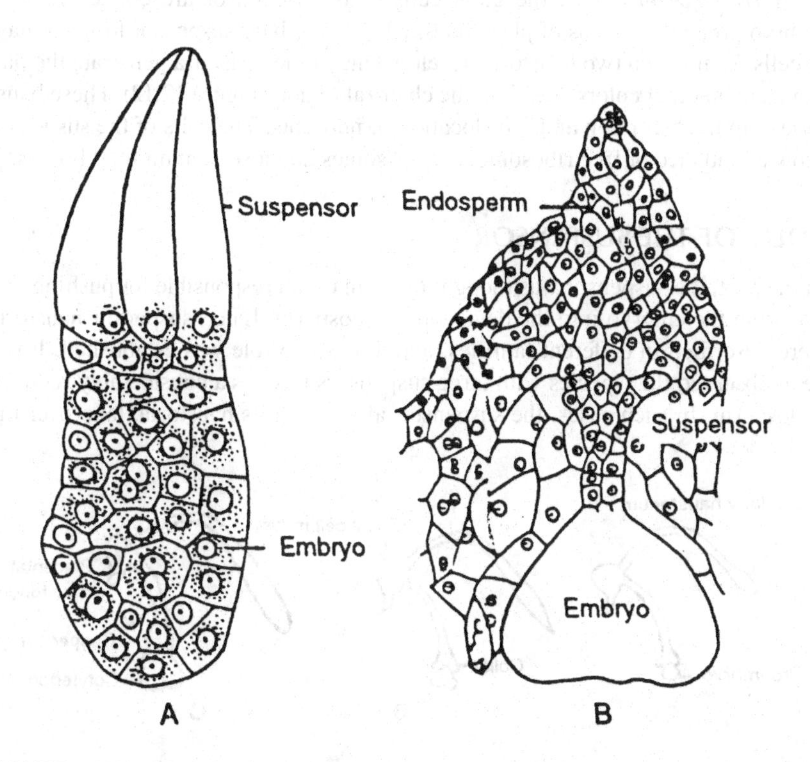


**FIGURE 12.11** Embryo and suspensor. A. *Rhynchostylis retusa* showing distinct suspensor (after Sood and Sham, 1987). B. *Ipomoea hispida.* Heart-shaped embryo showing finger-like cells of suspensor (after Kaur and R.P. Singh, 1987).

suspensor cells. The wall ingrowths of the long, tapering basal cell are abundant at the micropylar pole. Wall ingrowths are usually associated with several large mitochondria. The suspensor cells contain rough ER, mitochondria, dictyosomes, plastids and ribosomes. The innumerable mitochondria associated with the wall ingrowths are probably designed for the supply of necessary energy for active absorption and transport of the solutes towards the embryo. Such a nutrient flow is further supplemented by the plasmodesmatal connections present between suspensor cells.

In *Phaseolus*, suspensor cells possess only a few small vacuoles. ER in mature cells consists of twisted, mainly smooth, tubular structures. Nagl (1973) suggested that the ER functions as a storage organ. The basal giant cells possess leucoplasts of varying shape and structure. The suspensor cells with leucoplasts apparently are more active in early stages of embryogeny and concerned with the synthesis of lipids. Besides filamentous forms, cup-shaped and irregular-shaped plastids are common in suspensor cells. These plastids vary in their internal structure, such as fibrils and regular and irregular thylakoid layer. The function of suspensor plastids in providing nutrition to the developing embryo is well established in several genera, e.g., *Pisum*, *Ipomoea*, *Stellaria*.

In *Tropaeolum majus*, the giant cells of the suspensor are connected with embryo proper by means of plasmodesmata. At the base suspensor forms a mass of cells from which two haustoria develop. One of the haustoria penetrate the placenta and another enters deep into the chalazal region (Figure 12.12). These haustoria help in absorption and translocation of nutrients. The cells of the suspensor show abundance of free ribosomes or polysomes and also contain lipid bodies.

## ROLES OF THE SUSPENSOR

In the past, the suspensor was thought to be an organ responsible for pushing and orienting the embryo into the developing endosperm. Later, suspensor haustoria were described in different families, and a nutritive role was established. It has been observed that species with large suspensors possess a considerably reduced endosperm. In such cases, the suspensor, along with its haustoria, take over the

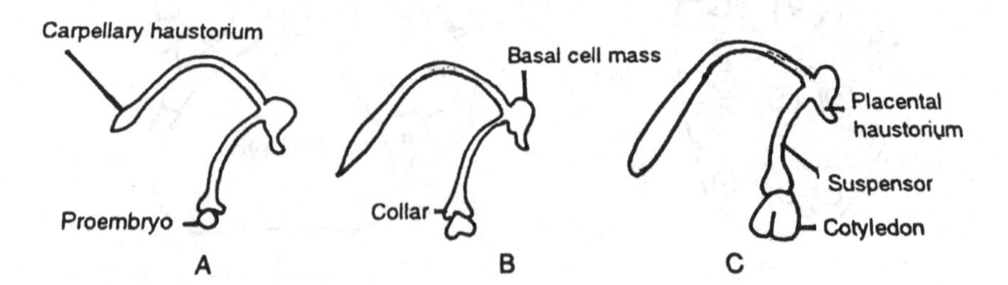


**FIGURE 12.12** *Tropaeolum majus.* A. Globular embryo. B. Heart-shaped embryo. C. Cotyledonary embryo. Note well-developed carpellary haustorium (after Bhalla et al., 1979).

function of endosperm. The following main functions have been attributed to the suspensor.

1. **Absorption of Nutritive Substances:** The suspensor plays an important role in the nutrition of the embryo. The suspensor, in most cases, grows more rapidly at a stage when the embryo is differentiating from globular to heart-shaped stage. During this period, the suspensor cells are metabolically very active. They contain abundant cell organelles. The basal cells of the suspensor show wall labyrinths which indicate that the main function of suspensor is absorption of nutrient material from maternal tissue.
2. **Synthesis of Phytohormones:** Alpi et al. (1975) extracted free and bound forms of gibberellic acid-like substances from cells of embryo and suspensor. The concentration of GA was reported to be greater in suspensor cells at the heart-shaped stage of embryo.
3. **Synthesis of Nutritive Substances:** Nagl and Kuhner (1976) reported that leucoplastids of suspensor cells synthesize lipids. In *Phaseolus* the suspensor cells show high DNA amplification and high protein synthesis. These materials may be transferred to the developing embryo when needed.
4. **Transport of Nutrients:** The cells of suspensor are interconnected with each other by means of plasmodesmata and help in transport of nutrients to the embryo.

## CHLOROPHYLLOUS EMBRYOS

Yakovlev and Zhukova (1980) reported the occurrence of chlorophyll in the embryos of 224 genera belonging to 72 families. According to Yakovlev and Zhukova (1975) the greening of embryos is a genetic characteristic. In many taxa the chlorophyll disappears in the mature embryo. Chlorophyllous embryos are common in Brassicaceae, Convolvulaceae, Fabaceae, Tiliaceae etc. In families like Anacardiaceae, Apocynaceae, Malvaceae, Scrophularaceae, both chlorophyllous and non-chlorophyllous embryos are reported. The monocotyledonous families rarely show chlorophyllous embryos.

## SUMMARY

- The process of syngamy results in the formation of a zygote. Zygote divides to form an embryo.
- The development of embryo is called embryogeny.
- Depending upon the species, zygotes give rise to either dicotyledonous or monocotyledonous embryos.
- The embryogeny patterns are of different types viz., Onagrad, Asterad, Solanad, Caryophyllad, Chenopodiad and Piperad types.

- The portion of the embryonal axis above the level of cotyledons is called epicotyl, and the portion below the level of the cotyledons is known as hypocotyl.
- The epicotyl terminates into the plumule (embryonic shoot), and the hypocotyl at its lower end bears radicle (embryonic root).
- The suspensor, which may or may not form a suspensor haustoria, helps in the absorption of nutrients from the surrounding tissues.

## SUGGESTED READING

Alpi, A., F. Tognoni and F. D'Amato. 1975. Growth regulator levels in embryo and suspensor of *Phaseolus coccineus* at two stages of development. *Planta* 127: 153–162.

Bhalla, P.L., M.B. Singh and C.P. Malik. 1979. Physiology of sexual reproduction VI. Embryogenesis in *Tropaeolum majus* L. enzyme changes. *Acta Bot. Indica* 7: 72–86.

Cave, M.S., R.J. Anolt and A.C. Stanton. 1961. Embryogeny of the *Californian peonies*, with reference to their systematic position. *Am. J. Bot.* 20: 358–364.

Chikkanaiah, P.S. 1961. Morphological and embryological studies in the Commelinaceae. In *Symp. Plant Embryology*, 23–26. New Delhi: Council of Scientific and Industrial Research.

Chikkanaiah, P.S. 1963. Embryology of some members of the Commelinaceae. *Phytomorphology* 13: 174–183.

Chopra, S. and R.P. Singh. 1974. Effect of gamma rays and 2, 4-D on ovule, female gametophyte, seed and fruit development in *Guizotia abyssinica. Phytomorphology* 26 (3): 240–249.

Chopra, S. and R.P. Singh. 1976. Effect of gamma rays and 2,4-D on ovule, female gametophyte, seed and fruit development in *Guizotia abyssinica. Phytomorphology* 26 (3): 240–249.

Crete, P. 1963. Embryo. In *Recent Advances in the Embryology of Angiosperms*, ed. P. Maheshwari, 171–220. Delhi: International Society of Plant Morphologists.

Deshpande, P.K. and Raju, P.S.G. 1979. Development of caryopsis and localization of DNA and proteins in Triticum species. *Phytomorphology* 29: 100–111.

Jensen, W.A. 1968. Cotton embryogenesis: The zygote. *Planta* 79: 346–366.

Jensen, W.A. 1974. Reproduction in flowering plants. In *Dynamic Aspects of Plant Ultrastructure*, ed. A.W. Robards, 481–503. London: McGraw Hill.

Johansen, D.A. 1945. A critical review of the present status of plant embryology. *Bot. Rev.* 11: 87–107.

Johansen, D.A. 1950. *Plant Embryology. Embryogeny of the Spermatophyta*. Waltham, MA: Chronica Botanica Co.

Johri, B.M. 1982. *Experimental Embryology of Vascular Plants*. Heidelberg: Springer Berlin.

Johri, B.M., K.B. Ambegaokar and P.S. Srivastava (eds.). 1992. *Comparative Embryology of Angiosperms*, Vols. 1 and 2. Heidelberg: Springer Berlin.

Kaur, H. and R.P. Singh. 1987. Development and structure of seed and fruit in some Convolvulaceae. *Phytomorphology* 37: 145–154.

Lakshmanan, K.K. 1972. Monocot embryo. In *Vistas in Plant Sciences*. Vol. 2., eds. T.M. Varghese and R.K. Grover, 61–110. Hissar: International BioScience Pub.

Maheshwari, P. 1950. *An Introduction to the Embryology of Angiosperms*. New York: McGraw-Hill.

Malik, C.P. and M.B. Singh. 1979. Embryo suspensor in angiosperms: Structural and functional aspects. In *Current Advances in Plant Reproductive Biology*, ed. C.P. Malik, 257–292. Ludhiana: Kalyani Publishers.

Mukkada, A.J. 1962. Some observations on the embryology of *Dicraea stylosa* Wight. In *Plant Embryology. A Symposium*, 139–145. New Delhi: Council of Scientific and Industrial Research.

Murgai, P. 1962. Embryology of Paeonia with a discussion on its systematic position. In *Plant Embryology-A Symposium*, 215–223. New Delhi: Council of Scientific and Industrial Research.

Nagl, W. 1973. The angiosperm suspensor and the mamrnalian trophoblast: Organs with similar cell structure and function. *Memoires Coll Morphologie. Bull. Soe. Bot. Fr.* 120: 289–301.

Nagl, W. and S. Kuhner. 1976. Early embryogenesis in *Tropaeolum majus* L.: Diversification of plastids. *Planta* 133: 15–19.

Natesh, S. and M.A. Rau. 1984. The embryo. In *Embryology of Angiosperms*, ed. B. M. Johri, 377–443. Berlin: Springer Verlag.

Rensing, S.A. and D. Weijers. 2021. Flwering plant embryos: How did we end up here? *Plant Reproduction*. https://doi.org/10.1007/s00497-021-00427-y.

Schnarf, K. 1929. *Embryologie der Angiospermen*. In: Handbook der Pflanzenanatomie (ed. K. Linsbauer), II/2. Berlin: Gebrueder Borntraeger, Berlin.

Sood, S.K. and N. Sham. 1987. Gametophytes, embryogeny and pericarp of *Rhynchostylis retusa* Blume (Epidendreae, Orchidaceae). *Phytomorphology* 37: 307–316.

Swamy, B.G.L. 1962. The origin of cotyledon and epicotyl in *Ottella alismoides*. *Beitr. Biol. Pflanzen* 39: 1–16.

Venturelli, M. and F. Bouman. 1986. Embryology and seed development in *Mayaca fluviatilis* (Mayacaceae). *Acta Bot. Neer.* 35 (4): 497–516.

Yakovlev, M.S. and M.D. Yoffe. 1957. On some peculiar features in the embryogeny of Paeonia L. *Phytomorphology* 7: 74–87.

Yakovlev, M.S. and G.Y.A. Zhukova. 1980. Chlorophyll in embryos of angiosperm seeds: A review. *Bot. Not.* 133: 323–336.

# 13 Polyembryony

The phenomenon of polyembryony was discovered by A. Leeuwenhoek in 1719 who found two seedlings developing from the same *Citrus* seed. Later, Strasburger (1878) demonstrated the nature of polyembryony in *Citrus*. He observed the formation of adventive embryos from nucellar cells of *Citrus* and called this phenomenon nucellar or adventive polyembryony. Since then, polyembryony has been recorded in a large number of angiosperms.

Maheshwari (1950) defined polyembryony as the occurrence of more than one embryo in a seed. According to Bouman and Boesewinkel (1969), polyembryony includes all cases in which there are clear indications of the potential or actual occurrence of two or several proembryos or embryos in a single seed.

## TYPES OF POLYEMBRYONY

Polyembryony can be broadly classified into "Simple" and "Multiple" depending upon the presence of one or more embryo sacs in the same ovule (Lakshmanan and Ambegaokar, 1984). In simple polyembryony the embryos develop within a single embryo sac, but in multiple polyembryony accessory embryos are produced from two or more embryo sacs in the same ovule (Figure 13.1). Simple polyembryony may be sexual or asexual.

On the basis of embryogenesis, Ernst (1918) divided polyembryony into two categories: (i) True polyembryony and (ii) False polyembryony. In true polyembryony, two or more embryos develop in the same embryo sac from the cells of nucellus (citrus, mango), integument or synergids. In false polyembryony, more than one embryo sac is formed in the same ovule (e.g., *Fragaria*) which results in the formation of multiple embryos.

Yakovlev (1967) classified polyembryony in to two main groups: gametophytic polyembryony and sporophytic polyembryony. In gametophytic polyembryony, the embryos develop from the gametic cells of the embryo sac (e.g., synergids, antipodals) before or after fertilization. In such cases, depending upon the fertilization, embryos may be haploid or diploid. In sporophytic polyembryony, multiple embryos develop from the sporophytic cells of the ovule (e.g., nucellus, integument) without fertilization. Such embryos are diploid in nature.

### NUCELLAR POLYEMBRYONY

In nucellar polyembryony (also known as adventive polyembryony) embryos develop from the cells of nucellus. The nucellar embryos are frequent in Anacardiaceae, Cactaceae, Rutaceae, Combretaceae, Poaceae, Malpighiaceae and Urticaceae. The nucellar cells destined to form embryos acquire dense

DOI: 10.1201/9781003260097-13

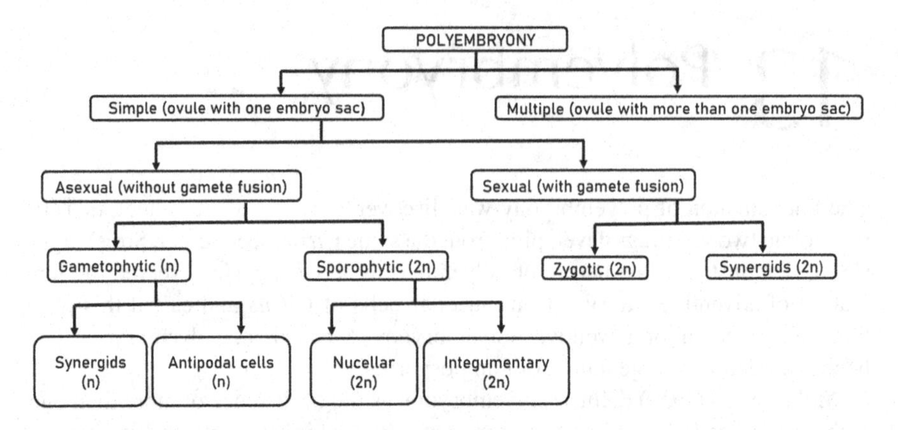


**FIGURE 13.1** Schematic representation of polyembryony in angiosperms.

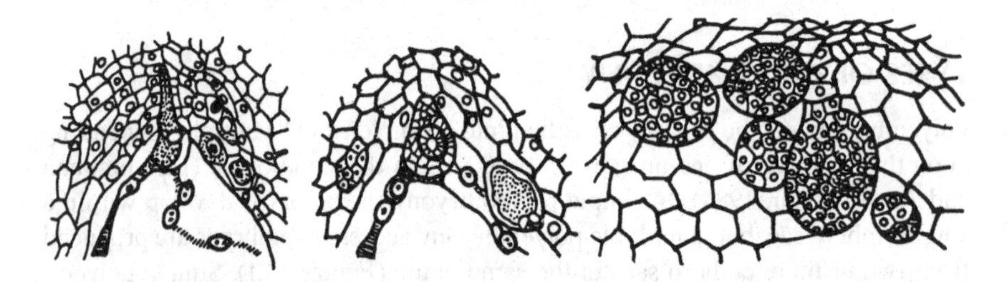

**FIGURE 13.2** Development of adventive embryos in *Citrus trifoliata*. A. Micropylar portion of embryo sac showing fertilized egg, pollen tube and endosperm nuclei; some of the nucellar cells have enlarged. B. Same, more advanced stage. C. Upper part of the embryo sac showing several embryos lying in endosperm; only zygotic embryo has suspensor (after Osawa, 1912).

cytoplasm and starch. These cells divide repeatedly to form embryonic masses. Some of these embryonic masses develop further to form embryos (Figure 13.2). They usually lack a suspensor except in a few cases (e.g., *Citrus macrocarpa*) where a well-defined suspensor has been observed (Rangaswamy, 1959). In nucellar embryos the cotyledons are usually unequal. Tricotyly and fasciation at the radicular side have also been observed. The origin of nucellar embryos may be autonomous, sporadic, induced or stimulated.

In *Aegle marmelos* 75–80% of the organized embryo sacs degenerate completely. In the remaining embryo sacs, the egg apparatus and antipodals degenerate but well-developed endosperm is formed due to the occurrence of triple fusion. Towards the micropylar end, a nucellar cell enlarges considerably, acquires dense cytoplasm and divides repeatedly to form an embryonal mass which finally gives rise to 2 or 3 embryos. Of these embryos, only one reaches maturity.

In *Citrus macrocarpa*, in a single embryo sac, 9–21 proembryos are formed, many of which attain complete differentiation but the cotyledons of these embryos are unequal in length and thickness. The nucellar embryos show a well-defined suspensor and therefore it is rather difficult to distinguish them from zygotic embryos.

In *Mangifera indica* as many as 50 embryos have been observed in a single seed. The nucellar embryos develop in the micropylar region and grow up to the middle of the embryo sac (Figure 13.3). Sometimes nucellar embryonal masses fuse with each other to varying degrees.

In *Opuntia dillenii* the egg degenerates, several nucellar embryos develop, and one of them reaches up to the mature stage.

## INTEGUMENTARY POLYEMBRYONY

Embryos may develop from the cells of the integument. The development of such embryos is not dependent on the fertilization of an egg. In *Euonymus*, the epidermal and subepidermal cells of the inner integument develop into adventive embryos.

In *Spiranthes cernua* the cells of inner integument develop into embryos. The terminal and a few subterminal cells enlarge and acquire dense cytoplasmic contents. The normally formed embryo sac degenerates at the four-nucleate stage and the cytoplasm rich cells divide to form 2–6 proembryos in the micropylar side. These proembryos, however, do not differentiate further.

In unitegmic, tenuinucellate ovules, the innermost layer of the integument differentiates as the endothelium. Sometimes some of the endothelial cells may become richly cytoplasmic and divide repeatedly to form several proembryos. This feature has been observed in *Melampodium divaricatum* (Figure 13.4) and

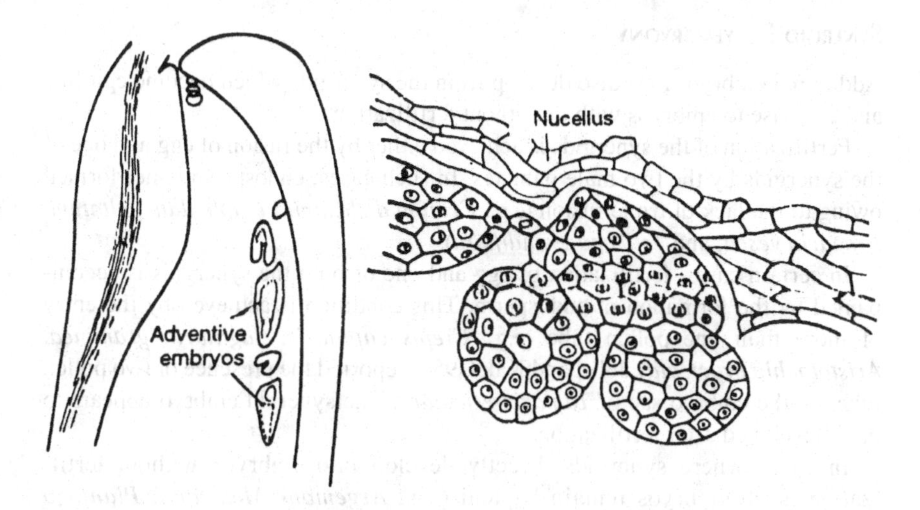


**FIGURE 13.3** A. *Mangifera indica* var. *paho*. Longitudinal section of the upper part of seed with eight embryos at different stages of development (after Sachar and Chopra, 1957). B. *Mangifera indica* var. *Higgins*. Part of nucellus with embryonic masses.

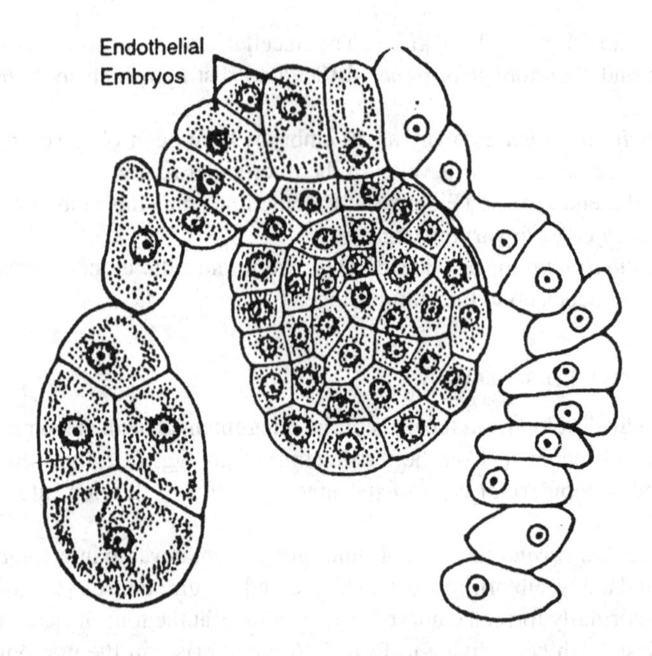


**FIGURE 13.4** *Melampodium divaricatum*, endothelial polyembryony (after Maheswari Devi and Pullaiah, 1976, 1977).

*Carthamus tinctorius* (Maheswari Devi and Pullaiah, 1976, 1977). These proembryos neither differentiate nor reach maturity.

## Synergid Polyembryony

Additional embryos may also develop from the synergids which become egg-like and give rise to embryos with or without fertilization.

Fertilization of the synergids is achieved either by the fusion of egg and one of the synergids by the two male gametes. In such cases, endosperm is not formed owing to the lack of triple fusion (e.g., *Crinum deflexum*, *Peltiphyllum peltatum*, *Fragaria vesca*, and *Tellisma grandiflora*).

In certain cases, fertilization of egg and one or both the synergids is accompanied by the formation of endosperm. This condition is achieved by the entry of more than one pollen tube (e.g., *Crepis capillaris*, *Sagittaria graminea*, *Aristolochia bracteata*). Johri and Kak (1954) reported the presence of two pollen tubes in the embryo sac of *Tamarix ericoides*. The synergid embryo appears to have developed after fertilization.

In cases where synergids directly develop into embryos without fertilization, such embryos remain haploid (e.g., *Argemone Mexicana*, *Plantago lanceolata*, *Phaseolus vulgaris*). In *Dioscorea composita*, the zygotic and synergid proembryos develop until a late stage (Figure 13.5A). They follow a uniform pattern of development up to organogenesis in *Nechamandra*

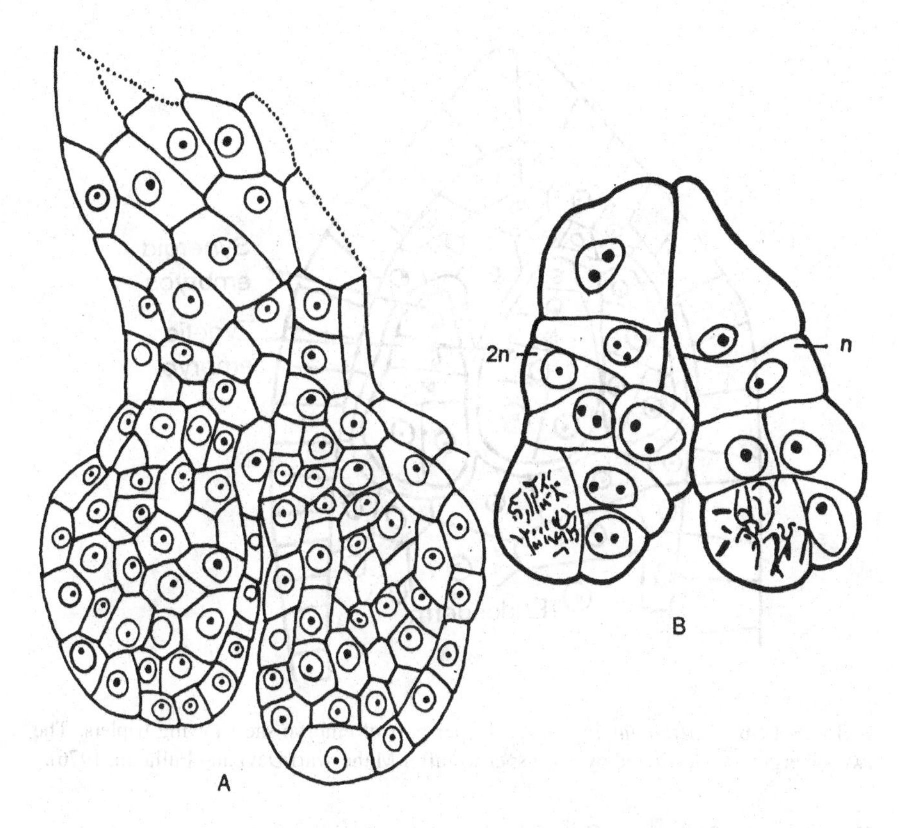


**FIGURE 13.5** Synergid polyembryony. A. *Dioscorea composita*. Zygote (left) and synergid (right) proembryos. B. *Lilium martagon*. Synergid n and zygotic 2n proembryo (A. after Thirumaran and Lakshmanan, 1982. B. after Cooper, 1943).

*alternifolia* (Lakshmanan, 1963). The development of the haploid embryo from the synergid along with the diploid zygote embryo is reported in *Lilium martagaon* (Figure 13.5B). In *Carthamus tinctorius*, the triplets develop due to fertilization of the egg and both synergids (Figure 13.6). The mechanism of stimulus that induces the unfertilized synergid to develop into an embryo is not yet clearly understood.

## Antipodal Polyembryony

The formation of antipodal embryos is not so common. It has been observed in *Alangium lamarkii*, *Paspalum scrobiculatum*, *Rudbeckia sullivantii*, *Ulmus americana* and *U. glabra* (Figure 13.7). According to Johri et al. (1992) the development of embryos from antipodals needs confirmation. When the antipodal cells divide, the configuration of cells may stimulate embryogenic patterns. Since advanced stages have never been recorded, it is not at all appropriate to interpret such cell assemblages as antipodal embryos.

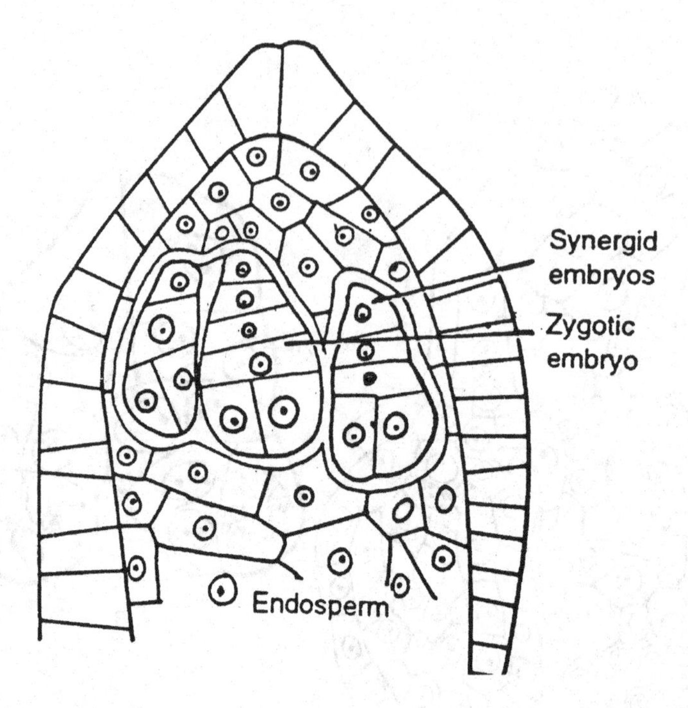


**FIGURE 13.6** *Carthamus tinctorius.* Upper part of embryo sac showing triplets. The two synergids are fertilized by extra sperms (after Maheswari Devi and Pullaiah, 1976).

## Zygotic and Suspensor Polyembryony

The simplest method of an increase in the number of embryos is a cleavage of the zygote or proembryo into two or more units (Figure 13.8). Among angiosperms, cleavage polyembryony is quite common in orchids. In *Cymbidium bicolor* (Orchidaceae) the zygote divides vertically or obliquely; the daughter cells get somewhat separated and divide further to form two independent proembryos. In *Erythronium americanum*, *Eulophia epidendraea* and *Hebenaria platyphylla* the zygote divides to form a small group of cells. Supernumerary embryos develop due to cleavage of the apical cells of the embryonic mass. The filamentous proembryo in *Eulophia* becomes branched, and an embryo may be formed at the tip of each branch (Figure 13.9).

Suspensor polyembryony is of common occurrence in members of family Acanthaceae and Solanaceae. In *Exocarpus*, as many as six embryos may develop simultaneously in an ovule by the proliferation of the suspensor cells (Figure 13.10).

## Causes of Polyembryony

Causes of polyembryony in nature are still unknown. Various theories have been put forward to explain the occurrence of polyembryony in angiosperms.

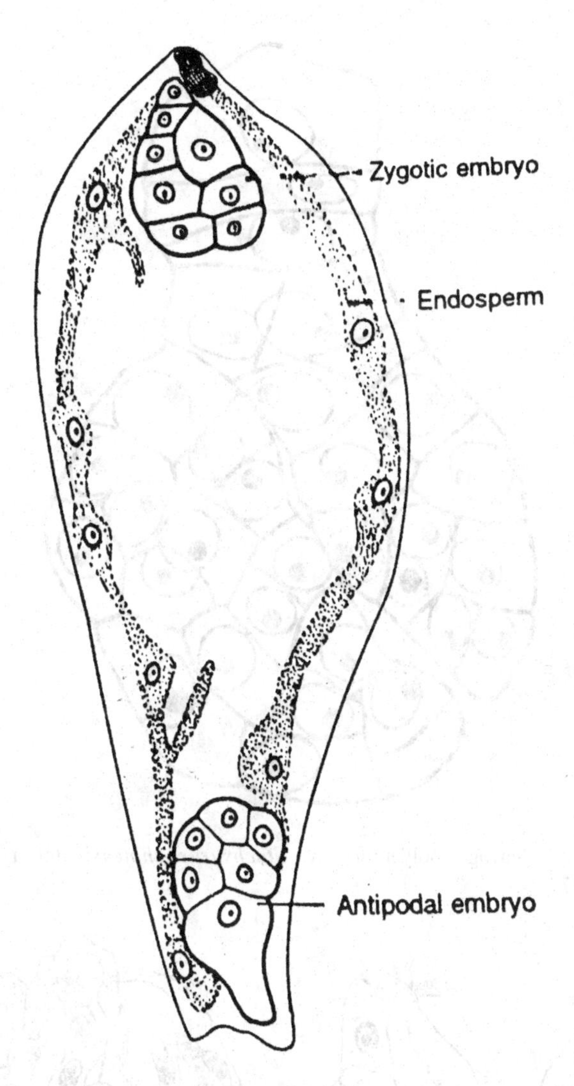


**FIGURE 13.7** Embryo sac of *Ulmus glabra* showing zygotic and antipodal embryos (after Ekdahl,1941).

1. According to Haberlandt (1921), the stimulus for the cell division and proliferation of nucellar embryo is furnished by certain substances which are secreted by the degenerating cells of the nucellus. On this basis, Haberlandt proposed the "necrohormone theory".
2. Kappert (1933) and Maheshwari and Rangaswamy (1958) opined that polyembryony is a recessive characteristic controlled by a series of multiple genes.
3. Leroy (1947) holds the view that in *Mangifera indica* adventive embryony is probably due to the effect of one of the more recessive genes.

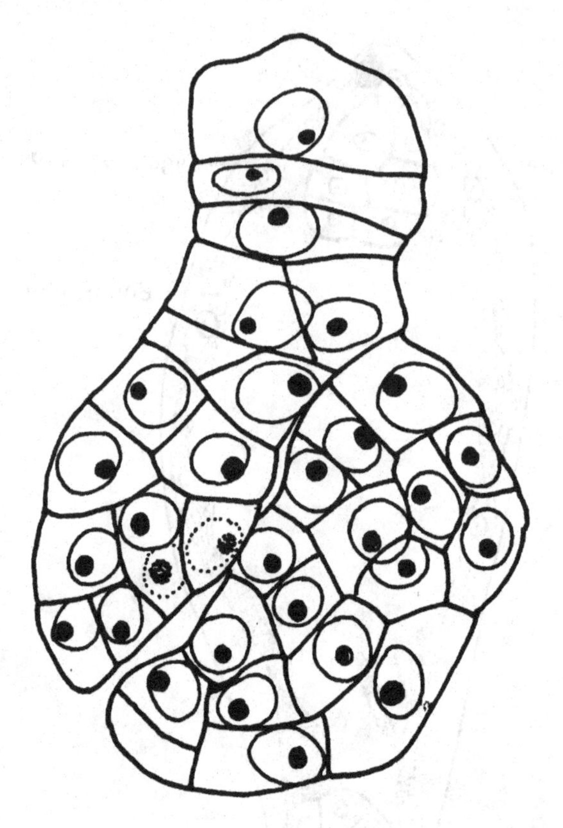

**FIGURE 13.8** Cleavage polyembryony. *Stachyurus chinensis* (after Mathew and Chaphekar, 1977).

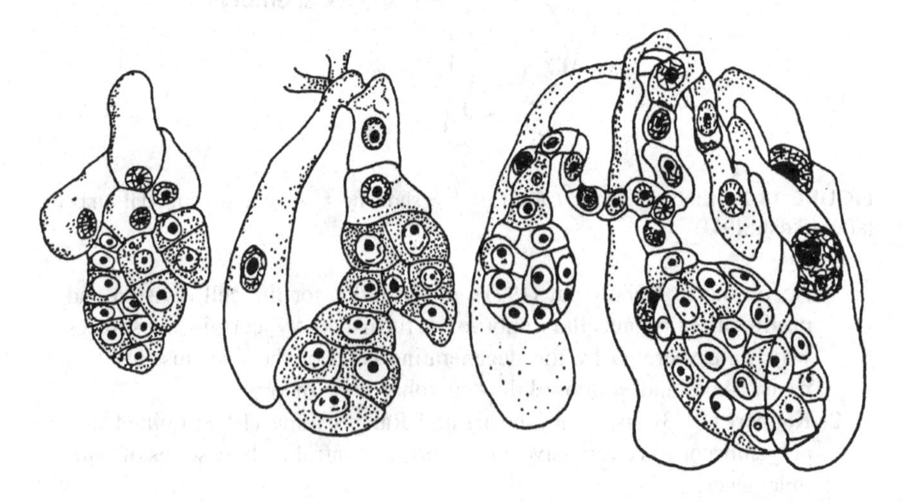

**FIGURE 13.9** Zygotic and suspensor polyembryony in *Eulophia epidendraea* (after Swamy, 1943).

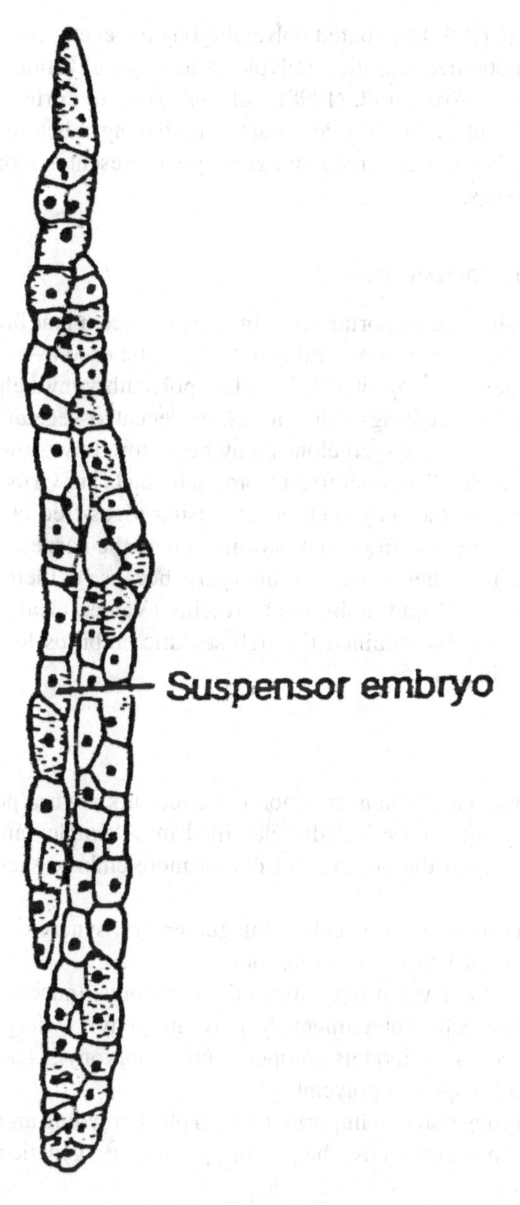


**FIGURE 13.10** Suspensor polyembryony in *Exocarpus sparteus* (after Ram, 1959).

4. Atabekova (1957) observed that certain races of the *Lupinus* show tendency towards production of the polyembryonic seeds and suggested that this tendency is hereditary.
5. Ernst (1918), Baker (1960) and Cesca (1961) believed that hybridization is responsible for polyembryony.

6. Reiser et al. (1993) attributed polyembryony to genetic causes as meiotic and/or mitotic irregularities, polyploidy and hybridization.
7. According to Aron et al. (1998), polyembryony in citrus and mango is generally controlled by a dominant gene having a heterozygous allele (Pp) while homozygous recessive gene (pp) is present in monoembryonic *Citrus* species.

### Significance of Polyembryony

Polyembryony plays an important role in cytogenetics, plant breeding and horticulture. Polyembryony is a natural trait that can be used as an alternative tool in the development of crop plants. Nucellar polyembryony helps in producing genetically uniform seedlings. The clones, by repeated vegetative propagation, deteriorate. The vigour of such clones may be restored by growing plants from nucellar embryos. Seedlings obtained from such plants are virus free. The nucellar polyembryony is the only method of raising virus-free clones of polyembryonate *Citrus* varieties. In polyembryonic maize, the kernels contain a higher nutritional value than their normal counterparts because of their enhanced quantity and quality in oil and embryonic proteins (Valdez et al., 2004). Desired genotypes may also be obtained through somatic embryos formed in cultures (for details see Chapter 16).

## SUMMARY

- Occurrence of more than one embryo in a seed is known as polyembryony.
- Polyembryony can be broadly classified into "simple" and "multiple", depending upon the presence of one or more embryo sacs in the same ovule.
- Polyembryony may be nucellar, integumentary, synergids, antipodal or zygotic and suspensor polyembryony.
- Embryos may develop from the cells of nucellus (nucellar polyembryony), integument (integumentary polyembryony), synergids (synergid polyembryony), antipodals (antipodal polyembryony), cleavage polyembryony and suspensor polyembryony.
- Polyembryony plays an important role in plant breeding and horticulture.
- Nucellar polyembryony helps in producing genetically uniform seedlings.

## SUGGESTED READING

Aron, Y., H. Czosnek, S. Gazit and C. Degani. 1998. Polyembryony in mango (Mangifera indica L.) is controlled by a single dominant gene. *Hort. Sci. Alexandria* 33(7): 1241–2.1.

Baker, H.G. 1960. Apomixis and polyembryony in Pachiraoleaginea (Bombacaceae). *American J. Botany*, 47: 296–302.

Bouman, F. and F.D. Boesewinkel 1969. On a case of polyembryony in Pteroearya fraxinifolia (Juglandaceae) and on polyembryony in general. *Acta Bot. Neer.* 18: 50–57.
Cesca G. 1961. Ricercheembryologischesu Rudbeckiamissouriensis. *Caryologia* 14: 129–139.
Chen, M., J.-Y. Lin, X.Wu, N.R. Apuya, K.F. Henry, B.H. Le, A Q. Bui, J.M. Pelletier, S. Cokus, M. Pellegrini, J.J. Harada and R.B. Goldberg 2021. Comparative analysis of embryo proper and suspensor transcriptomes in plant embryos with different morphologies. *PNAS* 118 (6). https://doi.org/10.1073/pnas2024704118.
Cooper, D.C. 1943. Haploid-diploid twin embryos in *Lilium* and *Nicotiana*. *Am. J. Bot.* 30: 408–413.
Ekdahl, I. 1941. Die entwicklung von embryosack und embryo bei *Ulmus glabra* Huds. *Sven. Bot. Tidskr.* 35: 143–156.
Ernst, A. 1918. Bastardierungals Ursache der Apogamie im Pflanzenreich: Ein Hypothese zu experimentellen Vererbungs- und Abstammungslehre. Fischer, Jena. 665.
Esen, A. and K. Soost. 1977. Advetive embryogenesis in *Citrus* and its relation to pollination and fertilization. *Am. J. Bot.* 64: 607–614.
Haberlandt. G. 1921. Ober experimentelle Erzeugung von Adventiv embryonen bei Genathera lamarckiana. Sit.,h. Frellssisch. A had. *Wiss. Berlin* 40: 695–725.
Johri B.M., Kak D. 1954. The embryology of Tamarix Linn. *Phytomorphology* 4: 230–247.
Johri, B.M., K.B. Ambegaokar and P.S. Srivastava (eds.) 1992. Comparative Embryology of Angiosperms, Vols. 1 and 2. Heidelberg: Springer Berlin.
Kappert, H. 1933. Erbliche Polyembryonie bei Linum usitatissimum. *Bioi. Zenlralbl.* 53: 276–307.
Lakshmanan, K.K. 1963. Embryological studies in the Hydrocharitaceae. III. Nechamandra alternifolia. *Phyton (Argentina)* 20: 49–58.
Lakshmanan, K.K. and K.B. Ambegaokar. 1984. Polymebryony. In *Embryology of Angiosperms*, ed. B.M. Johri, 445–474. Berlin: Springer-Verlag.
Leroy J.F. 1947. La polyembryonie chez les Citrus son intérêtdans la culture etamélioration. *Rev. Intern. Bot. Appl. Paris*, 27: 483–495.
Maheshwari, P. 1950. *An Introduction to Embryology of Angiosperms*. New York: McGraw-Hill Book Co.
Maheshwari P, Rangaswamy N.S. 1958. Polyembryony and in vitro culture of embryos of Citrus and Mangifera. *Indian J. Hortic.* 15: 275–282.
Maheshwari Devi, H. and T. Pullaiah. 1976. Embryological investigations in *Melampodium divaricatum*. *Phytomophology* 26: 77–86.
Maheshwari Devi, H. and T. Pullaiah. 1977. Embryological abnormalities in *Carthamus tinctorius* Linn. *Acta Bot. Indica* 5: 8–15.
Maheshwari, P. and R.C. Sachar. 1963. Polyembryony. In *Recent Advances in the Embryology of Angiosperms*, ed. P. Maheshwari, 265–296. Delhi: International Society of Plant Morphologists.
Mathew, C.J. and M. Chaphekar. 1977. Development of female gametophyte and embryogeny in *Stachyurus chinensis*. *Phytomorphology* 27: 68–79.
Naumova, T.N. 1981. Nucellar and integumentary embryony in angiosperms. *Acta. Soc. Bot, Polon* 50: 213–216.
Osawa, I. 1912. Cytological and experimental studies in *Citrus*. *J. Col. Agr. Imp. Univ. Tokyo* 4: 83–116.
Ram, M. 1959. Morphological and embryological studies in the family Santalaceae. II. *Exocarpus*, with a discussion on its systematic position. *Phytomorphology* 9: 4–19.
Rangaswamy, N.S. 1959. In vitro studies on the ovules of Citrus microcarpa Bunge. *Ph. D. Thesis, Univ.* Delhi.

Sachar, R.C. and R.N. Chopra. 1957. A study of the endosperm and embryo in *Mangifera* L. *Indian Agric. Sci.* 27: 219–228.
Swamy, B.G.L. 1943. Gametogenesis and embryogeny of *Eulophia epidendraea. Proc. Natl. Inst. Sci., India B.* 9: 59–65.
Thirumaran, K. and K.K. Lakshmanan. 1982. Synergid embryo in *Dioscorea composita* Hemsl. *Curr. Sci.* 57: 428–429.
Wakana, A. and S. Uemoto. 1987a. Adventive embryogenesis in *Citrus.* The occurrence of adventive embryos without pollination or fertilization. *Am. J. Bot.* 74: 517–530.
Wakana, A. and S. Uemoto. 1987b. Adventive embryogenesis in *Citrus* (Rutaceae) II. Post-fertilization development. *Am. J. Bot.* 75: 1033–1047.
Yakovlev, M.S. 1967. Polyembryony in higher plants and principles of its classification. *Phytomorphology* 17: 278–282.

# 14 Apomixis

Flowering plants have two modes of reproduction: amphimixis and apomixis. Amphimixis (sexual reproduction) involves fusion of male and female gametes, whereas apomixis (asexual reproduction) may be achieved by means of asexual production of seeds. Sexual reproduction is a means to develop new genetic combinations, whereas apomixis can help in the maintenance of desired plant phenotypes (Hojsgaard and Hörandl, 2019).

Apomixis may be defined as the phenomenon in which normal sexual reproduction is completely replaced by asexual reproduction. The plants exhibiting such a phenomenon are called apomicts. Winkler (1908), who coined the term, defined apomixis as the substitution of the usual sexual reproduction by a form of reproduction which does not involve meiosis and syngamy. He included vivipary, pseudovivipary and other modes of vegetative reproduction such as bulbils, runners, rhizomes, etc.

According to Solntzeva (1969), apomixis is a special and peculiar mode of seed production. This mode of reproduction implies the use of morphological structures inherent in seed reproduction – flower, ovule, embryo sac, embryo, seed and fruit. In this mode of reproduction, disturbances may occur both in sporogenesis and in the sexual process. The production of a new organism takes place through the formation of an embryo.

Apomictic reproduction has been detected in 78 out of 460 families of the flowering plants – mainly in Rosaceae, Asteraceae and Poaceae. Within the Poaceae, apomixis has been reported in more than 125 species, including members of the economically important genera *Bothriocloa*, *Brachiaria*, *Dichanthium*, *Eragrostis*, *Paspalum*, *Pennisetum*, *Poa*, *Tripsacum*, etc.

Apomixis is possibly an ancient feature in some major clades particularly in commelinids, fabids, and lamiids (Hojsgaard et al., 2014). The analysis of apomixis in current phylogenetic trees leads to the hypothesis that this trait evolved independently multiple times among plant families.

## CLASSIFICATION

Maheshwari (1950) divided apomixis into four classes: (1) non-recurrent apomixis, (2) recurrent apomixis, (3) adventive embryony, and (4) formation of vegetative propagules. Battaglia (1963) defined apomixis as the production of sporophyte from the gametophyte without sexual fusion. He classified apomixis into two types: (1) recurrent apomixis and (2) non-recurrent apomixis.

Apomixis may be classified into three different categories: (i) apomeiosis, the generation of a cell capable of forming an embryo without prior meiosis; (ii) parthenogenesis, the independent development of the embryo; and (iii) the

DOI: 10.1201/9781003260097-14

capacity to either produce endosperm autonomously or to use an endosperm derived from fertilization (Figure 14.1).

## AGAMOSPERMY

When embryos are produced without fertilization, the apomictic process is referred to as agamospermy (*a*= without; *gamo*= gametes; *sperm*= seed).

The agamospermy has several advantages: (1) it allows a sexual reproduction hybrid genotype which may possess adaptively valuable gene combinations, (2) it facilitates the perpetuation of a successful maternal genotype by allowing the hybrid to reproduce its own kind by seeds, and (3) seeds have dispersal and adaptive advantage over vegetative reproduction.

Agamospermy includes two different modes of reproduction: adventitious embryony and gametophytic apomixis.

### ADVENTITIOUS EMBRYONY (SPOROPHYTIC APOMIXIS)

In this type of agamospermy, the new sporophyte develops directly from a somatic cell of the ovule (nucellus or integument). Sporophytic apomixis differs from the gametophytic in that no alternation of generations intervenes prior to embryo development. Adventitious embryony leads to the development of more than one embryo in a seed, e.g., *Citrus*, Mango (*Mangifera indica*) and in orchids. In the mature embryo sac only endosperm is produced. The cells destined to form adventive embryos become richly protoplasmic and actively divide to form small groups of cells, which eventually push their way into the embryo sac and grow further to form true embryos. Frequently the zygotic embryo also develops at the

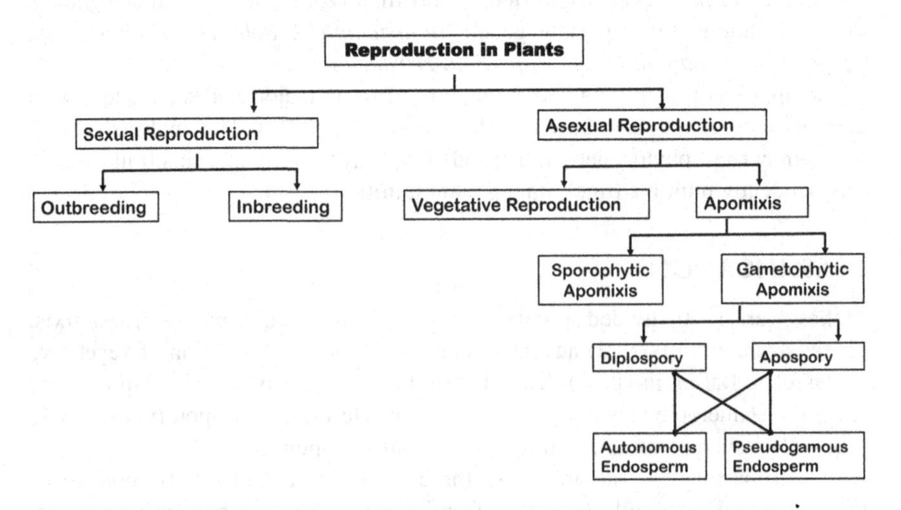


**FIGURE 14.1** Schematic representation of different categories of apomixis (after Sharma and Bhat, 2020).

same time and is distinguishable from the adventive embryos only by the somewhat lateral position and lack of a suspension in the latter. In *Citrus* (Rutaceae) the number of adventive embryos may be four or five. Sometimes as many as 13 viable embryos can be found in the same seed.

Adventitious embryony may be completely autonomous, e.g., independent of pollination and fertilization; or it may be induced by one or both of these factors. Adventitious embryony is of considerable significance (see Chapter 13).

## Gametophytic Apomixis

Gametophytic apomixis includes those apomictic pathways where the maternal embryo develops from a diploid egg cell differentiated in an unreduced embryo sac. In this type the female gametophyte develops from an unreduced generative cell (sporogenous or megaspore mother cell) directly by meiosis or indirectly by modified meiosis resulting in an unreduced restitution nucleus. An unreduced embryo sac can also develop from somatic cells of the nucellus.

In *Cenchrus ciliaris* (Poaceae), both sexual and apomictic mode of reproduction has been reported. During sexual reproduction, the megaspore mother cell divides to form a tetrad, out of which one megaspore becomes functional and gives rise to a seven-celled, eight-nucleate embryo sac (Figure 14.2). In apomictic mode, a diploid aposporous initial originates from nucellar tissues around

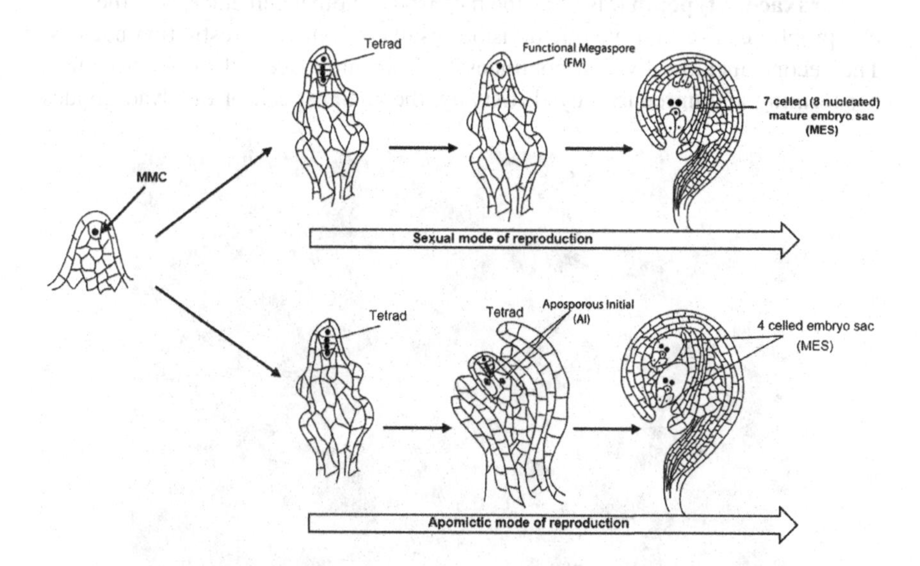


**FIGURE 14.2** *Cenchrus ciliaris.* Sexual and apomictic mode of reproduction. Megaspore mother cell divides to give rise to a tetrad which leads to the formation of a seven-celled mature embryo sac through the sexual mode of reproduction. The tetrad may take an apomictic mode of reproduction through the formation of aposporous initial (AI) followed by the development of four-celled unreduced MES (after Rathore et al., 2020).

the megaspore mother cell and produces an unreduced aposporous embryo sac (Ross et al., 2004). Aposporous initial may completely displace sexual structures, or sexual development may persist in the same ovule along with the apomictic embryo sac (Figures 14.2 and 14.3). There may be multiple embryo sacs in one ovule. Gametophytic apomixis has been mainly reported in Asteraceae, Poaceae and Rosaceae. The majority of gametophytic apomicts are polyploids.

In gametophytic apomixis, many apomictic taxa are facultative. A single individual can produce seeds through both sexual and apomictic pathways.

Gametophytic apomixis is classified into two categories: diplospory and apospory. In diplospory and apospory there is no meiosis, and fertilization of the egg does not occur. The reduced egg develops parthenogenetically. The formation of endosperm depends on the fertilization of polar nuclei/secondary nucleus.

## Diplospory (Generative Apospory)

The unreduced embryo sac develops from a generative cell, i.e., either from female archesporium or megaspore mother cell. Diplospory may be divided into three main groups:

**Antennaria type:** The megaspore mother cell does not divide meiotically and directly forms the unreduced eight-nucleate embryo sac, e.g., *Antennaria, Erigeron, Eupatorium* (Figure 14.4).

**Taraxacum type:** In this type, the megaspore mother cell enters into the meiotic prophase. The first meiotic division results in a meiotic restitution nucleus. The second meiotic division continues with an unreduced chromosome number forming and unreduced dyad. Usually, the chalazal cell of the dyad divides

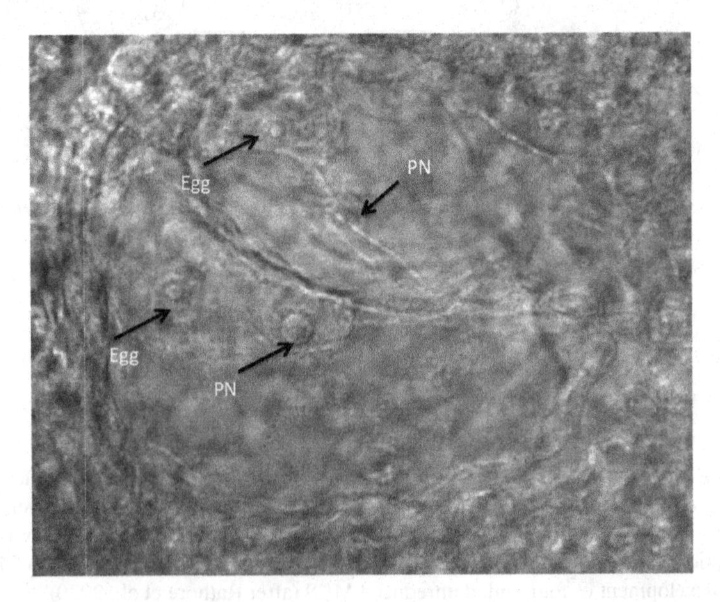


**FIGURE 14.3** *Cenchrus ciliaris*. Dual embryo sac (courtesy: Dr Anuj Dwivedi, Delhi).

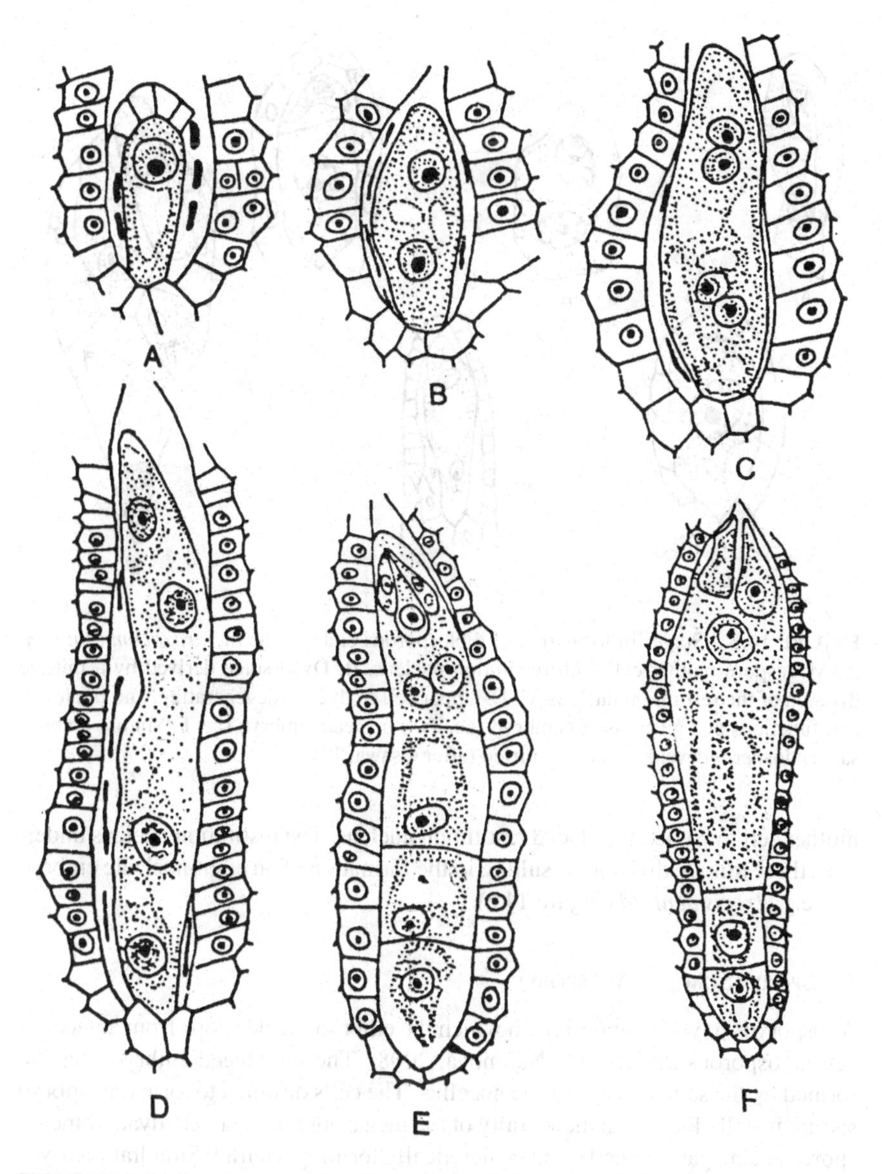


**FIGURE 14.4** Development of embryo sac in *Eupatorium triplinervis*. A. Megaspore mother cell. B. Two-nucleate embryo sac. C, D. Four-nucleate embryo sac. E. Organized embryo sac. F. Embryo sac showing degenerating egg apparatus and secondary nucleus (after Deasi and Mukherjee, 1983).

mitotically to form an eight-nucleate embryo sac. The micropylar cell of the dyad, however, degenerates, e.g., *Taraxacum* (Figure 14.5), *Chondrilla*, *Paspalum*.

**Ixeris type:** This type of diplospory occurs in species which show tetrasporic type of embryo sac as in the Taraxacum type, where the nucleus of the megaspore

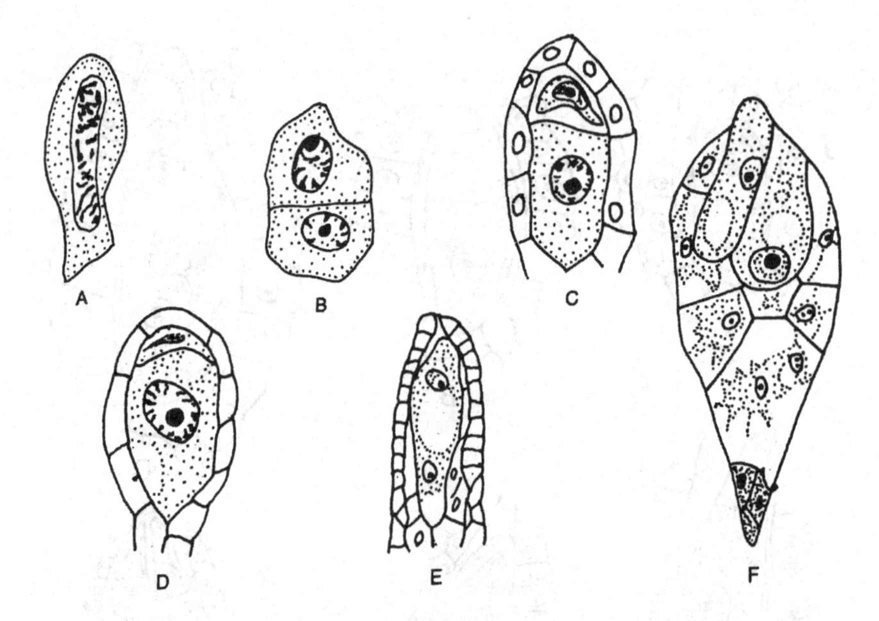


**FIGURE 14.5** Stages in formation of diplosporous embryo sac in *Taraxacum albidum*. A. Megaspore mother cell with restitution nucleus. B. Dyad stage derived by a mitotic division of the restitution nucleus. C, D. Upper cell of dyad is degenerating, and the lower cell functions and gives rise to embryo sac. Two-nucleate embryo sac. F. Mature embryo sac, endosperm development has started (after Osawa, 1913).

mother cell forms an unreduced restitution nucleus. The restitution nucleus undergoes three mitotic divisions resulting in the formation of an eight-nucleate embryo sac, e.g., *Ixeris dentata* (Figure 14.6).

### Apospory (Somatic Apospory)

Apospory is a type of apomixis in which the embryo sac develops from a nucellar cell (aposporous initial cell) (Naumova, 2008). The unreduced embryo sacs are formed by the somatic cells of the nucellus. The cells destined to form the embryo sac are usually located in the vicinity of the megaspore mother cell, dyad or megaspore. The megaspore cell divides meiotically forming a tetrad. Simultaneously, a cell of nucellus enlarges and develops into an embryo sac (aposporic embryo sac) (Figure 14.7). The megaspores that formed earlier due to the meiotic division of the megaspore mother cell gradually degenerate. Aposporic embryo sac development is found in *Hieracium*, *Crepis* (Asteraceae), *Panicum*, *Paspalum* (Poaceae).

Voronova and Babro (2019) reported the occurrence of apospory in *Helianthus ciliaris* (Asteraceae). Aposporous embryos are the clones of the mother plant. Formation of aposporous plants leads to genetic heterogeneity (the presence of both maternal and mixed inheritance) even among offspring of the same plant (Batygina and Vinogradova, 2007).

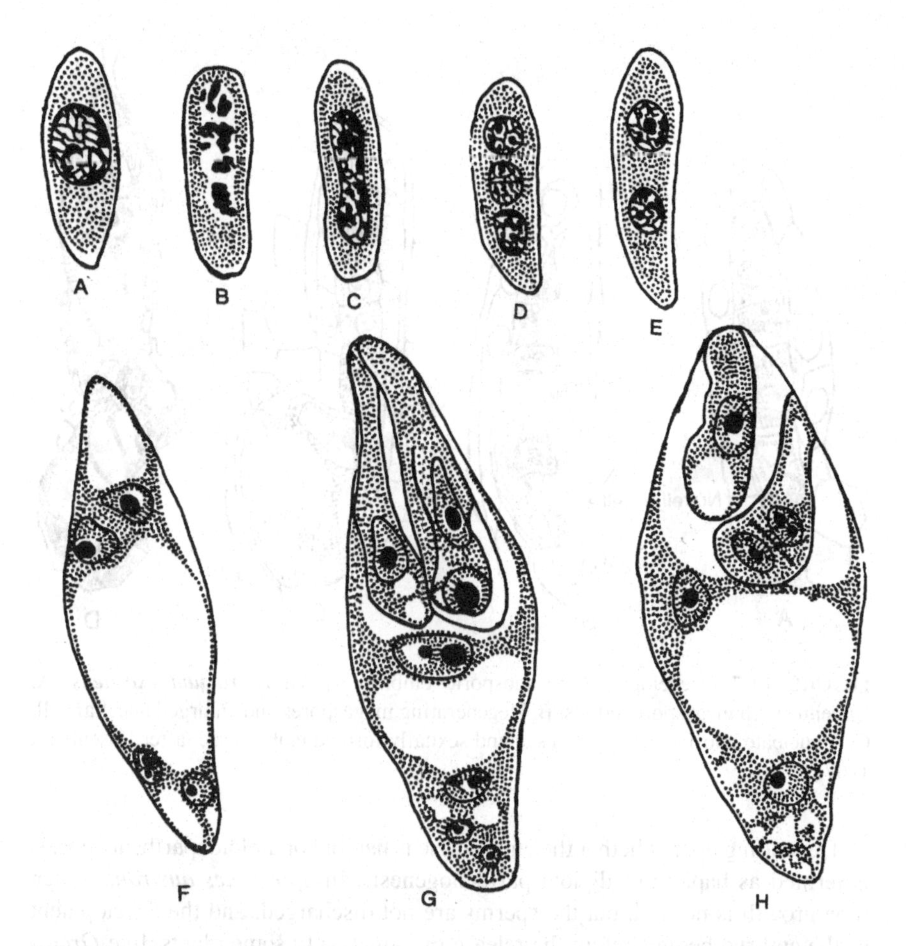


**FIGURE 14.6** Development of diplosporous embryo sac in *Iberis dentata*. A. megaspore mother cell showing prophasic nucleus. B. Later stages of meiosis with univalents. C. Restitution nucleus, D, E. 2-nucleate embryo sac. F. 4-nulceate embryo sac. G. Organized female gametophyte. H. Later stage showing 2-celled proembryo (after Okabe, 1932).

## PARTHENOGENESIS

When an embryo develops directly from an unfertilized egg, the phenomenon is known as parthenogenesis. Kerner (1876) observed that in *Antennaria alpina*, male plants were extremely rare in nature, but still there was development of seeds. Juel (1900) made a detailed study of the development of embryo in *A. alpina* and reported that even when staminate flowers are present, fertile pollen are very few. In this plant, the megaspore mother cell directly develops into the embryo sac without meiosis, and the diploid egg produces an embryo without fertilization.

Parthenogenesis may be classified into four categories: (1) haploid parthenogenesis, (2) diploid parthenogenesis, (3) facultative parthenogenesis and (4) induced parthenogenesis.

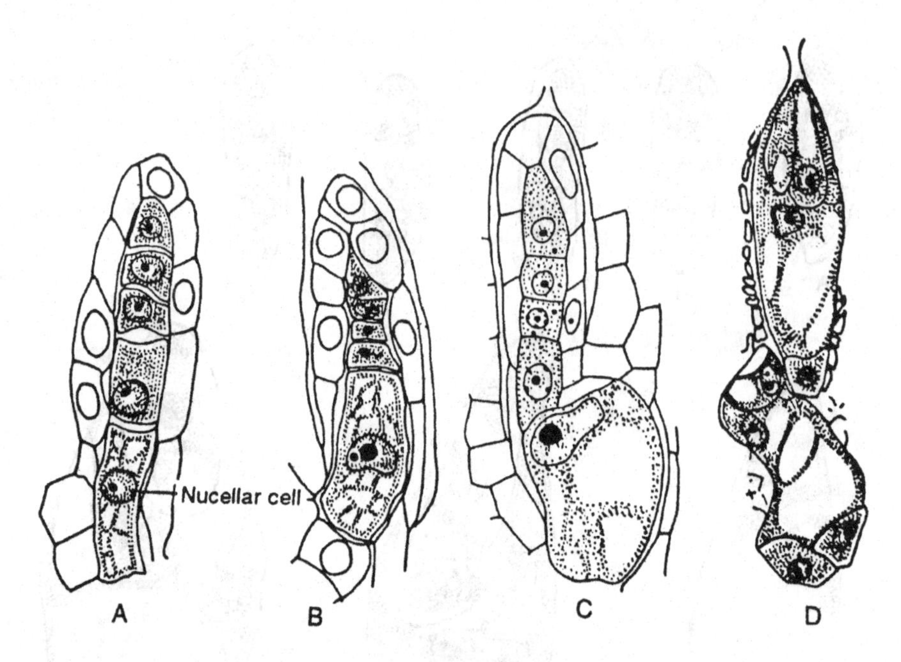


**FIGURE 14.7** Development of aposporic embryo sac in *Hieracium excellens*. A. Nucellus with megaspore tetrads. B. Degenerating megaspores and enlarged nucellar cell. C. 4-nucleate aposporous embryo sac and sexually formed embryo sac (after Rosenberg, 1907).

Depending upon whether the embryo sac is haploid or diploid, parthenogenesis is termed as haploid or diploid parthenogenesis. In *Spiranthes australis* pollen tube growth is normal, but the sperms are not discharged, and the development of the embryo begins before the release of gametes. In some plants, like *Orchis maculata*, *Listera ovata* and *Platanthera chlorantha*, some ovules receive more than one pollen tube while others receive none, or the pollen tube may arrive too late. The former develops into normal embryos, while the latter gives rise to haploid embryos (facultative parthenogenesis).

If the chromosome number of the gamete has been reduced in the normal manner at meiosis, and chromosome doubling of the unfertilized gamete does not occur, the apomictically produced embryo and the plant developing from it will be haploid.

Parthenogenesis can be induced by: (i) exposing the ovary to very high or very low temperatures soon after pollination, (ii) use of x-rayed pollen on stigma, (iii) use of foreign pollen or of delayed pollination and (iv) chemical treatment.

Male gametes, besides imparting genes, also provide stimulus for fruit development. If stimulus can be provided without discharge of sperms, it would be easier to produce a homozygous true breeding type, which otherwise requires a long and labourious process of self-fertilization.

Parthenogenetic development of haploid plants in the cultures of unfertilized ovaries has been reported in *Hordeum vulgare, Nicotiana tabacum and Triticum aestivum.*

## ORIGIN OF APOMIXIS

The origin of apomixis still represents an unsolved problem, as it may be either evolved from sex or the other way around. Apomixis has evolved several times during evolution of angiosperms and a number of interrelated factors, such as hybridization, polyploidization and environmental factors, seem to play an important role (Hojsgaard and Hörandl, 2019). It is hypothesized that apomixis across the eukaryote kingdoms evolved from sexual reproduction by genetic or epigenetic modifications causing deregulation of events associated with sexual reproduction (Neiman et al., 2014). Several theories have been proposed for the origin of apomixis.

1. **Polyphyletic origin:** Apomixis is believed to have originated independently on several occasions (Hojsgaard et al., 2014). This assumption gets support from the fact that two main variants of apomixis, i.e., apospory and diplospory, do not appear to have a common ancestor (Albertini et al., 2019).
2. **Mutation theory:** The theory assumes that a combination of pre-existing tendencies for features like haploid parthenogenesis and relaxed 2 m:1p requirement for endosperm development along with dominant apomeiotic mutations would lead to apomixis (Carman, 2001). This theory falls short, however, in explaining the variability in the facultative nature of most apomicts.
3. **Hybridization-derived floral asynchrony theory:** According to this theory (Carman, 2007), the apomixis could result from temporal deregulation of alleles involved in the demarcation of reproductive fate of cells. This deregulation may occur due to the hybridization of divergent alleles. This theory has been supported through transcriptomic profiling of apomixis in *Boechera*. This study revealed that several genes conserved across both sexual and apomictic systems are heterochronously expressed in apomictic reproduction and could also be linked to parent of origin effects (Tucker and Koltunow, 2009). According to this theory, asynchrony in megasporogenesis and embryo sac development in hybrids are the main cause of apomixis.

Due to multiple independent origins of apomixis in different taxa and varied pathways known for apomictic reproduction, it is difficult to provide a single theory for the origin of apomixis in angiosperms. However, we can conclude that hybridity, ploidy, and environmental adaptations are some of the factors affecting the process of evolution of apomixis.

## GENETIC BASIS OF APOMIXIS

The inheritance of apomixis has been keenly debated for about one century. The genetic and molecular basis of apomixis is still not well known. This asexual mode of reproduction was first considered to be determined by a complex balance of recessive genetic factors. But since the mid-1960s, genetic analyses, mainly in tropical grasses, have showed that apomixis is simply inherited. Such analyses have generally focused on the expression of apomeiosis, apparently determined by one dominant gene. Because apomeiosis and apomixis commonly coseggregate, the one dominant gene hypothesis has been widened to the overall apomictic development.

The work of Leblanc et al. (1995) on maize-*Tripsacum* hybrids confirmed the one dominant gene hypothesis for apomeiosis, but also suggested that apomixis might rather be determined by a complex of genes mimicking a single genetic factor.

Apomixis is reported to be genetically controlled by the apospory-specific genomic region in *Cenchrus ciliaris* (Ozias-Akins et al., 2003).

Apomixis is widespread in the plant kingdom. It is often associated with polyploidy, whereas sexuality is usually found at lower ploidy levels and associated with regular chromosome pairing. Evidences exist that apospory and diplospory are genetically controlled. Early hypotheses assumed several genes were involved based on the logic that apomixis, like any mode of reproduction, is a complex physiological process. But the success of apomixis across plant species, genera and families weakens the model of several genes, since any single mutation could condemn a plant to sterility.

Contradicting the early hypothesis, a simple genetic determinism (mono- or oligogenic) was proposed for some grasses of the Andropogoneae and Panicae. Genetic control of apomixis was studied in *Brachiaria* using completely sexual biotypes in crosses with compatible apomictic species. Valle and Savidan, 1996 suggested a simple, monogenic control. Recent evidence shows that female gametophyte development and seed formation are controlled by epigenetic mechanisms that distinguish sexual and apomictic development (Rodriguez-Leal et al., 2015).

In *Potentilla puberula*, genetic contribution from at least one apomictic parent has been found to be necessary for the formation of new apomictic genotypes, and penta- and hexaploids derived from sexual background did not show apomixis and hence do not support de novo origin of apomixis (Nardi et al., 2018).

Usually, apomixis has been considered an evolutionary dead end due to lack of genetic variation and adaptability to changing environments and also absence of recombination and interbreeding. However, substantial variation among apomictic species has been demonstrated. The following attributes point to a rather diverse and promotive role of apomixis in evolution:

1. Facultative apomicts have the evolutionary advantages of both sexual and asexual reproduction.
2. Alternation of generations occur without change of ploidy.

3. New apomictic species avoid meiosis and fertilization, which breaks sterility barriers induced by interspecific hybridization.
4. Apomixis maintains heterogenous and heterogenomic genotypes.
5. Fixed heterosis enhances competitive ability and survival of new apomictic species.
6. Apomixis provides 2n gametes for BIII hybrid formation, and it obstructs the rapid increase of ploidy in progeny by avoiding fertilization.
7. The cell cycles responsible for megaspore and megagametophyte development may be shortened, which may improve fitness while invading new or disturbed sites.
8. Genes for apomixis may be transmitted in wide crosses, thus enlarging apomictic populations.
9. Facultative apomixis, somatic variation and BIII hybrid formation produce new variation which is maintained by apomixis.

Each of these attributes contribute to the distribution and persistence of apomictic species in nature. Such species often exhibit various ranges of ploidy and chromosome numbers, seed-set in male sterile individuals, and a wide distribution in high elevations and disturbed regions (Cai et al., 1995). Through NGS sequencing, Lovell et al. (2017) found increased sequence diversity in apomictic populations. More information is, however, needed on the "apomixis gene" and its possible effect.

## EMBRYO–ENDOSPERM RELATIONSHIP IN APOMICTS

When pollination does not occur, the endosperm develops from the polar nuclei without fertilization. This mode of endosperm development in apomicts is known as "autonomous" and in such cases the embryo develops parthenogenetically (e.g., many Asteraceae, *Cortaderia jubata*). In *Commiphora wightii*, a sporophytic apomict, pollination stimulus is not required for the development of an apomictic embryo, and the development of endosperm is autonomous.

When fertilization of polar nuclei takes place, the development of endosperm is called "pseudogamous". The development of an embryo in pseudogamous forms begins only after the formation of endosperm as is usually seen in sexually reproducing plants (e.g., *Panicum maximum*, *Hypericum perforatum*).

The endosperm is primarily a nutritive tissue. In normal sexual forms, the embryo depends upon endosperm for its nutrition, at least in the earlier stages of its development. As the species are predominantly aposporous, the dependence of the embryo on the endosperm tissue is not obvious. This fact gets support by the frequent occurrence of embryo sacs with embryos but without endosperm tissue. This led to the suggestion that the dependence of the embryos on endosperm has been relaxed in the case of apomicts. Brink and Cooper (1944) opined that the absence of endosperm in the earlier stages of development indicates that either the embryos in such plants do not need the nutritive environment of the endosperm or that some other mechanism exists which has taken over the role of endosperm.

This indicates that the embryo–endosperm relationship in the apomicts may be different from that observed in the sexually reproducing species. Presence of a well-developed endosperm with an undivided egg cell suggests that the development of the one does not influence the development of the other and that the development of the one does not influence the development of the other and that no functional relationship exists between the embryo and endosperm (Narayanan, 1962).

The presence of normal healthy embryos without any endosperm raises the question as to how such embryos are nourished in the early stages of their development. Cooper and Brink (1949) observed one ovule possessing an embryo without endosperm in *Taraxacum officinale* (an autonomous apomict). They believed that nutrition to such embryos was made available by the ovules. In *Panicum* and *Pennisetum* (Narayanan, 1962), several aposporous embryo sacs developed in the same ovule, thus depleting the ovule tissue and rendering the ovule nutritionally deficient. The presence of as many as seven well-developed embryo sacs in one ovule, in three of which there were well-developed embryos without any trace of endosperm, supports such a view. The presence of occasional apospory in diploid *Panicum antidotale* (2n = 18) and its predominance in a tetraploid race (2n = 36) of the species affords an example in support of the views expressed by Gustafsson (1947) that "the action of many of these apomixis-influencing genes is stronger on the polyploid level than it is on the diploid level".

## PRACTICAL APPLICATIONS OF APOMIXIS

Apomixis gives rise to clonal progeny with a maternal genotype through the seed. Apomixis is important in plant breeding to fix hybrid vigour and could significantly reduce the costs of producing high yielding hybrid seeds (Koltunow et al., 1995). Apomictic reproduction would allow one to fix any desired homozygous gene combination and to maintain new heterozygous varieties with valuable agronomical properties such as disease resistance. Apomixis possesses a significant role in crop improvement because of its potential role in fixing heterosis and facilitating hybrid seed production without the need of maintaining parental inbred lines. It will also contribute to the understanding of many fundamental processes in the biology of reproductive systems, such as genetic control of meiosis, development of the female gametophyte, double fertilization, embryo and endospermogenesis (Fiaz, 2020).

The traditional breeding approach for incorporating apomixis in crop plants through interspecific hybridization resulted in the production of unviable germplasm (Bicknell and Koltunow, 2004). Hence, efforts are being made to harness apomixis through molecular genetics by identifying the key genes associated with regulating apomixis in plants (Kumar, 2019).

Difficulties in breeding apomictic grasses are numerous but are typically associated with the limited number of sexual plants available within an apomictic species. Apomixis has been transferred from *Pennisetum squamulatum* (2n=6x=54) to pearl millet (2n=4x=28) (Hanna, 1995).

Apomixis is a nuisance when the breeder desires to obtain sexual progeny, selves or hybrids. But it is of great help when the breeder desires to maintain varieties. Thus, in breeding of apomictic species, the breeder has to avoid apomictic progeny while making crosses or producing inbred lines.

Apomixis involves perfect conservation of any heterozygosity and is therefore expected to simplify the development of hybrid cultivars and production of commercial hybrid seeds.

## SUMMARY

- The phenomenon in which normal sexual reproduction is completely replaced by asexual reproduction is called apomixis.
- Apomixis may be classified into three different categories: (i) apomeiosis, the generation of a cell capable of forming an embryo without prior meiosis; (ii) parthenogenesis, the independent development of the embryo; and (iii) the capacity to either produce endosperm autonomously or to use an endosperm derived from fertilization.
- Gametophytic apomixis includes those apomictic pathways where the maternal embryo develops from a diploid egg cell differentiated in an unreduced embryo sac.
- When embryo develops directly from an unfertilized egg, the phenomenon is known as parthenogenesis.
- When embryos are produced without fertilization, the apomictic process is referred to as agamospermy.
- Apomictic reproduction would allow one to fix any desired homozygous gene combination and to maintain new heterozygous varieties with valuable agronomical properties such as disease resistance.

## SUGGESTED READING


Albertini, E., G. Barcaccia, J.G. Carman and F. Pupilli. 2019. Did apomixis evolve from sex or was it the other way around? *J. Exp. Bot.* 70 (11): 2951–2964.

Barcaccia, G. and E. Albertini. 2013. Apomixis in plant reproduction: A novel perspective on an old dilemma. *Plant Reprod.* 26: 159–179.

Barcaccia, G., F. Palumbo, S. Sgorbati, E. Albertini and Pupilli, F. 2020. A reappraisal of the evolutionary and developmental pathway of apomixis and its genetic control in angiosperms. *Genes* 11: 859. https://doi.org/10.3390/genes11080859.

Bataglia, E. 1963. Apomixis. In *Recent Advances in the Embryology of Angiosperms*, ed. P. Maheshwari, 221–264. Delhi: International Society of Plant Morphologists.

Batygina, T.B. and G. Vinogradova. 2007. [Phenomenon of polyembryony. Genetic heterogeneity of seeds]. *Ontogenez* 38 (3): 166–91. [In Russian]

Bhat, V., K.K. Dwivedi, J.P. Khurana and S.K. Sopory. 2005. Apomixis: An enigma with potential applications. *Curr Sci.* 89 (11): 1879–1893.

Bicknell, R.A. and A.M. Koltunow. 2004. Understanding apomixis: Recent advances and remaining conundrums. *Plant Cell* 16 (Suppl): S228–S245. https://doi.org/10.1105/tpc.017921.

Brink, R.A. and D.C. Cooper. 1944. The antipodals in relation to abnormal endosperm behavior in *Hordeum jubatum* x *Secale cereale* hybrid seeds. *Genetics* 29 (4): 391–406. https://doi.org/10.1093/genetics/29.4.391.

Carman, J.G. 2001. The gene effect: Genome collisions and apomixis. In *Flowering of Apomixis: From Mechanisms to Genetic Engineering*, eds. Y. Savidan, J.G. Carman and T. Dresselhaus, 95–110. Mexico: CIMMYT, IRD, European Commission DG VI.

Carman, J.G. 2007. Do duplicate genes cause apomixis? In *Apomixis: Evolution, Mechanisms and Perspectives*, eds. E. Hörandl, U. Grossniklaus, P.J. van Dijk and T.F. Sharbel, 63–91. Liechtenstein: A. R. G. Gantner, Rugell.

Cooper, D.C. and R.A. Brink. 1949. The endosperm-embryo relationship in an autonomous apomict, *Taraxacum officinale*. *Bot. Gaz.* 111: 139–153.

Corner, H.E. and M.I. Dawson. 1993. Evolution of reproduction in *Lamprothyrus* (Arundineae: Gramineae). *Ann. Missouri Bot. Gard.* 80: 512–517.

Fiaz, S., W. Xiukang, A. Younas and B. Alharthi. 2020. Apomixis and strategies to induce apomixis to preserve hybrid vigor for multiple generations. *GM Crops and Food* 12 (1): 57–70.

Gustafsson, A. 1947. Apomixis in higher plants. II. The causal agent of apomixis. *Lund Univ. Arsskr. N.F. Avd. II* 43 (2): 71–179.

Hanna, W.W. and E.C. Bashaw. 1987. Apomixis: Its identification and use in plant breeding. *Crop Sci.* 27: 1136–1139.

Hanna, W.W. 1995. Use of apomixis in cultivar development. *Adv. Agron.* 54: 333–544.

Hojsgaard, D., J. Greilhuber, M. Pellino, O. Paun, T.F. Sharbel and E. Hörandl. 2014. Emergence of apospory and bypass of meiosis via apomixis after sexual hybridisation and polyploidisation. *New Phytol.* 204: 1000–1012. https://doi.org/10.1111/nph.12954.

Hojsgaard, D. and E. Hörandl. 2019. The rise of apomixis in natural plant populations. *Front. Plant Sci.* 10: 358. https://doi.org/10.3389/fpls.2019.00358.

Hörandl, E. and Hojsgaard, D. 2012. The evolution of apomixis in angiosperms: A reappraisal. *Plant Biosystems* 146: 681–693.

Jankun, A. 1993. Evolutionary significance of apomixis in the genus Sorbus (Rosaceae), In Polish, English Summary. *Fragm. Flor. Geobot.* 38 (2): 627–686.

Juel, H.O. 1900. Vergleichende Untersuchungen uber typische und parthenogenetische Fortpflanzung bei der Gattung Antennaria. *K. Svenska Vet. Akad. Handl.* 33 (5): 1–59.

Kerner, A. 1876. Parthenogenesis bei einer angiospermen Pflanzc. – *Sitzb Math. Nat. Classe Akad. Wiss. Wien, Abt. I, Bd.* 54: 469.

Khokhlov, S.S. 1967. Apomixis: Classification and distribution (in Russian). In *Uspekhi Sovremennoy Genetiki*, ed. Dubinin, H.P., 1: 43–105. Moskva: Nauka.

Koltunow, A.M., R.A. Bicknell and A.M. Chaudhury. 1995. Apomixis: Molecular strategies for the generation of genetically identical seeds without fertilization. *Plant Physiol.* 108: 1345–1352.

Kumar, S. 2017. Epigenetic control of apomixis: A new perspective of an old enigma. *Adv Plants Agric. Res.* 7: 15406.

Kumar, S., S. Saxena, A. Rai, A. Radhakrishna and P. Kaushal. 2019. Ecological, genetic, and reproductive features of Cenchrus species indicate evolutionary superiority of apomixis under environmental stresses. *Ecol. Indic.* 105: 126–136.

Leblanc, O., D. Grimanelli, D. Gonzaléz de Léon and Y. Savidan. 1995. Detection of the apomictic mode of reproduction in maize–*Tripsacum* hybrids using maize RFLP markers. *Theor. Appl. Genet.* 90: 1198–1203.

Lovell, J.T., R.J. Williamson, S.I. Wright, J.K. McKay and T.F. Sharbel. 2017. Mutation accumulation in an asexual relative of *Arabidopsis*. *PLoS Genet.* 13 (1): e1006550. https://doi.org/10.1371/journal.pgen.1006550.

Maheshwari, P. 1950. *An Introduction to Embryology of Angiosperms*. New York: Mc-Graw Hill Book Company.

Narayan, K.N. 1962. Apomixis in some species of *Pennisetum* and in *Panicum antidotale*. In *Plant Embryology—A Symposium*. Council of Scientific and Industrial Research, 55–61.

Nardi, F.D., C. Dobeš, D. Müller, T. Grasegger, T. Myllynen, H. Alonso-Marcos and A. Tribsch. 2018. Sexual intraspecific recombination but not *de novo* origin governs the genesis of new apomictic genotypes in *Potentilla puberula* (Rosaceae). *Taxon* 67: 1108–1131.

Naumova, T. 1992. *Apomixis in Angiosperms: Nucellar and Integumentary Embryony*. Boca Raton, FL: CRC Press, Inc.

Naumova, T.N. 2008. Apomixis and amphimixis in flowering plants. *Cytol. Genet.* 42: 179–188. https://doi.org/10.3103/S00954527080300.

Neiman, M., T.F. Sharbel and T. Schwander. 2014. Genetic causes of transitions from sexual reproduction to asexuality in plants and animals. *Journal of Evolutionary Biology* 27: 1346–1359.

Nogler, G.A. 1984. Gametophytic apomixis. In *Embryology of Angiosperms*, ed. B.M. Johri, pp. 475–518. Berlin: Springer.

Okabe, S. 1932. Parthenogenesis bei *Ixeris dentata*. *Bot. Mag. (Tokyo)* 46: 518–532.

Osawa, J. 1913. Studies on the cytology of some species of *Taraxacum*. *Arch. Zellforsch.* 10: 450–469.

Ozias-Akins, P., Y. Akiyama and W.W. Hanna. 2003. Molecular characterization of the genomic region linked with apomixis in *Pennisetum/ Cenchrus*. *Funct. Integr. Genomics* 3: 94–104.

Rathore, P., S.N. Raina, S. Kumar and V. Bhat. 2020. Retro-element Gypsy-163 is differentially methylated in reproductive tissues of apomictic and sexual plants of *Cenchrus ciliaris*. *Front. Genet.* 11: 795. https://doi.org/10.3389/fgene.2020.00795.

Rodríguez-Leal, D., G. León-Martínez, U. Abad-Vivero and J. Vielle-Calzada. 2015. Natural variation in epigenetic pathways affects the specification of female gamete rrecursors in *Arabidopsis*. *The Plant Cell* 27 (4): 1034–1045. https://doi.org/10.1105/tpc.114.133009.

Rosenberg, O. 1907. Cytological studies on the apogamy in *Hieracium*. *Bot. Tidsskrift* 28: 143–170.

Ross, A., A. Bicknell and A.M. Koltunow. 2004. Understanding apomixis: Recent advances and remaining conundrums. *Plant Cell* 16 (Supplement): S228–S245. https://doi.org/10.1105/tpc.017921.

Sharma, R. and V. Bhat. 2020. Role of apomixis in perpetuation of flowering plants: Ecological perspective. In *Reproductive Ecology of Flowering Plants: Patterns and Processes*, eds. R. Tandon, K. R. Shivanna and M. Koul, 275–297. https://doi.org/10.1007/978-981-15-4210-7_13.

Solntseva, P.M. 1969. Principles of embryological classification of apomixis in angiosperms. *Rev. Cytol. Biol. Veg.* 32: 371–377.

Tucker, M.R. and A.M.G. Koltunow. 2009. Sexual and asexual (apomictic) seed development in flowering plants: Molecular, morphological and evolutionary relationships. *Funct. Plant Biol.* 36 (6): 490–504. https://doi.org/10.1071/FP09078.

Valle, C.B. do and Y.H. Savidan. 1996. Genetics, cytogenetics and reproductive biology of Brachiaria. In *Brachiaria: Biology, Agronomy and Improvement*, eds. J.W. Miles, B.L. Maass and C.B. do Valle, 164–177. Cali, Colombia: Centro Internacional de Agricultura Tropical.

Voronova, O.N. and A.A. Babro. 2019. Apospory in *Helianthus ciliaris* DC. (Asteraceae). *Int. J. Plant Reprod. Biol.* 11 (1): 66–69. https://doi.org/10.14787/ijprb.2019 11.1.

Winkler, H. 1908. *Uber Parthenogenesis und Apogamie im Pflanzenreiche*. Jena.

# 15 Seed

The seed develops from an ovule. It contains a tiny plantlet technically known as an embryo. The embryo has a tiny embryonic root (radicle), a small shoot (plumule) and one or two leaves (cotyledons) (Figure 15.1). The outer and inner integuments undergo structural modifications, and are termed testa and tegmen, respectively. They together constitute the seed coat.

The nucellus is usually consumed during development of the seed. But if it persists, it is known as perisperm (e.g., *Eriospermum capense*, *Orygia decumbens*) (Figure 15.2). A storage tissue, either of the endosperm or perisperm, or both, may or may not be present in the seed. In many parasitic angiosperms (e.g., members of Santalales), where the ovules do not show differentiated integument(s), a protective covering of integumentary origin is not found.

A true seed is a fertilized mature ovule that possesses an embryonic plant, stored food material (rarely missing) and a protective coat or coats. Application of the term "seed" is seldom restricted to this morphologically accurate definition. Rather, seed is usually used in a functional sense, viz., as a unit of dissemination, a disseminule. In this sense, the term seed includes dry, one-seeded (rarely, two- to several seeded) fruits, such as caryopsis (Poaceae), cypsela (Asteraceae), cremocarp and mericarp (Apiaceae), achenes as well as true seeds.

As the seed gets detached from the funicle, a scar is left on its surface, known as hilum. In the family Fabaceae the funicular scar persists as the hilum and the micropyle, as a more or less inconspicuous cleft, often visible with only moderate magnification (Figure 15.1). The lens, a circular to variously elongate area of modified seed coat, lies somewhat along the sagittal midline of the seed, usually near the hilum on the opposite side from the micropyle (Figure 15.1). Seed morphology (external) refers to the dynamics of form and surface patterns, whereas anatomy reveals the structural arrangements in its components and their spatial relationships.

## SEED MORPHOLOGY

Seeds may be spherical, subspherical, oblong, oval, ellipsoid, triangular, lenticular, sublenticular, pyramidal, subpyradimal, cuboid, subcuboid, reniform etc. Seeds may be turgid or compressed and vary greatly in size from dustlike particles (Orchidaceae, Orobanchaceae) to large coconuts and double coconuts. The weight of many of the minutest seeds is about 0.000003–0.000005 g. In Orobanchaceae, each seed weighs only 0.000001 g.

Seeds may be monochromed or marked by points, mottles, streaks etc. Brown and brown derivatives are by far the most common colours of seeds. Browns and

DOI: 10.1201/9781003260097-15

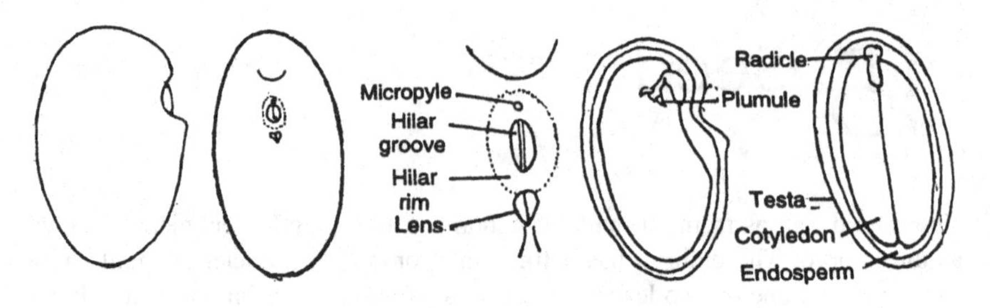


**FIGURE 15.1** Seed morphology in Fabaceae.

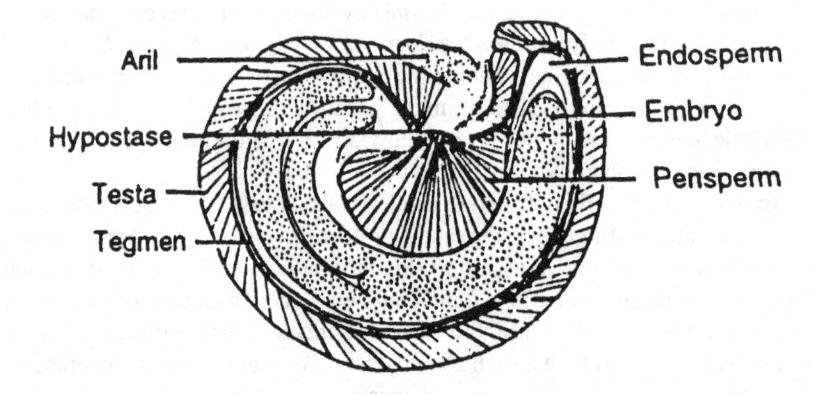


**FIGURE 15.2** *Orygia decumbens.* Longisection of mature seed. Note copious perisperm (after Narayana, 1962).

blacks make up over half of the seed colours. Conspicuous colours such as red, green, yellow and white are infrequent. Different species of *Abrus* may be identified on the basis of their typical colours. Seeds are dichromatic, scarlet red and jet black in *Abrus precatorius*; reddish brown, mottled in *A. laevigatus* and olive green in *A. fruticulosus*. The surface of the seed coat varies from highly polished to markedly roughened or sculptured. Various types of sculpturing occur, namely, wrinkled, striate, ridged, furrowed, reticulate, punctate, tuberculate, alveolate, hairy, spinescent etc. (Figure 15.3). Some of the common testa ornamentations are shown in Figure 15.4.

## Spermoderm

The seed coat morphologically has provided valuable information with regard to taxonomy and identification of seed because seed characteristics are considered fairly conservative. Tobe et al. (1987) have emphasized that seed coat structures can be useful for assessing relationships and delimiting taxa. Scanning electron microscopy (SEM) has proved to be an exciting and important additional tool to taxonomists for solving systematic problems. In many genera and families, testa topography is characteristic to each species.

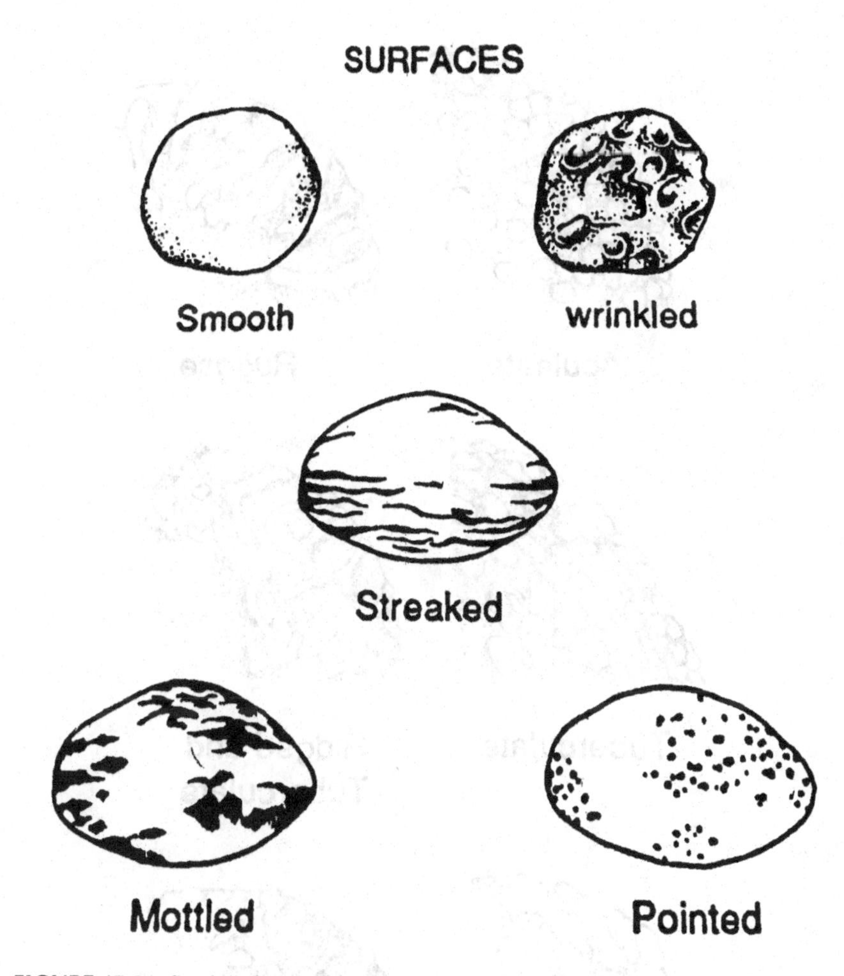


**FIGURE 15.3** Seed surface and marks.

Pandey (1991) studied spermoderm pattern in all the 6 genera of Trifolieae (Fabaceae) and observed tuberculate (*Medicago*, *Melilotus*, *Trifolium*, *Trigonella*), reticulate (*Ononis*) to levigate (*Parochetus*) spermoderm patterns (Figure 15.5, Figure 15.6). In *Vigna unguiculata* (tribe Phaseoleae, Fabaceae), spermoderm shows reticulate patterns with exaggerated walls. Seeds of *Lotus* species show tuberculate (*Lotus angustissimus*), foveolate (*Lotus corniculatus*) or distinct irregular ridges all over the surface (Figure 15.7). In *Phaseolus* species, four types of testa topography are observed (Table 15.1).

## SPECIAL FEATURES

During the development of the seed in some taxa, special structures arise from various parts of the ovule. They include aril, caruncle, operculum, elaiosome, etc.

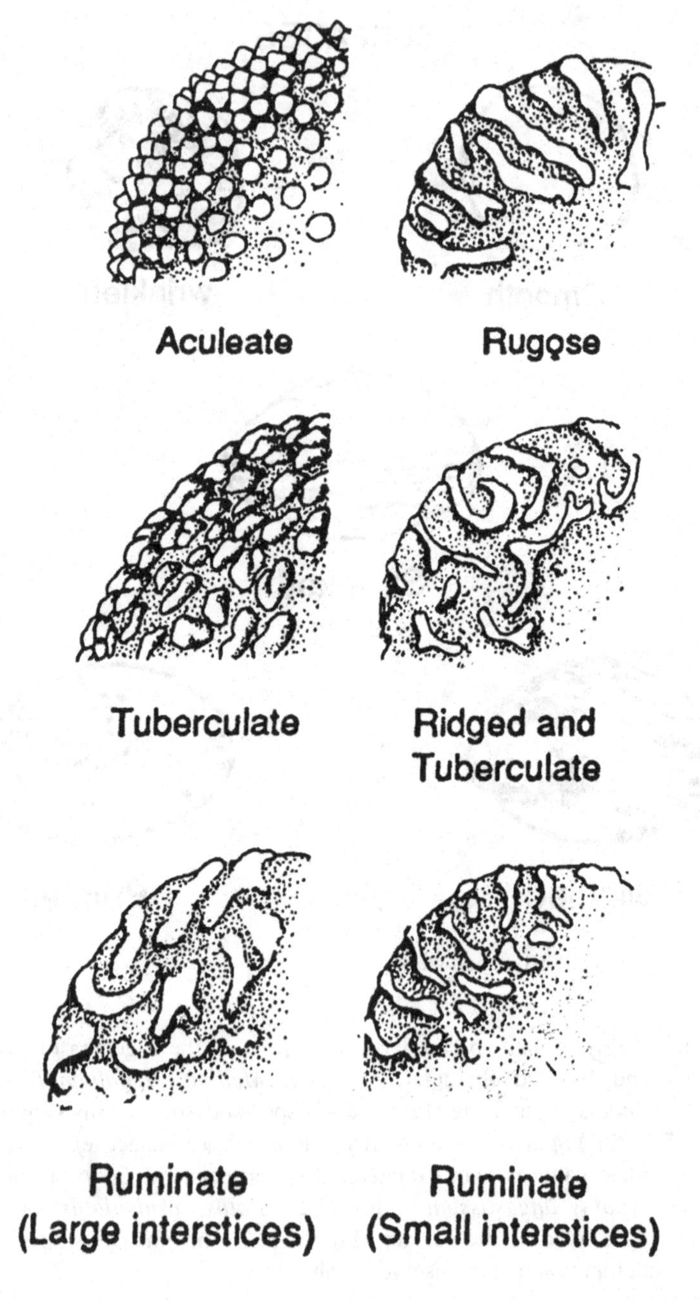


**FIGURE 15.4** Common testa ornamentation in Vicieae (after Lersten and Gunn, 1982).

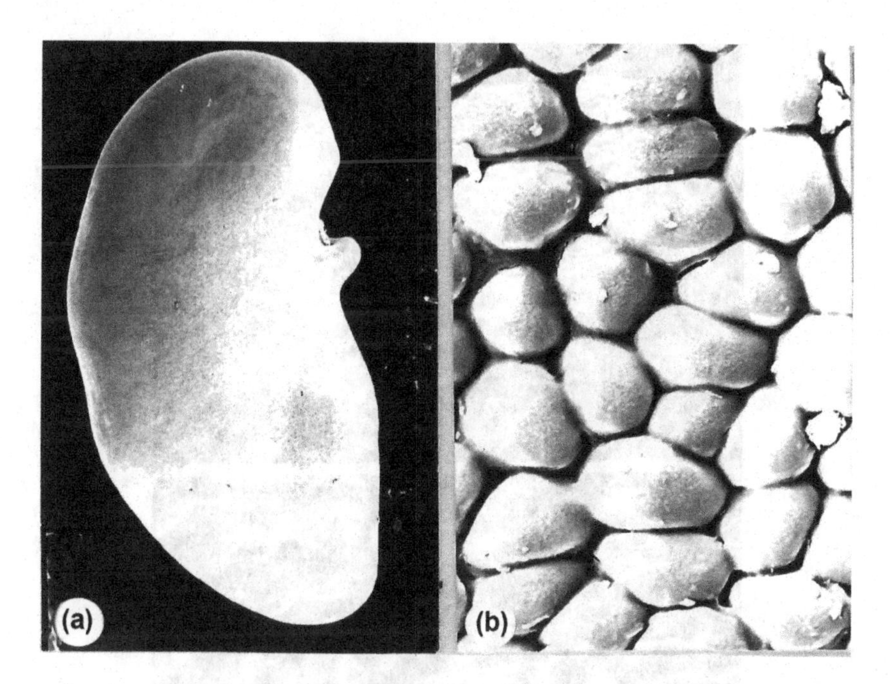


FIGURE 15.5 *Medicago arabica.* A. Seed. B. Spermoderm (after Pandey, 1995).

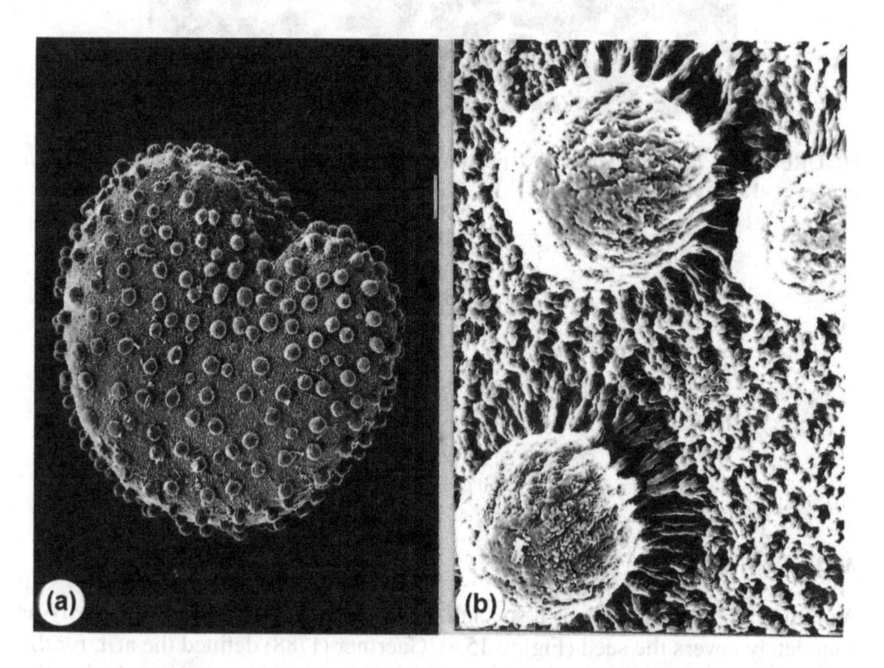


FIGURE 15.6 *Ononis reclinata.* A. Seed. B. Spermoderm (after Pandey, 1995).

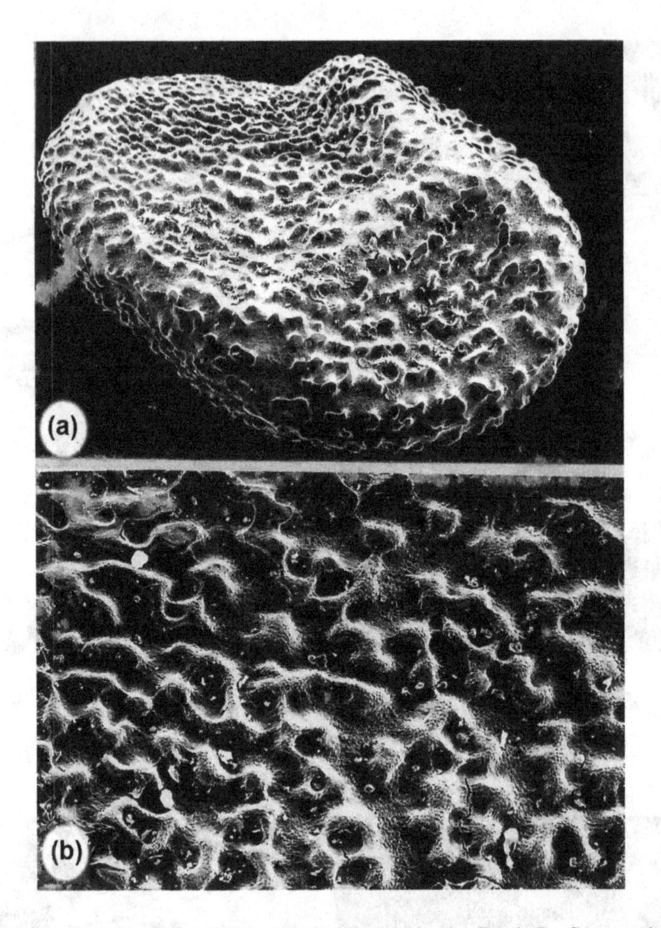


**FIGURE 15.7** *Lotus edulis.* SEM photomicrograph. A. Seed. B. Spermoderm (after Pandey and Jha, 1991).

**TABLE 15.1**
**Spermoderm Pattern in *Phaseolus* Species**

| Spermoderm Pattern | Species |
|---|---|
| Rugose | *Phaseolus calcaratus, P. ricciardianus* |
| Reticulate | *P. atropurpureus, P. trilobus, P. lathyroides* |
| Foveolate | *P. aconitifolius* |
| Tuberculate | *P. acutifolius* |

## ARIL

Aril, also called an arillus, is a specialized outgrowth from a seed that partly or completely covers the seed (Figure 15.8). Gaertner (1788) defined the aril, for the first time, as an accessory integument attached to the hilum and enveloping the

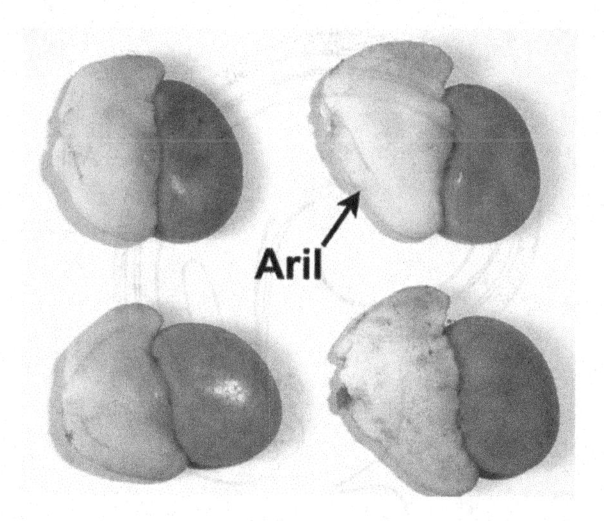


**FIGURE 15.8** Aril on seeds of weeping boer-bean (*Schotia brachypetala*) (Credit: JMK, via wikimedia commons, CC BY-SA 3.0).

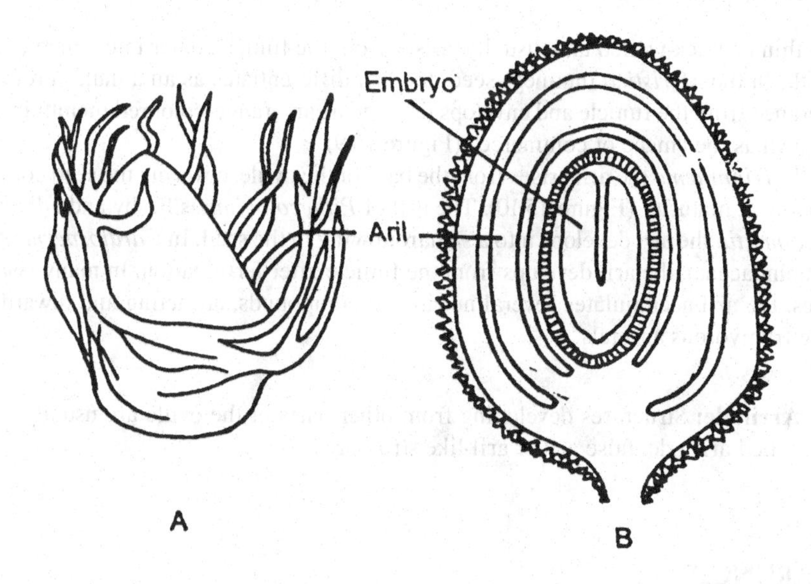


**FIGURE 15.9** A. *Myristica* seed with aril. B. Longitudinal section of mature seed of *Litchi chinensis* showing aril

seed completely or incompletely. Corner (1976) designates the term aril as a general name for pulpy structures which grow from some part of the ovule or funicle after fertilization and envelopes the seed partially or completely. According to Van der Pijl (1972) the aril originates from the funiculus.

Arils are fleshy seed appendages, often with vivid colours to attract animals e.g., *Myristica*, *Litchi* (Figure 15.9), *Asphodelus* and *Trianthema*. The aril may

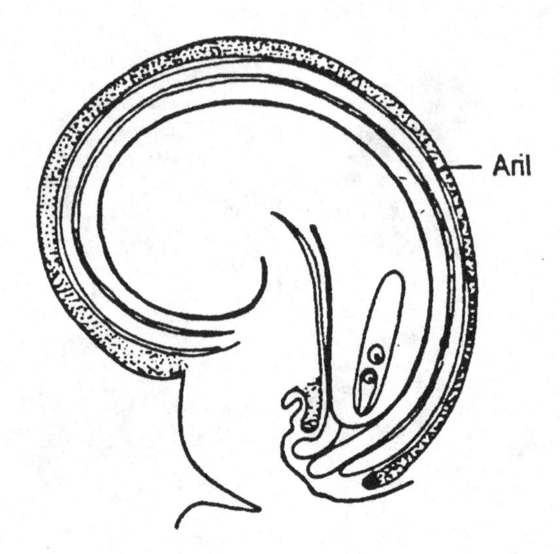


**FIGURE 15.10** *Trianthema monogyna.* Longitudinal section of ovule showing aril (after Bhargava, 1935).

be thin or thick-walled and, usually, arises from the funicle, outer integument, or both. In the *Myristica* (nutmeg) seed, the aril differentiates as an annular protuberance from the funicle and envelops the seed as an orange-coloured membrane, known as the "mace of commerce" (Figure 15.9).

In *Trianthema* the aril arises from the base of the ovule, covering the seed completely at maturity (Figure 15.10). The aril of *Pithecolobium* is fleshy and edible. In *Lomatia* the aril develops into a samaroid wing in the seed. In *Cardiospermum* (Sapindaceae), the aril develops from the funicle after fertilization. In many species, the aril accumulates several nutritional compounds, attracting and rewarding frugivorous animals.

**Arrilode:** Structures developing from other parts of the ovule are usually called arrilode, false aril or aril-like structures.

## CARUNCLE

The tip of the outer integument forms a reflexed outgrowth leading to the formation of a massive or small structure known as the caruncle (Figure 15.11). The caruncle is present in the micropylar region of Euphorbiaceae seeds. This is commonly seen in the seeds of castor (*Ricinus communis*) and several other Euphorbiaceae (Figure 15.12). Sometimes the caruncle becomes more pronounced and directed backwardly, thus resembling an aril. The caruncle accumulates up to 40% oil by weight in the form of triacylglycerol (TAG) with a highly contrasting fatty acid composition compared to the seed oil.

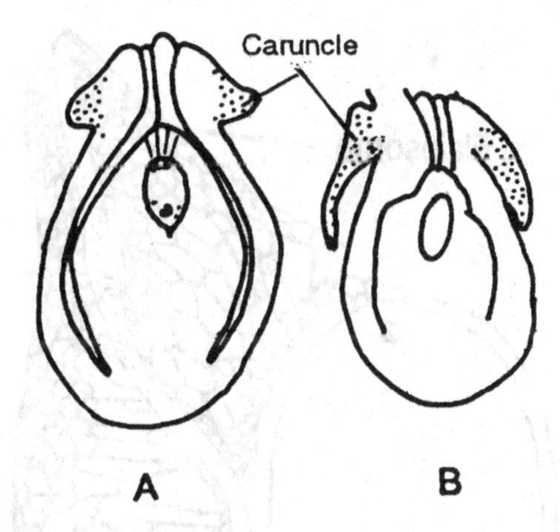


**FIGURE 15.11** Ovules showing caruncle. A. *Brachychilum horsfieldii*. B. *Burbidgea schizocheila* (after Mauritzon, 1936).

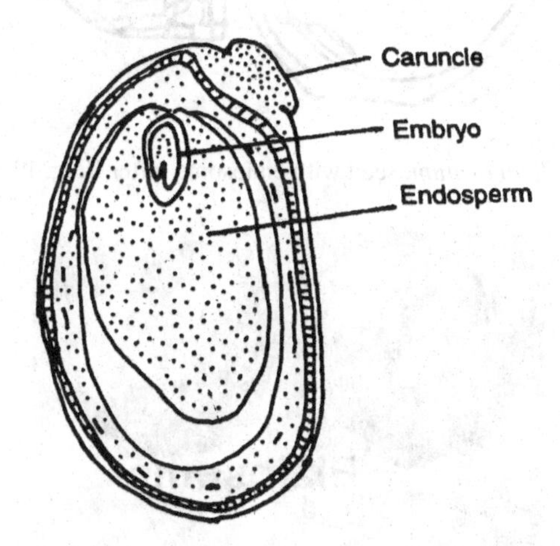


**FIGURE 15.12** *Ricinus communis.* Longitudinal section of seed showing caruncle formed by the outer integument.

### Functions

1. The caruncle promotes the dispersal of seed by ants (myrmecochory).
2. Being hygroscopic in nature, it absorbs water from the soil and passes it on to the embryo at the time of germination.

## ELAIOSOME

Sernander (1906) introduced the ecological term elaiosome for all fleshy and edible parts of seeds dispersed by ants. The elaiosome arises as an outgrowth of raphe, or hilum (e.g., *Trillium ovatum*, Liliaceae) (Figure 15.13). The elaiosome is

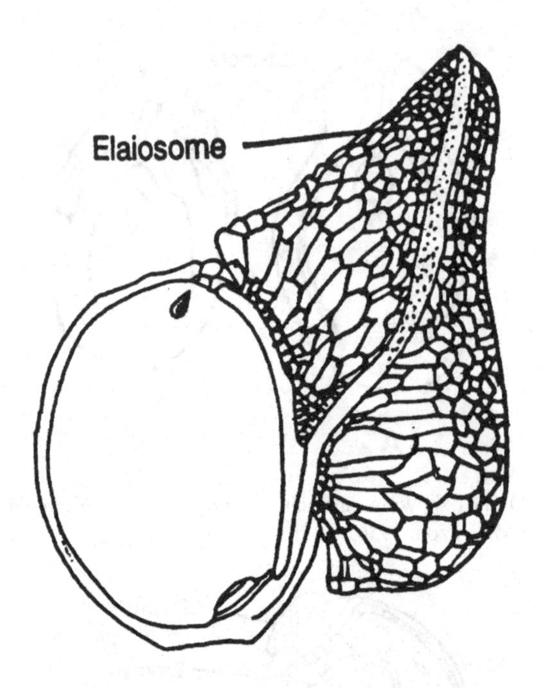


**FIGURE 15.13** *Trillium ovatum*, seed with elaiosome (after Berg, 1958).

**FIGURE 15.14** Elaiosomes on seeds of prickly seeds bur (*Datura innoxia*) (credit: Stefan. Lefnaer, via wikimedia commons, CC BY-SA 4.0).

rich in lipids and proteins. It attracts ants which help in seed dispersal. The seeds get ants to move them away from the mother plant by offering up the fatty, nutritious elaiosomes. The colour of the elaiosome is white or yellow and differs from that of the rest of the seed which is generally darker (Figure 15.14). According to Bresinky (1963), the ants are attracted due to the presence of an unsaturated fatty acid, rincinolid acid. The elaiosome also contains nutritive substances like proteins, lipids, starch and vitamins.

## OPERCULUM

The operculum (plural Opercula) is a cap-like structure in some flowering plants. In angiosperms, the operculum, also known as calyptra, is formed by the micropylar and hilar region, exo- and endostome, endostome alone, or by the endosperm haustorium. The operculum is generally present in monocotyledonous families. In *Lemna paucicostata* the cells of tegmen in the micropylar region undergo excessive expansion and develop into the operculum (Figure 15.15). In *Plantago* the capsule has an opening covered by an operculum. When the operculum falls, the seed is sticky and is easily carried by animals that come in contact with it. The opercula facilitate germination and also provide extra protection to the micropylar region of the seed.

## JACULATOR

A curved outgrowth from the funiculus, on the side away from the micropyle, is termed jaculator or retinaculum (Figure 15.16). It is commonly found in the members of family Acanthaceae. In *Ruellia*, the funicle of the seed forms a hooklike projection. These hooklike jaculators make the fruits burst and the seeds

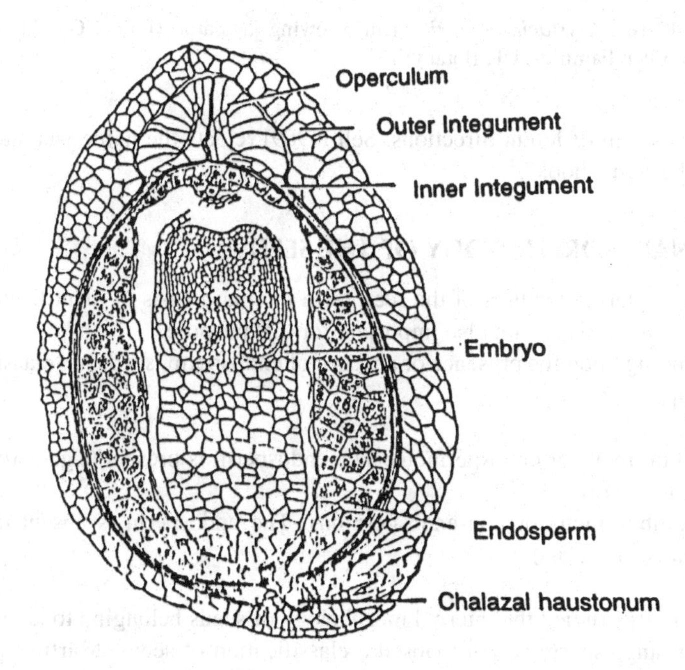


**FIGURE 15.15** *Lemna pausicostata.* Longisection of seed to show operculum (after S.C. Maheshwari and Kapil, 1963).

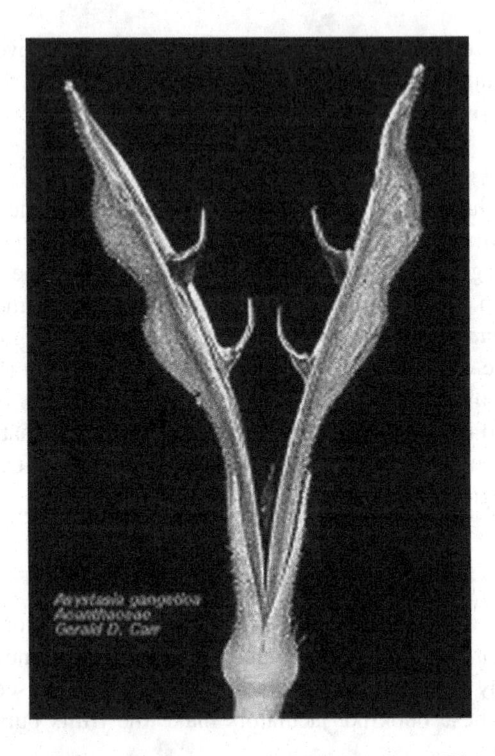

**FIGURE 15.16** *Asystacia gangetica* fruit showing jaculator. (Photo: Gerald D. Carr; Flowering Plant Families, UH Botany)

are dispersed in different directions. Sell (1969) rejects the name jaculator, and prefers the term "hook".

## INTERNAL MORPHOLOGY OF THE SEED

While the external features of the seed form the main basis of identification, its internal characteristics are also important for this purpose.

Depending upon the presence or absence of endosperm, seeds are classified in two groups:

1. Albuminous or endospermous: The endosperm is present (e.g., cereals, castor bean).
2. Ex-albuminous or non-endospermous: The endosperm is absent (e.g., beans, cucurbits).

Martin (1946) studied the internal morphology of seeds belonging to about 1287 genera of angiosperms and proposed a classification of seeds. Martin's plan of classification takes into account (i) size of the embryo in relation to the endosperm, and (ii) differences in size, shape and position of the embryo in the seed.

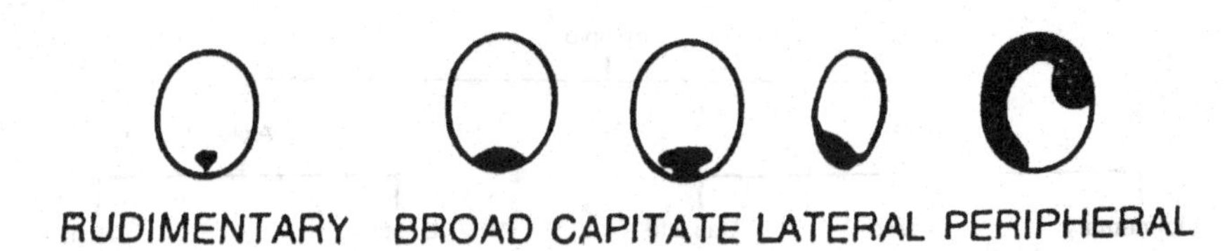


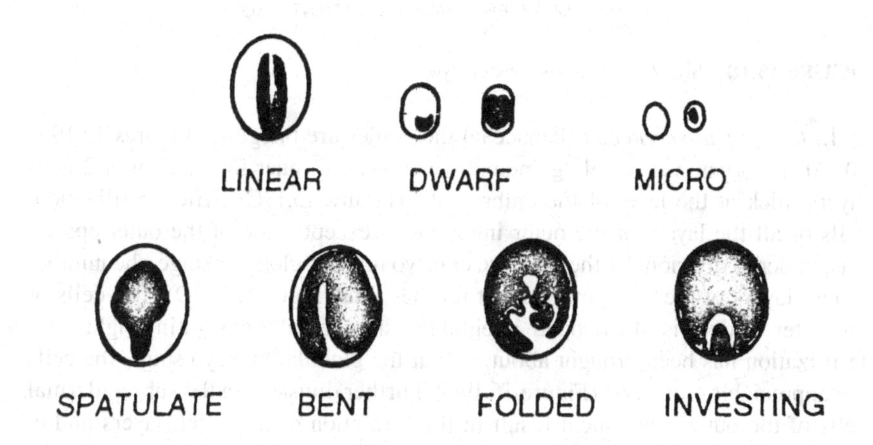


**FIGURE 15.17** Classification of embryo (after Martin, 1946).

He has given five size-designations represented volumetrically in quarter-units (Figure 15.17). The units are small (when the embryo is smaller than one-quarter), quarter (one-quarter), half (more than two-quarters), dominant (more than three-units) and total (when embryo occupies entire seed).

The embryo in a mature seed is also classified according to its position and shape (Figure 15.18). R. Dahlgren and Clifford (1982) classified embryos as follows:

## Development of the Seed Coat

The seed coat arises by the modification, including destruction, sclerification lignification as well as other chemical impregnations and variations in the distribution of the mechanical tissues of the integument or integuments.

During the development of a seed, the integuments mature into a seed coat. In bitegmic ovules, the seed coat may be formed by both the integuments (e.g., *Gossypium* species), or the inner integument may degenerate and seed coat is formed by the outer integument alone (e.g., members of family Fabaceae).

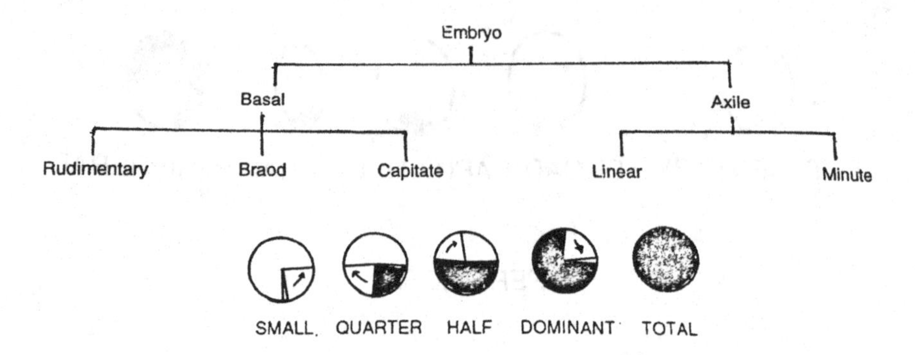


**FIGURE 15.18** Size designations for embryo.

In *Crotalaria verrucosa* (Fabaceae) the ovules are bitegmic (Figures 15.19A, B). At the organized female gametophyte stage, the inner integument is 2 cell-layers thick at the level of the embryo sac (Figure 15.19C). After fertilization, cells of all the layers of the outer integument, except those of the outer epidermis, undergo division; by the time the embryo reaches globular stage, the number of cell-layers of the outer integument reaches four (Figure 15.19D). The cells of the outer epidermis of the outer integument start radially elongating right after fertilization has been brought about, and at the globular embryo stage, the cells are appreciably elongated (Figure 15.19D). Further divisions in the sub-epidermal cells of the outer integument result in the formation of more cell layers and at the stage at which the cotyledons are well differentiated in the embryo, and the number of cell layers goes up to six (Figure 15.19E). At this stage the cells of the outer epidermis of the outer integument show greater radial elongation and slight thickening of their walls (Figure 15.19E). These cells ultimately differentiate into palisade-like, highly lignified macrosclereids in the mature seed (Figures 15.19F, G). The cells lying just below the outer epidermis of the outer integument organize in a definite manner and show rich cytoplasmic contents and biconcavity of their radial walls, at a stage when the cotyledons are well differentiated in the embryo (Figure 15.19E). These cells are destined to form hourglass-shaped, truncate, thick-walled osteosclereids in the mature seed (Figure 15.19H). The remaining cells of the outer integument remain parenchymatous and the inner few cell layers degenerate in the course of seed development.

At the organized female gametophyte stage, the inner integument is two-layered except towards the micropylar end where the number is greater (Figures 15.19B, C). After fertilization, the cells of both the layers of the inner integument get compressed and the cells show tangential elongation in the plane along the long axis of the ovule, and by the time embryo reaches globular stage, the cells are fully stretched (Figure 15.19D). The generation of the inner integument begins from the central region of raphe and antiraphe sides and gradually progresses towards the chalazal and micropylar ends. As the seed reaches maturity no trace of inner integument is seen (Figure 15.19F, I).

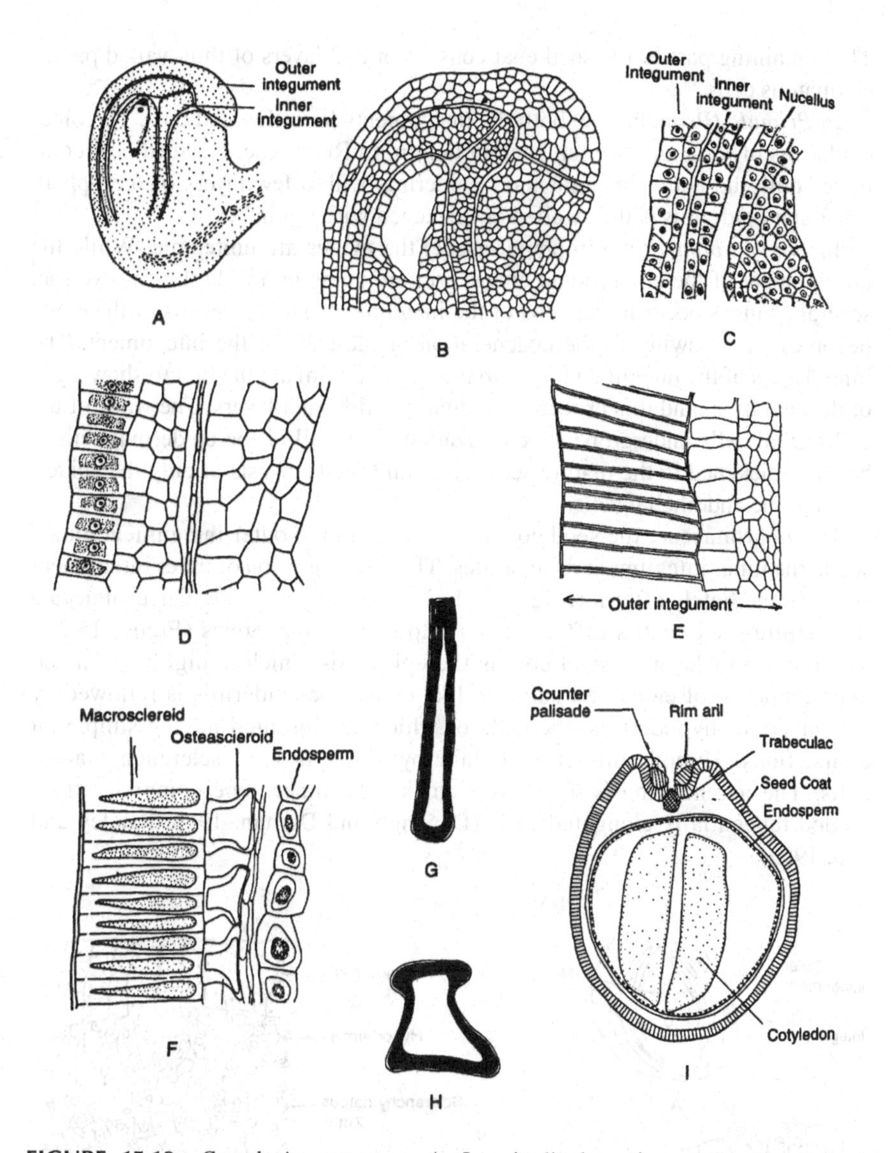


**FIGURE 15.19** *Crotalaria verrucosa.* A. Longitudinal section ovule at organized female gametophyte stage. B, C. L.s. part of ovule at organized female gametophyte stage. D, E, F. L.s. part of ovule at globular, at well-differentiated cotyledonary stage and mature seed stages. respectively. G, H. Macrosclereid and osteosclereid in surface view. I. Cross section mature seed (after Pandey and Jha, 1988).

The seed coat develops from the outer integument alone. The mature seed coat consists of five layers of cells. The outermost layer is composed of radially elongated, palisade-like, lignified cells, the macrosclereids (Figures 15.19G, H). The layer of macrosclereids is underlain by a layer of osteosclereids (Figures 15.19F, H).

The remaining part of the seed coat consists of 2–3 layers of thin-walled parenchymatous cells.

In *Primula* (Primulaceae) both the integuments form the seed coat. The outer epidermis of testa and tegmen contain tannin. In Rubiaceae, the seed coat comprises only the outer integumentary epidermis with a few layers of mesophyll. The outer epidermis is thin-walled or thickened with lignin.

In *Linaria bipartita* (Scrophulariaceae) the ovules are unitegmic. While the development of the embryo and endosperm are going on inside the embryo sac, several changes occur in the cells of the integument. The epidermis of the ovule becomes wavy, owing to the degeneration of the cells of the integument. The inner layers of the integument (next to the endothelium) are the first to show signs of degeneration, and this proceeds gradually to the outer layers. The walls of the endothelial cells undergo heavy cutinization. The wall of the epidermal cells is heavily thickened in the mature seed. A mature seed consists of only the epidermis and the endothelial cells.

In Cucurbitaceae, the seed coat develops from the outer integument alone, while the inner integument degenerates. The disintegration of inner integument starts immediately after fertilization. In *Trichosanthes dioica* (Cucurbitaceac) the mature seed coat is differentiated into five distinct zones (Figure 15.20). The outermost layer of seed coat is the epidermis which is highly cutinized and composed of elongated palisade-like cells. The epidermis is followed by 7–8 layers of hypodermis, the cells of which are lignified having simple pit connections. Hypodermis is underlain by single-layered sclerenchymatous cells. The aerenchyma is 4–6 layers thick. The inner zone comprises thin-walled tangentially elongated cells (D. Singh and Dathan, 1976; Pandey and Jha, 1992).

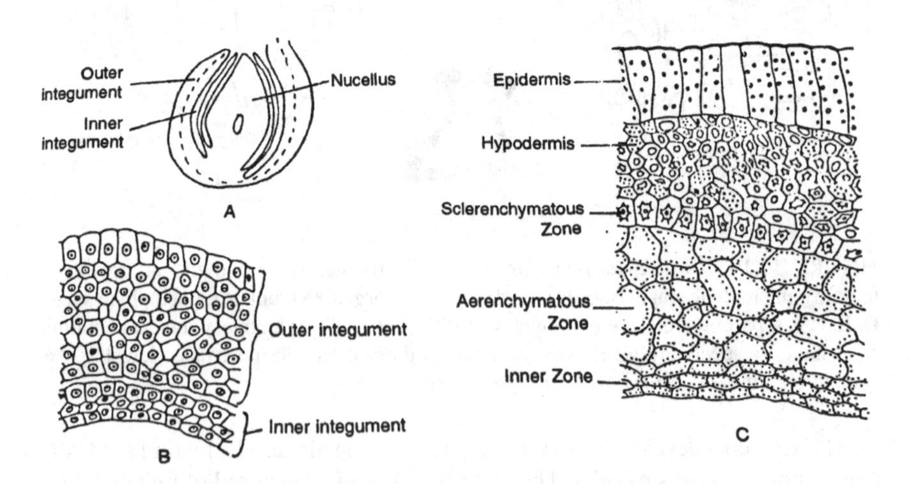


**FIGURE 15.20** *Trichosanthes dioica.* A. Longitudinal section ovule before fertilization. B. L.s. part of ovule at organized female gametophyte stage. Note massive outer integument and thin inner integument. C. L.s. part of mature seed coat.

## LABYRINTH SEEDS

Seeds that show an irregular internal structure when cut in any plane are called labyrinth seeds (van Heel, 1970). This may be due to folding of the cotyledons (Burseraceae, Dipterocarpaceae) or to the presence of testa tissue within the seed. The testa may be located between portions of folded and lobed, mostly flat, cotyledons. The seed coat first encroaches on the endosperm, and later the cotyledons, and in this way, the labyrinth structure is formed. The labyrinth becomes more pronounced if the lobing of the cotyledons occurs in combination with the intrusion of the testa between the folds and lobes, e.g., *Kingiodendron pinnatum* (Figure 15.21) and *Erysibe tomentosa*. In *Harnandia peltata* and *Mangifera* species, the seed coat finds its way in many crevices of the massive cotyledons.

In mature seeds of *Argyreia ridleyi* (Convolvulaceae), there develops a large plate of testa tissue in the median plane around which the halves of the cotyledons are seen bent and cap-shaped (Figure 15.22). The remaining space between the plates and the cotyledonary folds is filled up with endosperm.

## IMPORTANCE OF SEEDS

Committed for over a million years to a nomadic life, man settled down about 10,000 years ago when he learned to satisfy his hunger by growing food, especially seed foods. Seeds, the great staple food of the world, feeds more people than does any other type of food. The endosperm or cotyledons with their rich food reserves for the developing embryo and seedling offer man and other animals a highly nutritious food that can be easily stored.

Cereals contribute the major part of the human diet. It is because of the high value of their seeds (rice, wheat, maize, oats, barley, sorghum, millet, rye and other edible seeds). They are rich in carbohydrates, proteins, minerals and

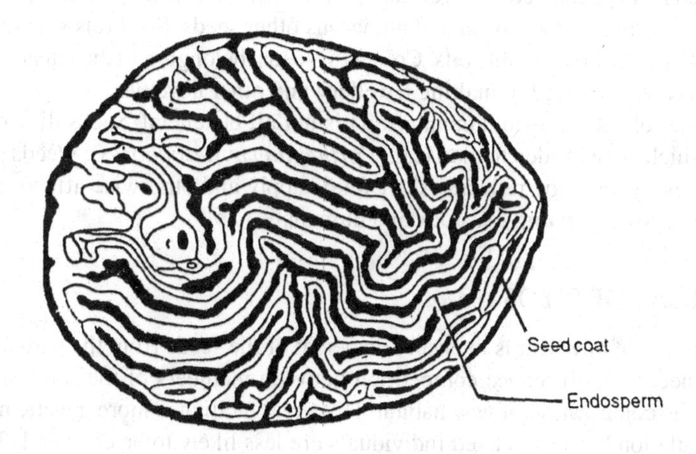


**FIGURE 15.21** *Kingiodendron pinnatum*. L.s. of labyrinth seed (after van Heel, 1970).

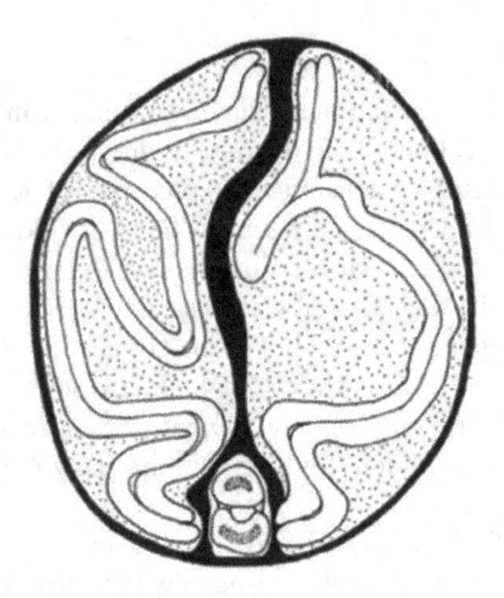

**FIGURE 15.22** *Argyreia ridleyi*, Cross-section of labyrinth seed (after van Heel, 1970).

vitamins. Being compact and dry, the cereals can be stored for long periods. The legume family Fabaceae is the second most important family that provides us with peas, gram, peanuts, soybeans, beans, lentils, chick-peas and other edible seeds. Besides the human diet, seeds constitute feed for livestock and poultry birds.

Man uses other seeds, such as spices, condiments and nuts, in his diet. Some of the popular beverages are derived from seeds: coffee (*Coffea arabica*) and chocolate (cocoa) made from coffee and cacao seeds, beers from barley, whisky and gins fermented from mashes of cereal grains. Seed and seed extracts are also used as medicines. Cotton (*Gossypium* species), a major fibre, is spun from the hairs of the cotton seed.

Another major contribution from seeds is the edible and industrial oils expressed from peanut, coconut, cotton, palm, mustard, sunflower, safflower, rape, flax, sesame, tung, castor bean and numerous other seeds. Seed oils supply about one-half of the world's edible oils. Colourful seeds, belonging to the Fabaceae and Palm families, are used in making jewellery and other novelties.

But the role of seeds in human affairs is not all beneficial. Seeds disseminate plants which are a burden to man (e.g., *Parthenium hysterophorus*). Weeds plague humans everywhere by reducing their food supply and otherwise affecting their health (e.g., by poisoning livestock and humans).

## DISPERSAL OF SEEDS

The function of any fruit is to ensure dispersal of the seeds from the parent plant. This is necessary to reduce competition between members of the same species, and for the colonization of new habitat. Dispersal facilitates more genetic mixing in a population because related individuals are less likely to be clustered close to one another. Fruits are adaptations that result in the protection and distribution

**TABLE 15.2**
**Agents of Dispersal**

| Agents | Descriptive Term |
|---|---|
| Dispersal by the plant itself | Autochory |
| Wind | Anemochory |
| Water | Hydrochory |
| Animals | Zoochory |
| Attachment to animal | Epizoochory |
| Eaten by animal | Endozoochory |
| Birds | Ornithochory |
| Mammals | Mammaliochory |
| Bats | Cheiropterochory |
| Ants | Myrmecochory |

of seeds. There are many different agents for the dispersal of seeds and fruits (Table 15.2).

### AUTOCHORY (SELF DISPERSAL)

The seeds of some plants are expelled explosively from the fruit by rapid movements caused by drying of the pericarp or, more rarely, by turgor pressure. These are sufficient to propel the seeds to various distances from their parent plants.

Many leguminous fruits like pea, pegionpea and beans burst along both the sutures to disperse the seeds. The ripe fruits of balsam (*Impatiens parviflora*) when touched, burst immediately. The valves roll up inwards and the seeds are thrown out with great force and scattered in different directions.

**Dispersal syndromes:** The relationship between particular seed traits and vectors is called dispersal syndrome. For example, wings are associated with wind-dispersal, whereas fleshy structures are associated with animal dispersal.

In *Ecballium elaterium* (squirting cucumber) the internal watery content during maturity exerts high osmotic pressure on the point of origin of the stalk. The basal part of the stalk acts as a cork or stopper in a bottle. After maturity of fruit the internal pressure is increased and the stalk is detached leaving the watery content with a great force to disperse the seeds at a long distance.

Seed dispersal in *Arceuthobium* (dwarf mistletoe) is accomplished by a remarkable explosive process. The seeds are forcibly ejected from the ripe single-seeded fruit (Figure 15.23). The seeds can be dispersed as far as 20 m from its source,

**FIGURE 15.23** Explosive seed discharge in *Arceuthobium americanum* (deBruyn, 2015).

with initial velocities approaching 100 km/h$^{-1}$. Viscin tissue, a mucilaginous region that forms a layer around the single seed within each fruit, accumulates the hydrostatic force needed for a water-driven explosive discharge. Dispersal is ultimately achieved by explosive fracture in the abscission layer at the pedicel (deBruyn et al., 2015).

## Anemochory (Wind Dispersal)

Anemochory is defined as seed dispersal by wind. Wind is one of the main agencies of seed dispersal. The seeds of dandelions (*Taraxacum officinale*), swan plants (*Gomphocarpus physocarpus*) and cottonwood trees (*Populus deltoides*) are light, have feathery bristles and can be carried long distances by the wind. Some plants, like Kauri and maple trees, have "winged" seeds.

In *Bigonia*, *Cinchona*, *Tecoma*, *Oroxylum* and *Moringa*, the seed coats form wings. Certain seeds have appendages which accumulate air and become light to give buoyancy to the seeds. The seeds of *Calotropis* bear tufts of hair called coma. The coma act like a parachute and provide support to the seeds to float easily in the air for dispersal.

Seeds of orchids are very small, dry and dusty, and they are easily carried away by wind. Some of the orchids, where seeds are dispersed by air, are said to produce about 700 million seeds per plant.

In legumes, the pericarp dries up and fruit splits open explosively (Figure 15.24A). The poppy (*Papaver*) produces spore-like seeds which are so light that they may be carried to long distances along with air current. The seeds are produced in a dry hollow fruit, called a capsule (Figure 15.24B). When mature, the seeds emerge through pores in the capsule as it is shaken by the wind.

## Hydrochory (Water Dispersal)

Hydrochory, or the passive dispersal of organisms by water, is an important means of propagule transport, especially for plants. The fruits, seeds and vegetative units of many plants, especially those growing in or near water sources, are adapted for floating, either because air is trapped in some part of the fruit or because the fruit contains buoyant tissue. The fruits of coconut are especially adapted for dispersal by ocean currents.

In mangroves, if seeds fall during low tide, it can begin to root in the soil. If the seeds fall in the water, they are carried away by the tide to grow somewhere else. Rain is also a common means of seed and fruit dispersal and is particularly important for plants that grow on hillsides or mountain slopes.

## Zoochory (Animal Dispersal)

Many animal-dispersed fruits are dispersed by vertebrates especially certain mammals and birds. Vertebrate dispersed seeds and fruits may be fleshy or may have fleshy coverings. Seed dispersal by ants is known as myrmecochory. Ant-dispersed seeds often have nutrient-rich appendages called elaiosomes (see Figure 15.12, 15.13). Zoochory is of two types: exozoochory and endozoochory.

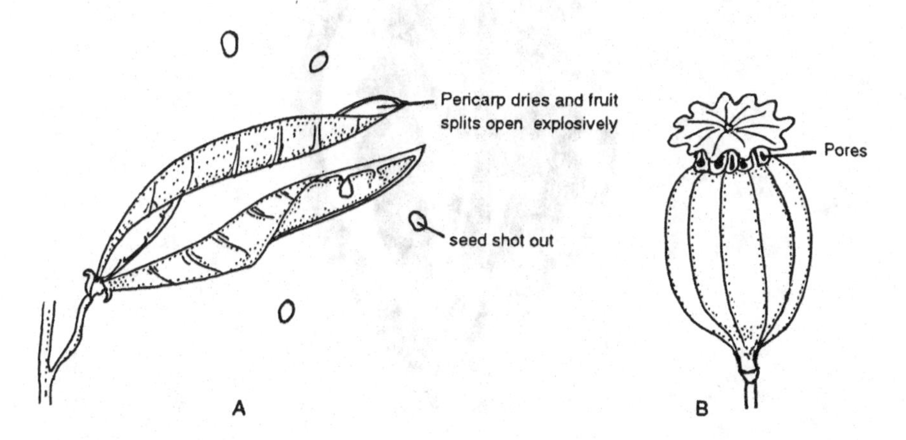


**FIGURE 15.24** Mechanism of seed dispersal by wind. A. Legume. B. Poppy.

1. **Exozoochory:** Some dried fruits and seeds have sticky or spiny projections designed to latch onto the bodies of passing animals. The fruits of *Martynia diandra* are woody and covered with hooked spines (Figure 15.25). The fruit terminates in two large, curved horns. These fruits are dispersed by animals to long distances. In *Tribulus*, the fruits bear sharp, rigid spines which enter the skin of animals and clothes of human beings and are carried away from the parent plant.

   In deserts, rodents are important dispersers of seeds and especially those species of rodents that scatter-hoard (scatter-hoarding is a behaviour of rodents that dig caches for seeds that are at varying distances from the home burrow). The rodents are the only dispersers of the seeds of *Myrcianthes coquimbensis.*
2. **Endozoochory:** Dispersal through transport in the gut of animals is called endozoochory. Fleshy fruits are attractive as food to a variety of animals, mainly birds. These fruits advertise their maturity by changing colour from an inconspicuous green to bright orange, red, blue or black. Such fruits are consumed by animals which later eject the seeds. Birds often fly away from the parent plant and disperse the seeds in their droppings. Plants like Pittosporum have sticky seeds that can be carried away by birds.

**Barochory:** Gravity, dispersal involving the seed falling below plant.
**Ombrohydrochory:** Seeds dispersed by rain drops.

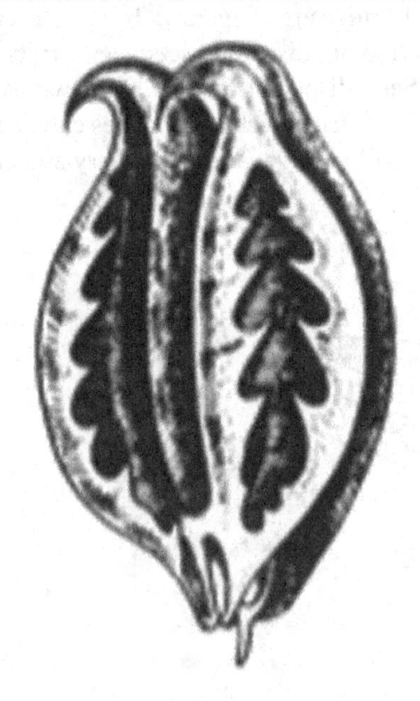

**FIGURE 15.25** *Martynia diandra.* Mature seed.

## SUMMARY


- A true seed is a fertilized mature ovule that contains an embryonal plant, stored food material (rarely missing) and protective coat(s).
- Seed surface viewed under SEM is known as spermoderm.
- Arils are fleshy seed appendages.
- The tip of the outer integument forms a reflexed outgrowth leading to the formation of a massive or small structure known as a caruncle.
- Elaiosomes are fleshy and edible part of seeds dispersed by ants.
- A curved outgrowth from the funiculus, on the side away from the micropyle is termed jaculator.
- Seed coat develops from the integument(s).
- Seeds that show an irregular internal structure when at in any plane are called labyrinth seeds.
- Depending upon the presence or absence of endosperm, seeds are classified in two groups: albuminous and non-albuminous.
- Seeds are dispersed by different abiotic and biotic agencies such as wind, water, animals. Dispersal also takes place by the plant itself.


## SUGGESTED READING


Bansal, S. 1995. Seed anatomy in *Lathyrus* (Papilionoideae, Fabaceae). In *Advances in Plant Reproductive Biology*, Vol. I., eds. Y.S. Chauhan and A.K. Pandey, 233–242. Delhi: Narendra Publishing House.

Berg, R.Y. 1958. Seed dispersal, morphology and phylogeny of Trillium. *Skr. Norske Vidensk.-Akad. Oslo* 1958: 1–36.

Bhargava, H.R. 1935. The life history of *Trianthma monogyna* Linn. *Proc. Ind. Acad. Sci.* 2: 49–58.

Boesewinkel, F.D. and F. Bouman. 1984. The seed structure. In *Embryology of Angiosperms*, ed. B.M. Johri, 567–610. Berlin: Springer-Verlag.

Corner, E.J.H. 1976. *The Seeds of Dicotyledons. I & II.* Cambridge: Cambridge University Press.

Dahlgren, R.M.T. and H.T. Clifford. 1982. The monocotyledons: A comparative study. *Brittonia* 34: 267–268.

deBruyn, R.A.J., M. Paetkau, K.A. Ross, D.V. Godfrey, J.S. Church and C.R. Friedman. 2015. Thermogenesis-triggered seed dispersal in dwarf mistletoe. *Nat. Commun.* 6: 6262. https://doi.org/10.1038/nComms7262.

Gaertner, J. 1788. *The Fruits and Seeds of Plants* (in Latin), 3 vols. Stuttgart Tübingen. Carpologia, III vol. Lipsiae 1805-1807.

Gunn, C.R. 1981. Seeds of Leguminosae. In *Advances in Legume Systematics. Part 2*, eds. R.M. Polhill and P.H. Raven, 913–925. In *Proceedings of International Legume Conference*, Kew.

Lersten, N.R. and C.R. Gunn. 1981. Testa characters in tribe Vicieae with notes about tribes Abreae, Cicereae and Trifolieae (Fabaceae). *Tech. Bull.*, U.S. Dept of Agriculture.

Lersten, N.R. and C.R. Gunn. 1982. Testa characters in the tribe Vicieae, with notes about tribes Abreae, Cicerae and Trifolieae (Fabaceae). *U. S. Dep. Agric. Tech. Bull.* 1667: 1–40.

Lersten, N.R., C.R. Gunn and C.L. Brubaker. 1992. *Comparative Morphology of the Lens on Legume (Fabaceae) Seeds, with Emphasis on Species in Subfamily Caesalpinoideae and Mimosoideae*, 1–44, Ist ed. Washington, DC: U.S. Dept. of Agriculture.

Maheshwari, S.C. and R.N. Kapil. 1963. Morphological and embryological studies on the Lemnaceae. II. The endosperm and embryo of *Lemna paucicostata*. *Am. J. Bot.* 50 (9): 907–914.

Martin, A.C. 1946. The comparative internal morphology of seeds. *The Am. Midl. Naturalist* 36: 513–660.

Martin, A.C. and W.D. Barkley. 1961. *Seed Identification Manual*. Calfornia: The Balckburn Press.

Mauritzon, J. 1936. Semenbau und Embryologie Einiger Scitamineen. *Acta Univ. London* 31: 1–31.

Narayana, H.S. 1962. Seed structure in Aizoaceae. *Proc. Summer School Bot.* 1960 (*Darjeeling*): 220–230.

Pandey, A.K. 1991. Spermoderm pattern in Trifolieae (Leguminosae-Papilionoideae). *Phytomorphology* 41: 293–297.

Pandey, A.K. 1992. Spermoderm pattern in Papilionoideae (Fabaceae). In *Current Concepts in Seed Biology*, eds. K.G. Mukerji, A.K. Bhatnagar, S.C. Tripathi, Manju Bansal and Manju Saxena, 31–44. Calcutta: Noya Prakash.

Pandey, A.K. 1995. Conservation of biodiversity: Present status and future strategy. In *Taxonomy and Biodiversity*, ed. A.K. Pandey, 44–51. New Delhi: CBS Publishers and Distributors.

Pandey, A.K. and S.S. Jha. 1988. Development and structure of seeds in Genisteae (Papilionoideae-Leguminosae). *Flora* 181: 415–424.

Pandey, A.K. and S.S. Jha. 1990. Development and structure of seeds in Phaseoleae (Papilionoideae-Fabaceae). *Flora* 184: 369–380.

Pandey, A.K. and S.S. Jha. 1991. SEM studies on spermoderm of some *Lotus* species (Fabaceae). *J.Indian Bot. Soc.* 70: 433–434.

Periasamy, K. 1962. The ruminate endosperm: Development and types of rumination. In *Plant Embryology: A Symposium* (ed. P. Maheshwari), 62–74. New Delhi: CSIR.

Sernander, R. 1906. Entwurf einer Monografie der Europäischen Myrmekochoren. *Kungliga Svenska vetenskapsakademiens handlingar* 41 (7): 1–410.

Singh, B 1964. Development and structure of angiosperm seed-I. Review of the Indian work. *Bull. Natl. Bot. Gdn.* 89: 1–115.

Singh, D. and A.S.R. Dathan. 1972. Structure and development of seed-coat in Cucurbitaceae. VI. Seeds of Cucurbita. *Phytomorphology* 22: 29–45.

Subramaniam, S., A.K. Pandey and S.A. Rather. 2015. Systematic and adaptive significance of seed morphology in *Crotalaria* L. (Fabaceae). *Int. J. Pl. Rep. Biol.* 7(2): 135–147.

Tobe, H., W.L. Wagner and H. Chin. 1987. A systematic and evolutionary study of *Oenothera* (Onagraceae): Seed coat anatomy. *Bot. Gaz.* 148: 235–257.

Van der Pijl, L. 1972. *Principles of Dispersal in Higher Plants*. Netherlands: Springer.

Van Heel, W.A. 1970. Some unusual tropical labyrinth seeds. *Proc. Kon Ned Akad Wetensch Amsterdam C* 73: 298–301.

# 16 Experimental Plant Reproductive Biology

Experimental plant reproductive biology (experimental embryology) is concerned with an imitation and a modification of the course of nature, with a view to understanding various processes that underlie the development and differentiation of the embryo.

The beginning of experimental embryology dates back to the observations of Massart (1902), who obtained swelling of the ovary by treating the ovaries of certain plants with spores of *Lycopodium*, dead pollinia, and aqueous extracts of pollen. Presently several formulations of nutrient media are available which support the *in vitro* growth of reproductive tissues, and very promising results have been obtained. Cell, tissue and organ culture is now a well-established discipline with a wide range of contacts with other disciplines including physiology, biochemistry and genetics. Cultural manipulations are now possible through tissue culture techniques listed in Figure 16.1.

## TECHNIQUES FOR CELL, TISSUE AND ORGAN CULTURE

### NUTRIENT MEDIUM

Cells, tissues and organs can only be cultured *in vitro* when all the required nutrients are available in the medium. A nutrient medium should contain inorganic nutrients, organic nutrients and growth hormones. The nutritional requirements of different species may vary, therefore, a wide range of culture media differing in their elemental compositions have been formulated (Table 16.1). A medium with these ingredients is referred to as the basal medium. Some growth hormones, like auxins, cytokinins and gibberellins may be added to the basal medium either alone or in various combinations.

Tissue cultures are maintained either on semi-solid medium or in liquid medium. Agar (a polysaccharide obtained from sea weeds) is used to solidify the medium. Agar is commonly used at a concentration of 0.8–1% (w/v). The advantages of growing tissues in agitated liquid media are that it facilitates gaseous exchange, reduces nutrient gradients and is amenable to manipulations in experiments where radioisotopes are employed.

As there is no single medium suitable for culturing all cells, tissues and organs, it is essential that the most suitable medium be determined to accommodate specific requirements of the plant material in question. For the medium preparation, appropriate quantities of agar-agar and sucrose are dissolved in distilled water. Required quantities of stock solutions, growth hormones and other supplements

DOI: 10.1201/9781003260097-16

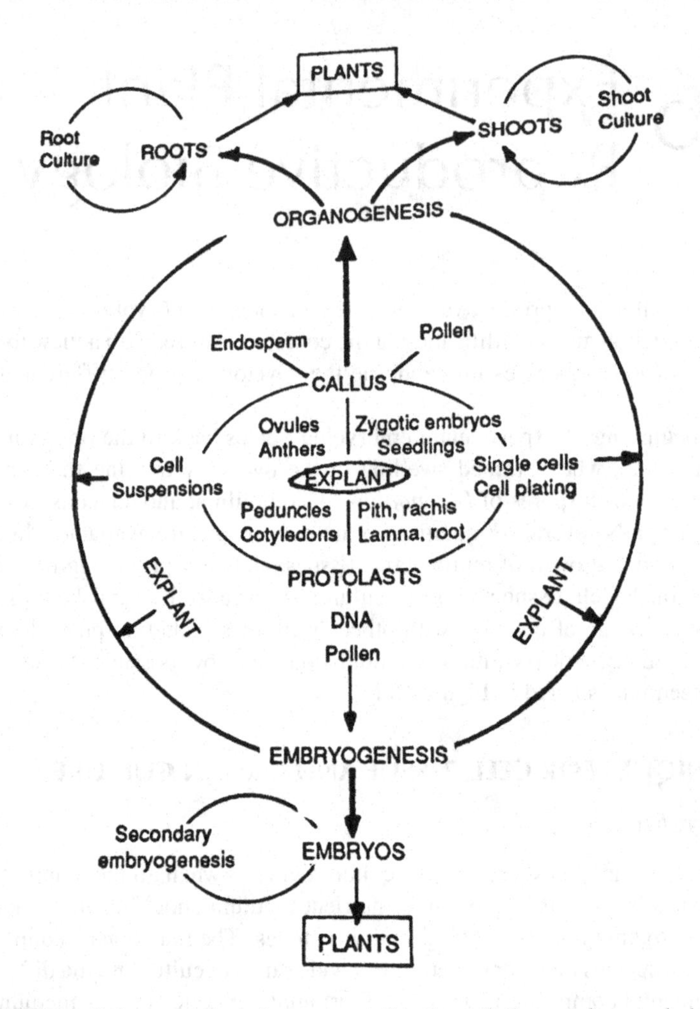


**FIGURE 16.1** Culture manipulation now possible through tissue culture technique.

are added. More distilled water is added to make the final volume of the medium. Stock solutions of varying strengths are prepared from different media components viz., inorganic salts, vitamins and other organic compounds and growth regulators. pH of the medium is adjusted between 5–5.8 by adding 0.1N NaOH or 0.1N HCl. The medium is poured in culture tubes, flasks, petri plates or any other containers. Culture vessels are plugged with non-absorbent cotton wool wrapped in cheese cloth. This allows free gaseous exchange but inhibits microbial contaminations. Culture vessels are autoclaved at 120°C (1.06 kg/cm²) for 15–20 min. Culture medium is allowed to cool at room temperature and stored at 4°C. While preparing a solid medium in culture tubes, it is desirable to make slants by keeping the tubes tilted during cooling. Such slants provide larger surface area for tissue growth.

## TABLE 16.1
## Constituents of Some Common Plant Tissue Culture Media (mg litre⁻¹)

| Constituents | | WM | MS | NM | NTM |
|---|---|---|---|---|---|
| **Inorganic Compounds** | | | | | |
| Potassium chloride | KCL | 65 | - | - | - |
| Potassium nitrate | $KNO_3$ | 80 | 1900 | 950 | 950 |
| Ammonium nitrate | $NH_4NO_3$ | - | 1650 | 720 | 825 |
| Calcium chloride | $CaCl_2.2H_2O$ | - | 440 | - | 220 |
| Calcium nitrate | $Ca(NO_3)_2$ | 300 | - | - | - |
| Magnesium sulphate | $MgSO_4.7H_2O$ | 750 | 370 | 185 | 1233 |
| Sodium sulphate | $Na_2SO_4.10H_2O$ | 200 | - | - | - |
| Potassium dihydrogen phosphate | $KH_2PO_4$ | - | 170 | 68 | 680 |
| Sodium dihydrogen phosphate | $NaH_2PO_4.7H_2O$ | 165 | - | - | 1233 |
| Manganese sulphate | $MnSO_4.4H_2O$ | 5 | 22.3 | 25 | 22.3 |
| Boric acid | $H_3BO_3$ | 1.5 | 6.2 | 10 | 6.2 |
| Zinc sulphate | $ZnSO_4.7H_2O$ | 3 | 8.6 | 10 | - |
| Potassium iodide | KI | 0.75 | 0.83 | - | 0.83 |
| Sodium molybdate | $Na_2MoO_4.2H_2O$ | - | 0.025 | 0.25 | 0.25 |
| Copper sulphate | $CuSO_4.5H_2O$ | 0.01 | 0.025 | 0.025 | 0.025 |
| Cobalt chloride | $CoCl_2.6H_2O$ | - | 0.25 | - | - |
| Cobalt sulphate | $CoSO_4.7H_2O$ | - | - | - | 0.03 |
| Disodium ethylene diamene tetraacetate | $Na_2EDTA.2H_2O$ | - | 37.3 | 37.3 | 37.3 |
| Ferrous sulphate | $FeSO_4.7H_2O$ | - | 27.8 | 27.8 | 27.8 |
| Ferric sulphate | $Fe(SO_4)_3$ | 2.5 | - | - | - |
| **Organic compounds** | | | | | |
| Inositol | | - | 100 | 100 | 100 |
| Nicotinic acid | | 0.05 | 0.5 | 5 | - |
| Pyridoxine HCL | | 0.01 | 0.1 | 0.5 | 1 |
| Thiamine HCL | | 0.01 | 0.1 | 0.5 | 1 |
| Glycine | | 3 | 2 | 2 | - |
| Ca-D-pantothenate | | 1 | - | - | - |
| Sucrose | | 2% | 3% | 2% | 1% |
| Mannitol | | - | - | - | 12.7% |
| **Growth Regulators** | | | | | |
| IAA | | - | 1.30 | - | - |
| NAA | | - | - | - | 3 |
| Kinetin | | - | 0.4–10 | - | - |
| 6 benzyl amino purine | | - | - | - | 1 |

WM, White's medium (1954);MS- Murashige and Skoog medium (1962); NM-Nitsch medium (Nitsch, 1969); NTM- Nagata and Takebe's medium (Nagata and Takebe, 1971)

### Callus Initiation

Callus is an organized mass of loosely arranged parenchymatous cells (Figure 16.2A). Callus initiation begins when segments of plant organs (explants) are cultured on solidified culture media. The initiation of the callus formation is visible when the surface of the explant begins to glisten and the texture become rough. Repeated cell divisions cover the entire explant with unorganized callus in 2–3 weeks. Plants can be regenerated from cultured materials by the process of organogenesis (Figure 16.2B).

## ASEPTIC CONDITIONS

The media used for plant tissue cultures also provide suitable environment for the growth of microorganism like fungi and bacteria. There are a number of sources through which medium may get contaminated. These include the medium itself, explant, culture vessels and instruments, environment of culture room and place of transfer.

1. **Medium:** In order to destroy microorganisms present in the medium, culture and vessels containing the medium are properly plugged and autoclaved.
2. **Explant:** The microorganisms may also be carried along with the tissue that is being cultured. To avoid this, plant tissue is sterilized before culturing. Table 16.2 lists the commonly used sterilizing agents.
3. **Culture Vessels and Instruments:** Glassware and metal instruments are sterilized by placing them in an oven at 160–180°C for 3–4 hours. The instruments such as scalpel, forceps, needles etc., dipped in 95% ethyl alcohol, are flamed before use.

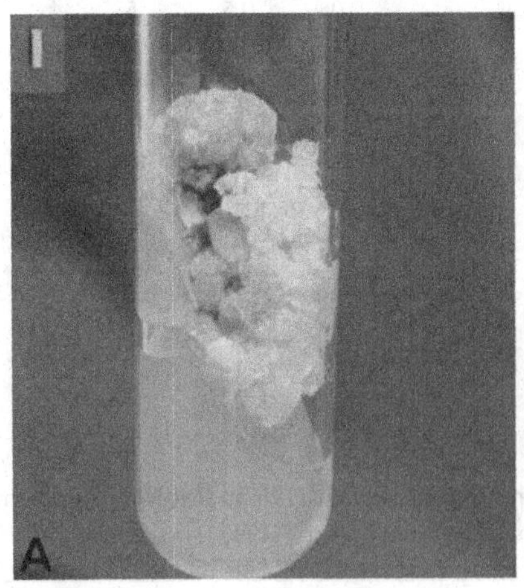

**FIGURE 16.2** *Tecoma instans.* A. Callus from explant; B. Regenerated plantlet (courtesy: Dr M. Anis, Aligarh).

**TABLE 16.2**
**Sterilizing Agents Commonly Used in Plant, Cell and Tissue Culture**

| Sterilizing Agent | Level Used | Time of Sterilization (minutes) |
|---|---|---|
| Calcium hypochlorite | 9–10% | 5–30 |
| Sodium hypochlorite | 1–2% | 5–30 |
| Hydrogen peroxide | 10–12% | 5–15 |
| Mercuric chloride | 0.1–1.0% | 2–10 |
| Antibiotics | 4–50 $mgl^{-1}$ | 30–60 |

4. **Transfer Area:** Aseptic manipulations are conducted in laminar air flow cabinets (Figure 16.3A). Air passes through ultrafilters which flow in the working areas at a constant speed. These filters eliminate microbes, dust and other particles. Thus, an aseptic environment is maintained at the time of inoculation. The cultures in culture tubes, flasks, etc. are kept in the culture room (Fig. 16.3B). The temperature of culture room should be maintained at ±25°C. However, a wider range of temperature may be required for specific experiments.

## ANTHER CULTURE

Anther culture has gained much importance mainly for two reasons. Firstly, it is an excellent approach for the quantitative production of true isogenic pure lines in an appreciable time period and thus has an edge over the traditional methods of obtaining pure lines. Secondly, it provides a valuable tool for mutation research. In anther cultures homozygous lines can be immediately produced after doubling the pollen-derived haploid plants. In haploids there is only one set of chromosomes; therefore, whenever there is gene mutation, it is immediately expressed. Haploids have been obtained from anther cultures of number of plant species (Table 16.3). The members of the family Solanaceae, Poaceae and Brassicaceae have been found to be more responsive to androgenesis.

The induction of haploid embryoids from pollen grains of *Datura innoxia* was first reported by Guha and Maheshwari (1964, 1966). Bourgin and Nitsch (1967) were the first to obtain mature flowering haploid plants of pollen origin in *Nicotiana tabacum* and *N. sylvestris*. Sharp et al. (1972) grew haploid clones of tomato plants by placing the isolated anther of the plant on small pieces of filter paper, kept over cultured anthers of the same species.

## METHODS OF CULTURE

Anther culture technique involves excising closed flower buds which have anthers containing uninucleate microspores. At this stage, microspores are most suitable

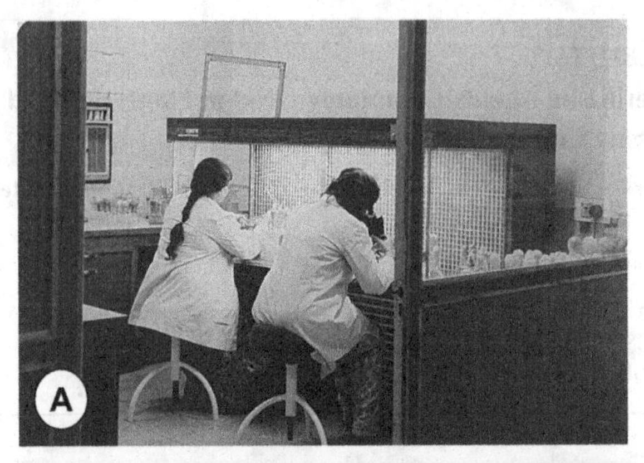

**FIGURE 16.3** A. Inoculation work being performed in a laminar air flow cabinet. B. A view of the culture room of the Tissue Culture Laboratory, CSIR-National Botanical Research Institute, Lucknow (courtesy: Dr Ashok Sharma, Lucknow).

for the induction of androgenesis. The excised flower buds are surface-sterilized with alcohol or chlorine water and anthers are carefully removed (Figure 16.4). Special care is taken to ensure that the anthers are not injured during operation. Anthers are detached from their filaments and placed horizontally on the medium. The cultures are incubated at 24–27°C using light of about 2000 lux for 14 h per day. Depending upon the plant species, it takes 3–8 weeks for pollen plantlets to regenerate from anthers.

## FACTORS AFFECTING ANDROGENESIS

### 1. Culture Medium

Culture medium plays an important role in inducing divisions in microspores. Basal media suggested by Murashige and Skoog (1962) and Gamborg et al. (1968) have

**TABLE 16.3**
**Species for Which Androgenic Haploids Have Been Raised Through Anther or Pollen Culture**

| Family | Species |
|---|---|
| Asteraceae | *Gerbera jamisonii* |
| Brassicaceae | *Arabidopsis thaliana, Brassica campestris, B. napus* |
| Chenopodiaceae | *Beta vulgaris* |
| Caricaceae | *Carica papaya* |
| Euphorbiaceae | *Hevea brasiliensis* |
| Solanaceae | *Atropa belladonna, Datura innoxia, Hyoscymus niger, Solanum lycopersicum, Nicotiana tabacum, Petunia hybrida, Solanum tuberosum* |
| Fabaceae | *Phaseolus aureus, Dolichos uniflorus, Arachis hypogea* |
| Poaceae | *Triticum aestivum* |

been found to be advantageous for many species. Surcrose is added in all anther culture media and is generally used at a concentration of 2–4%. Iron is indispensable for the production of plantlets from microspores. Incorporation of activated charcoal also stimulates androgenesis (development of plants from the male gametophyte).

The nutritional requirements vary greatly from species to species. Kinetin (KN) added alone to the medium induces pollen division, but incorporation of yeast extract (YE), casein hydrolysate (CH) and 2–4 dichlorophenoxyacetic acid (2,4-D) triggers callus growth from anther wall tissues. In rice, genotype and media play a significant role on both callus induction frequency and green plantlet regeneration efficiency (Ali et al., 2021).

2. **Physiological State of Donor Plant**

Age and growth conditions of the donor plants greatly affect the embryogenic potential of the cultured anthers. Generally, the anthers from young plants are more responsive. Anthers derived from older inflorescences show decline in the frequency of haploid production due to reduced pollen viability. In *Oryza sativa* increased anther response was observed when plants were given pre-treatment with ethereal for 84 hrs at 10°C. In *Nicotiana* an increase in embryoid formation was obtained when anther donor plants were subjected to nitrogen starvation.

3. **Developmental Stage of the Microspore**

The developmental stage of microspores at culture is critical for the induction of division. Usually, pollen cultured immediately before or after first mitosis give better response. In *Nicotiana tabacum* the yield of embryoids was increased when

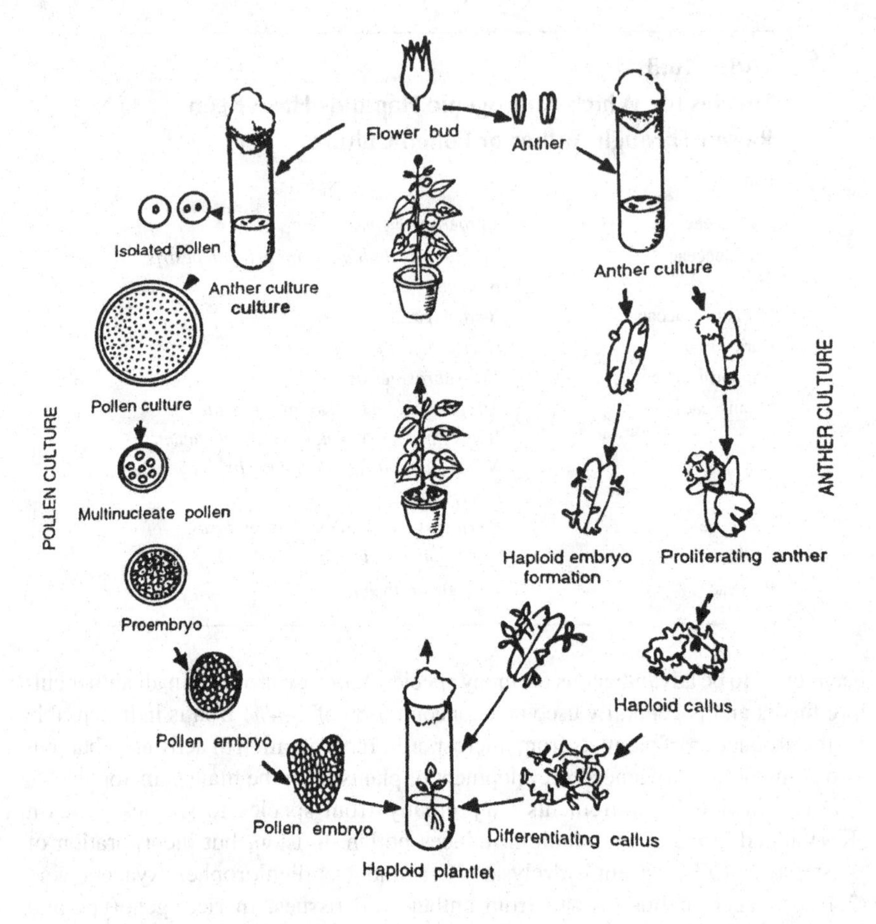


**FIGURE 16.4** Diagrammatic illustrations showing various modes of androgenesis and haploid plant formation by anther and isolated pollen culture.

anthers were cultured at the free microspore stage. In *Brassica* and *Petunia* pollen grains cultured at bi-celled stage gave better response.

4. **Anther Tissues**

The anther wall plays an important role in embryoid formation. The degeneration of anther tissue has been observed in all successful cultures and release of hormones by senescing cells is known. Sharp et al. (1972) reported nursing effects of whole anthers in tomato. Occasionally, the embryoids developed suspensor-like structures attached to the anther wall which indicated a nutritive role for the diploid tissue.

5. **Temperature and Light**

Temperature and light affect the embryogenic potential of the cultured anthers. Usually, a temperature range from 23 to 28°C is most suitable for pollen divisions.

In *Datura innoxia* embryoids are not formed at all at 20°C or below. In *Oryza sativa* the highest frequency of pollen callusing occurred when the excised panicles were treated at 13°C for 10–14 days.

In float cultures of *Nicotiana* initial incubation of anthers in darkness favoured embryoid formation. In *Vitis vinifera* anthers responded better when they were kept in continuous light for 24 hrs. In ornamental kale (*Brassica oleracea* L. var. *acephala*), cold treatment significantly improved plant regeneration from microspore-derived embryos to a rate of up to 79.0% under 4°C for 2 days or 5 days (Wang et al., 2011).

## ONTOGENY OF POLLEN EMBRYOIDS

Normally the uninucleate microspores divide unequally forming a small generative cell and a large vegetative cell. The generative cell further divides to form two sperm cells. In cultures, uninucleate microspores divide to form multicellular pollen grains (Figure 16.5) by any of the following pathways:

(i) **Pathway I:** The microspores divide by an equal division forming two cells. Both the cells contribute towards the formation of the sporophyte.
(ii) **Pathway II:** The microspores divide by an unequal division forming a small generative cell and a large vegetative cell. The generative cell either does not divide further or divides once or twice before degeneration.
(iii) **Pathway III:** The microspore divides unequally forming a small generative cell and a large vegetative cell. The pollen embryos are formed from the generative cell only.
(iv) **Pathway IV:** The microspores divide by an unequal division resulting in the formation of a generative cell and a vegetative cell. Both these cells participate in the development of the sporophyte.

Irrespective of the early pattern of divisions, the responsive pollen grains become multicellular and burst open to release a cellular mass having an irregular shape. This tissue gradually becomes globular and undergoes the normal stages of post-globular embryogeny (e.g., *Atropa*, *Datura*, *Nicotiana*). In this way the haploid plant is formed from one microspore. In *Arabidopsis* and *Asparagus* where, androgenic sporophytes have been obtained through anther culture, the multicellular mass liberated from the bursting of pollen grain undergoes further proliferation and forms a callus. This callus later on differentiates forming sporophytes (Figure 16.6).

Haccius and Bhandari (1975) studied ontogeny of pollen embryos of tobacco and observed and about 82% of the well-developed pollen embryos had their basal end adhering to a supporting tissue. These studies have revealed that, for establishment of initial polarity and for normal development of pollen embryos, a temporary support is essential.

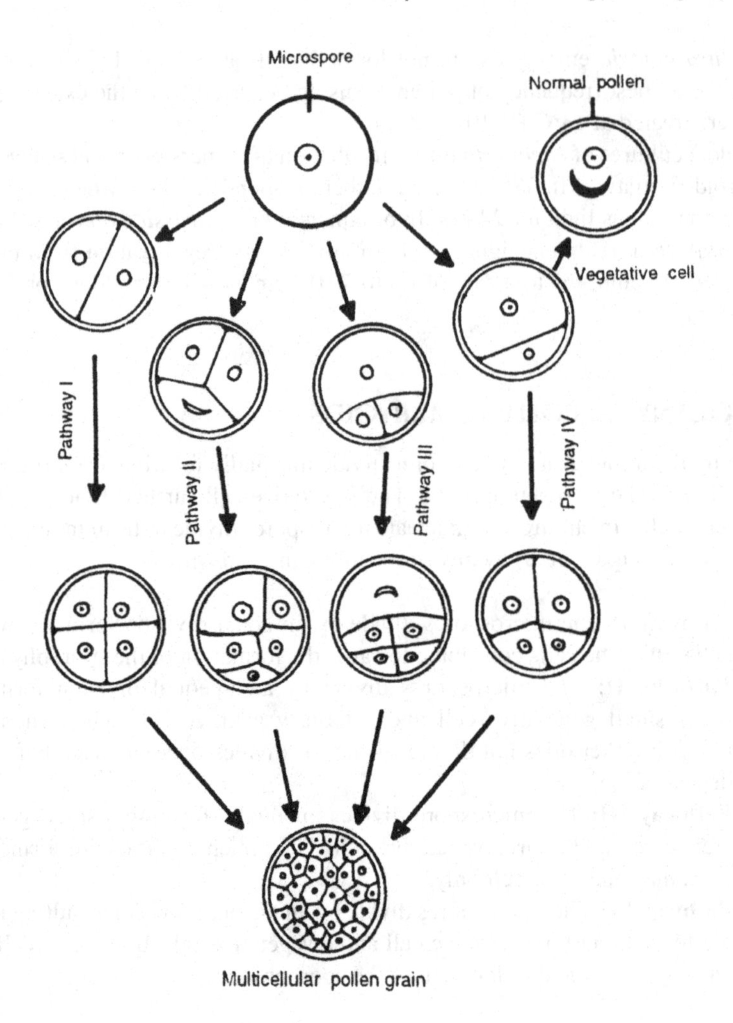


**FIGURE 16.5** Diagram showing development of multicellular pollen grains.

## POLLEN CULTURE

Pollen culture (microspore culture) is a technique in which haploid plants are obtained from isolated pollen grains by the induction of embryogenesis. Pollen culture provides a system with haploid single cells and thus is a promising haplophase tool for basic and applied biotechnological breeding programmes. C. Nitsch and Norreel (1973) were the first to demonstrate the induction of embryoids from isolated pollen grains of *Datura innoxia*. Since then, embryoids have been obtained by culturing isolated pollen grains in *Petunia*, *Nicotiana*, *Atropa*, *Solanum* and many other taxa.

The pollen grains within an anther are genetically heterogenous. The plants thus arising from an anther constitute a heterogenous population. In cases where

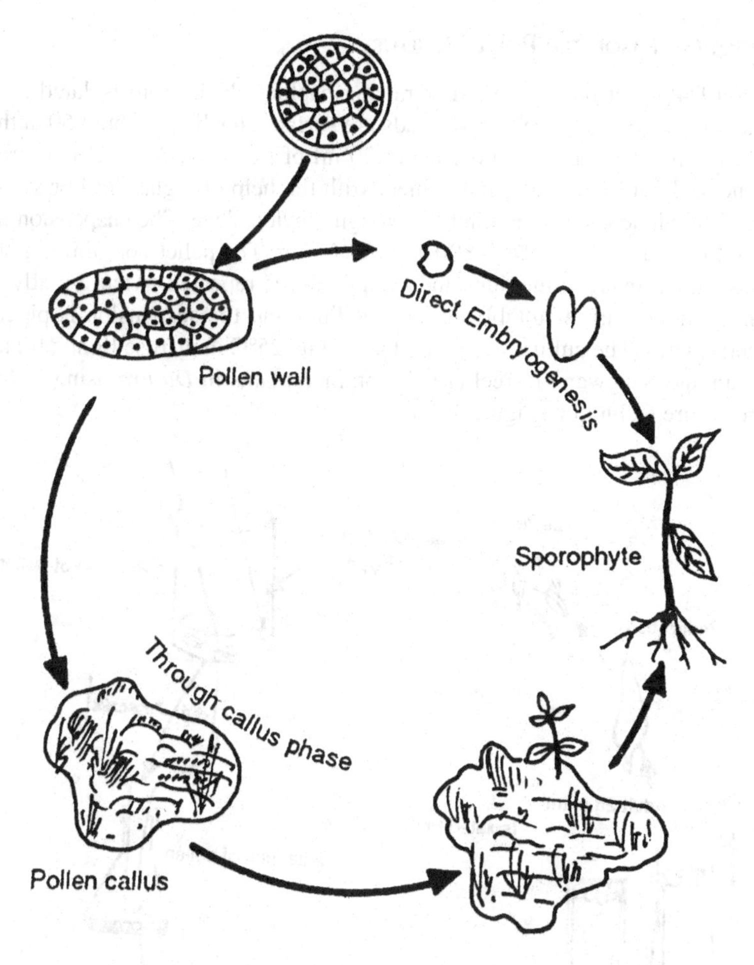


**FIGURE 16.6** Diagram showing origin of sporophyte from pollen grain (after Bhojwani and Rajdan, 1983).

development of haploid plants is preceded by callusing of pollen grains, the tissue thus formed would be a chimera. This happen due to the mixing of tissues derived from several pollen grains. Isolated microspore or pollen grain cultures can be a solution to avoid chimera formation. Another problem which can be overcome by isolated pollen culture is the completion between a large number of pollen grains for growth in the limited space available within the anther locule.

Shmykova (2021) induced embryogenesis in isolated microspore cultures of carrot (*Daucus carota*). The first embryogenesis was observed in microspores at the late vacuolated stage when the nucleus moved from the centre to one pole following the long axis. In pollen cultures, it is important to correctly determine the phase of development of microspores in buds for embryogenesis in a culture of isolated microspores *in vitro*.

## Techniques of Isolated Pollen Culture

Nitsch (1974) described a method of raising haploid plants from isolated pollen grains of tobacco. Unopened flower buds are surface sterilized. About 50 anthers are placed in a beaker containing about 20 ml of liquid medium. Pollen grains are squeezed out by pressing the anthers with the help of a glass rod or syringe piston. The whole solution is filtered through a nylon sieve. The suspension thus obtained is centrifuged at 500–800 rpm for 5 min. The pellet-containing pollen is resuspended in fresh medium and the process is repeated twice. Finally, the pollen is mixed with a suitable culture medium and the suspension is pipetted into petri plates. The cultures are stored at around 25°C in diffuse light (500 lux). Sangwan and Sangwan-Norreel (1987) obtained plants in *Datura* using isolated pollen culture technique (Figure 16.7)

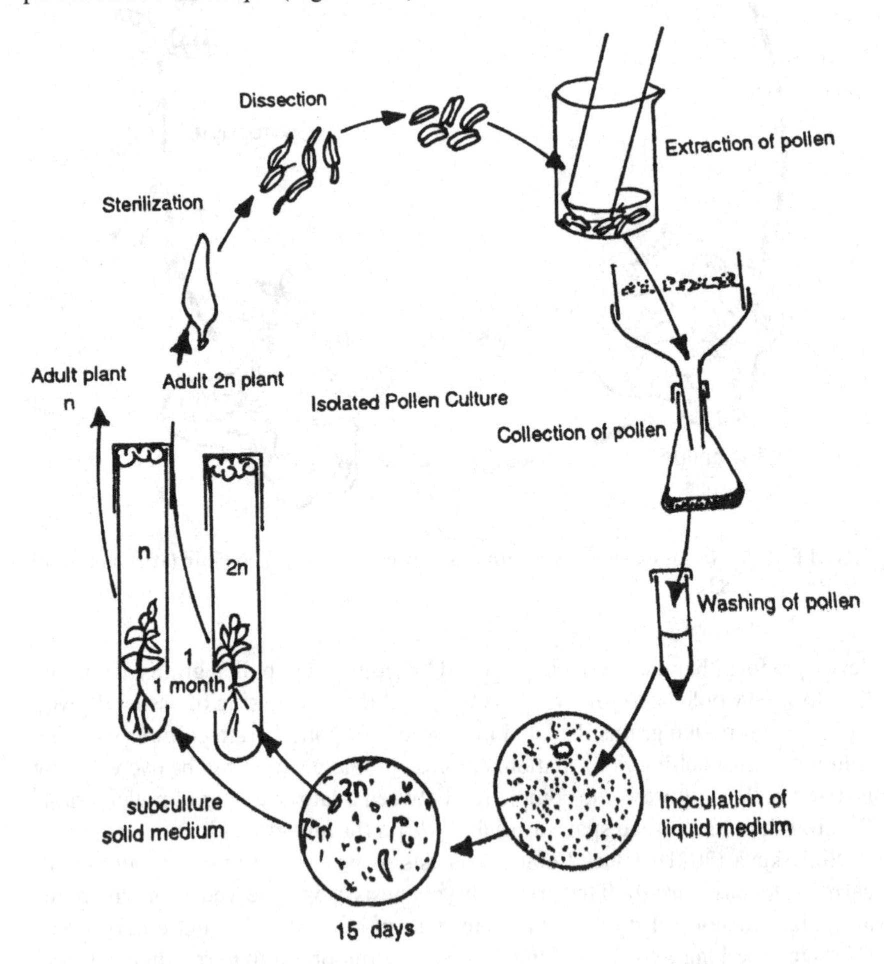


**FIGURE 16.7** Diagram showing *in vitro* androgenesis in cultured isolated pollen grains of *Datura* (after Sangwan and Sangwan-Norreel, 1987).

Wenzel (1980) used the technique of gradient centrifugation in order to improve the efficiency of isolated pollen culture for production of haploid plants. Sunderland and Roberts (1977) proposed a modified method of pollen culture. Flower buds were given adequate chill treatment. Anthers were then excised and floated on a liquid medium. After a few days, anthers were dehisced and discharged batches of pollen grains which also included grains at various stages of development. This method has been successfully employed in producing haploid plants in *Datura*, *Nicotiana* and *Hordeum*.

## DIPLOIDIZATION OF HAPLOIDS

The haploid plants obtained by anther and pollen culture are sterile as they lack homologous chromosomes. To obtain fertile homozygous diploids, a chromosome complement of the haploid must be duplicated. Diploidization of haploids can be achieved by colchicine treatment of the sporophyte, haploid cell suspension culture and culture of the vegetative parts of haploid plants (Figure 16.8). Since colchicine acts on dividing cells, it is ideal to use this chemical during the early stages of development when the cells are actively dividing.

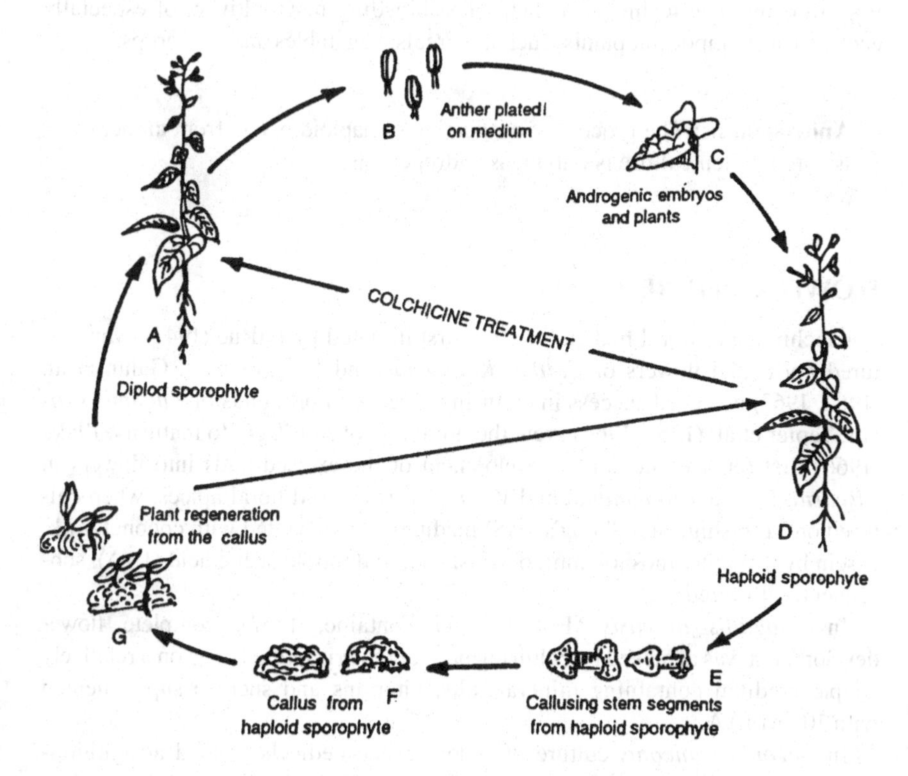


**FIGURE 16.8** Regeneration of diploid plants from haploid sporophyte.

## SIGNIFICANCE AND USES OF HAPLOIDS

Anther culture technique is the most viable and efficient method of producing homozygous doubled haploid plants within a short period. Homozygosity can be achieved quickly through haploids and doubled haploids have been utilized for crop improvement. Nakamura et al. (1974) raised three promising lines of tobacco through anther culture of the hybrids between lines MC-1610 and Cocker-139. The new lines exhibited higher resistance to bacterial wilt and black shank. Production of haploids and subsequent chromosome doubling enables development of true breeding recombinants in one generation (Stringam et al., 1995).

In the production of new mutant forms, haploids provide excellent material for experimentation. A large number of mutants have been obtained in *Nicotiana tabacum* and *Brassica napus* by subjecting haploid plants to x- or gamma irradiation. Free cells of the haploids in suspension could be used as an experimental material for mutagenic studies and to evolve biochemical mutants resistant to pathogens, cold, salinity or drought conditions.

Culture of isolated microspores provides a novel experimental system for study of factors controlling pollen embryogenesis of higher plants. *In vitro* culture techniques of haploid production not only shorten the breeding cycle but also recover more types of gene combinations and provide more valuable materials for plant improvement. The technique is helpful in breeding new cultivars of especially economically important plants, such as cereals, vegetables and tree crops.

**Androgenesis:** The process of production of haploid plants from anther or isolated pollen culture is known as androgenesis.

## FLOWER CULTURE

The technique of floral bud culture was first initiated by LaRue (1942), who cultured pollinated flowers of *Caltha*, *Kalanchoe* and *Lycopersicon*. Galun et al. (1962, 1963) achieved success in culturing detached floral buds of *Cucumis sativus*. Tepfer et al. (1963, 1966) grew the floral bud of *Aquilegia* to maturity. Blake (1966) first reported complete development of the excised buds into flowers in *Viscaria*. In *Viscaria candida* and *V. cardinalis* excised floral apices, when cultured on Murashige and Skoog's (MS) medium supplemented with coconut milk, casein hydrolysate, inositol, amino acids, $GA_3$ and indole acetic acid (IAA), subsequently flowered.

In *Anagallis arvensis* (Brulfert and Fontaine, 1967), complete flower development was achieved by culturing buds at the primordial stage on a relatively simple medium containing mineral salts, vitamins and sucrose supplemented with $10^{-6}$M IAA.

In *Begonia franconis* culture of inflorescence pedicels yielded adventitious flower buds; and for complete development, liquid medium proved better than solid medium. After callus induction, auxin generally decreased the number of flower buds.

In *Haworthia* (Majumdar, 1970) the inflorescence segments bearing flower buds were inoculated on White's medium supplemented with CM, IAA and KN. Some flower buds shrivelled, whereas others developed into mature flowers.

In *Ranunculus sceleratus* (Konar and Natraja, 1969), the cut ends of undifferentiated flower buds proliferated rapidly forming a mass of white, friable callus. Optimal callus formation was observed on White's Medium (WM) containing 2% sucrose, 10% coconut milk, 500 ppm casein hydrolysate (CH) and 1 ppm 2, 4-D. Callus further differentiated into roots, shoots and embryoids leading to the formation of plantlets. In *Begonia franconis* when young floral primordials were cultured on a medium containing BA, normal floral buds developed.

In *Swertia chirayita* (Sharma et al., 2014) the MS medium supplemented with BAP 1.0 $mgL^{-1}$ and adenine sulphate 70.0 $mgL^{-1}$ was found optimum for production of multiple shoots from axillary bud explants. The incubation of cultures on BAP-supplemented medium was found necessary for flowering.

## OVARY CULTURE

LaRue (1942) was the first to attempt culture of ovaries of a few angiosperms. He obtained rooting of the pedicel and a limited growth of the ovaries. Jansen and Bonner (1949) grew the ovaries of *Lycopersicon pimpinellifolium* on a medium supplemented with casein hydrolysate, IAA and a mixture of B vitamins.

Maheshwari (1958) cultured the ovaries of *Iberis amara* (Brassicaceae) excised from flowers one day after pollination. *In vitro* formed fruits were even larger than the *in vivo* formed fruits when IAA was added to the mineral salt–sugar–vitamin medium.

In *Tropaeolum majus*, the ovaries excised two days after pollination were cultured on Nitsch Medium (NM) containing, vitamins, glycine and growth promoters. The ovaries grew normally but the size of test tube fruits was smaller than those developed *in vivo* (Sachar and Kanta, 1958).

In *Althea rosea* (Chopra, 1962), the unpollinated ovaries cultured on modified NM + IBA (20/mg/l) and NM + KN (0.5 mg/I) + IAA (μmg/I) produced parthenocarpic fruits (Table 16.4).

In *Hyoscyamus niger* ovaries excised three days after pollination (with calyx intact developed into fruits within four weeks on WM. When the calyx was removed, the fruits remained much smaller, seedless and malformed.

Nitsch (1963) reported that roots have a stimulatory effect on fruit growth. He cultured the ovaries of tomato with a small pedicel and induced rooting of the pedicel by adding tomato juice in the medium.

**Gynogenic haploids:** Originate from unfertilized egg or occasionally other cells of the embryo sac from cultured unpollinated ovaries/flowers.

**TABLE 16.4**
**Effect of Calyx Removal on Cultured Ovaries of *Althea rosea* (after Chopra, 1962)**

| Characteristics | Calyx Intact | Calyx Removed |
|---|---|---|
| Fruit diameter | 19 mm | 12 mm |
| Endosperm | Cellular | Nuclear |
| Embryo | Dicotyledonous | Heart-shaped |

**Stenospermocarpy:** When pollination and fertilization trigger ovary development, but the ovule/embryo aborts without producing mature seed.

## OVULE CULTURE

White (1932), for the first time, cultured the ovules of *Antirrhinum* and observed callus formation from the cells of integument. LaRue (1942) cultured the ovules of *Erythronium americanum* and observed an increase in size but the seeds failed to develop. N. Maheshwari (1958) obtained normal, mature seeds of *Papaver somniferum* by culturing the fertilized ovules on NM. In *Zephyranthes* (Sachar and Kapoor, 1959) reported that the embryo failed to develop beyond globular stage when ovules were cultured on NM alone. In such cases endosperm formation was also inhibited.

Poddubnaya-Arnoldi (1959, 1960) successfully grew the ovules from pollinated ovaries of many orchids on a medium containing only 10% sucrose. Rangaswamy (1961) cultured the ovules of *Citrus microcarpa* and observed that the nucellar embryos showed development and differentiation on WM containing 5% sucrose. However, the embryos dissected out from such ovules showed normal growth only on addition of casein hydrolysate (400 mg/l) to the medium. In *Abelmoschus esculentus* (Bajaj, 1964), the ovules bearing proembryos developed into seeds and even germinated *in situ*, on a medium containing CH (500 mg/l) + coconut milk (10–25%).

Guha and Johri (1966) cultured the ovules of *Allium cepa* before and after pollination on NM + IAA + GA + 2% sucrose and observed that ovules failed to develop. They further observed that ovules cultured after 13 days after pollination on NM developed normally and germinated *in situ* after 20 days. The ovules of *Trifolium repens* cultured at zygote (or the two-celled proembryo stage) developed into mature seeds only if the medium was supplemented with the juice prepared from the young fruits of cucumber or watermelon (Nakajima et al., 1969). In *Gossypium* (Joshi and Johri, 1972) six-day old ovules when cultured on medium containing KN developed folded cotyledons in embryos. Beasley (1971, 1973) cultured the unfertilized ovules of cotton on kinetin-enriched medium. The immature ovules enlarged and produced fibres if the medium was supplemented with IAA or GA or both.

Somatic embryogenesis has been observed in cultured ovules of *Ribes rubrum* and *Carica papaya*. Placental tissue plays a significant role on the development and maturation of seeds. This effect may be either due to increased surface absorption or the influence of some growth regulators present in it. In *Papaver somniferum* normal development of ovules occurred when cultured with placenta.

When *Gossypium arboreum* was crossed with *G. hirsutum*, hybrid embryos failed to develop. Pundir (1972) excised the ovules three days after pollination and cultured them on MS medium containing 50 mg/l of inositol. After seven weeks, 70–80% of ovules developed hybrid seedlings. In *Gossypium hirsutum*, excised ovules on the day of anthesis or a day later either failed to develop or developed a few fibres, but the ovules cultured on the second day of anthesis produced more fibres. Ovule culture assumes great significance in plant breeding. The hybrids which usually fail to develop either due to abortion of embryo at an early stage or premature abscission of fruits, could be obtained by ovule culture.

Uchimiya et al. (1971) cultured unfertilized ovules of *Solanum melongena*. They obtained vigorous callus formation on a medium supplemented with IAA and kinetin. In *Gerbera hybrida* (Li et al., 2020), haploid induction was obtained via unpollinated ovule culture. During the process of ovule induction, three distinct stages were identified from ovules to regenerates: (i) gradual expansion of the ovule (Figure 16.9A), (ii) callus formation (Figure 16.9B), and (iii) adventitious bud induction (Figure 16.9C). A small number of ovules are directly induced

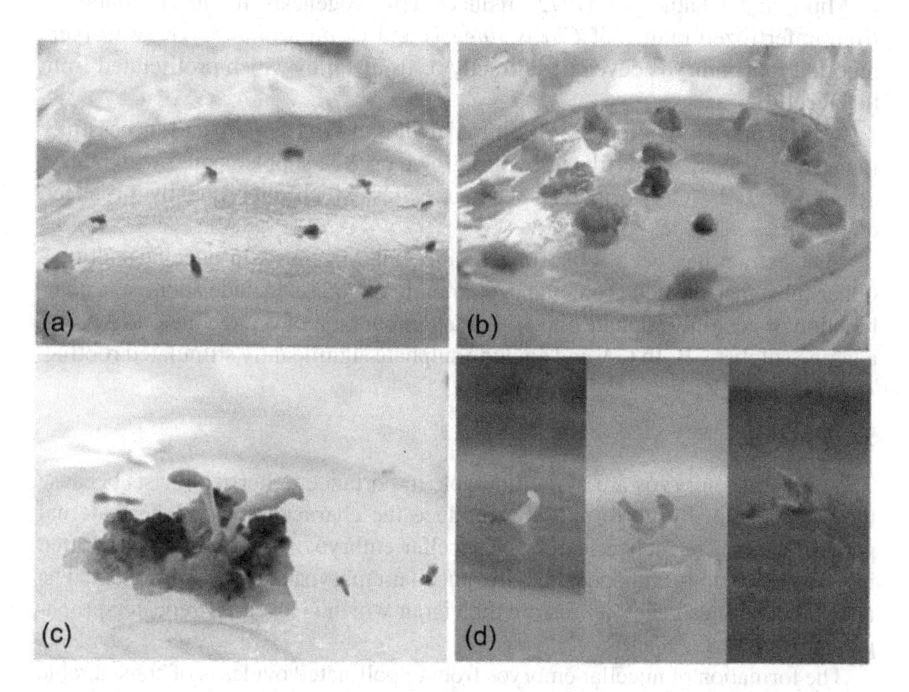


**FIGURE 16.9** The stages of ovule induced to form adventitious buds in *Gerbera hybrida* (after Li et al., 2020).

to form adventitious buds (Figure 16.9D). When adventitious buds formed in this manner, the induction time was shorter than that of buds formed via callus formation.

## NUCELLUS CULTURE

Rangaswamy (1958) excised nucellus from pollinated ovules of *Citrus microcarpa* and cultured it on a modified WM supplemented with CH. The nucellus proliferated and produced embryo-like structures (termed pseudobulbils) which finally gave rise to plantlets. Sabarwal (1963) obtained similar results in *Citrus reticulata* but opined that the pseudobulbils developed from nucellar embryos. Rangan et al. (1969) demonstrated that adventive embryos were successfully initiated in nucellus cultures of three monoembryonic species of *Citrus* (*Citrus grandis, C. limon* and *C. reticulata* x *C. sinensis*). Besides *Citrus, in vitro* nucellar embryogenesis is also reported in *Aegle marmelos, Agrostemma githago, Cynanchum vincetoxicum, Malus domestica, Poncirus trifoliata,* and *Vitis vinifera.*

Button and Bornman (1971) observed that excised nucellus tissue from unfertilized ovules can also form embryoids under suitable culture conditions. These embryoids may form plantlets if excised embryoids were planted individually on a medium supplemented with $GA_3$. Button and Bornman could successfully transplant these plantlets to soil.

Mitra and Chaturvedi (1972) induced embryogenesis in nucelli obtained from unfertilized ovules of *Citrus sinensis* and *C. aurantifolia.* These workers reported that embryos developed directly or from callus which proliferated from the nucellus.

Mullins and Srinivasan (1976) cultured ovules of *Vitis vinifera* and successfully induced embryony. Unfertilized ovules grown on NM supplemented with benzyl adenine and B-NOA formed a nucellar callus which subsequently produced embryoids.

Malt extract has been found to promote embryogenesis in nucellar cultures. Other substances which promote nucellar embryogenesis include adenine, a combination of adenine sulphate and KN or a combination of orange juice, NAA and adenine sulphate. Both GA and adenine sulphate significantly stimulated rooting.

### SIGNIFICANCE

The adventive embryos are of considerable importance to horticulturists because they are genetically uniform and reproduce the characteristics of the maternal parent. Another attractive feature of nucellar embryos is that cuttings are often infected by various pathogens, but the nucellar embryos are free from them. The nucellar seedlings generally restore the vigour which is lost after repeated propagation by cuttings.

The formation of nucellar embryos from unpollinated ovules is of considerable interest in that it can be employed to import seedless *Citrus* varieties without the risk of introducing new viral diseases.

## ENDOSPERM CULTURE

The first attempt to grow endosperm tissue was by Lampe and Mills (1933). They cultured the endosperm of maize after 10–25 days of pollination in a medium containing extract of potato or young corn kernels. The endosperm callused on the side towards the embryo. However, the credit for the first elaborate and systematic study of endosperm tissue under cultural conditions goes to LaRue (1949), who succeeded in obtaining callus from immature maize endosperm. LaRue used several media with different combinations of nutrients for culturing the maize endosperm, but only in a few instances was he successful in inducing roots – and only in one instance both root and shoot. Since then, several workers have cultured the endosperm of cereals (maize, wheat, barley) but have failed to obtain organogenesis (Johri and Rao, 1984).

### Culture of Immature Endosperm

The cereal endosperm proliferates only if excised during a proper period of development (Table 16.5). Endosperm younger than 7 days after pollination (DAP), or older than 11 DAP, failed to proliferate. It is suggested that, in cereals, the endosperm undergoes certain changes after 12 days after pollination and is incapable of responding to treatments in culture. Sturani and Cocucci (1965) observed that there is marked decrease in almost all enzyme activities in the endosperm of maturing seeds.

### Culture of Mature Endosperm

The mature endosperm of several taxa has now been grown successfully in culture, but taxa belonging to families Euphorbiaceae, Loranthaceae and Santalaceae respond better in cultures. Rangaswamy and Rao (1963) established tissue cultures from the mature endosperm of *Santalaun album*. This report was followed by similar observations on *Ricinus communis*, *Codium variegatum*, *Putranjiva roxburghii* and several other plants.

**TABLE 16.5**
**Proliferation of Endosperm Tissue**

| Species | Endosperm Proliferation (days after pollination) |
|---|---|
| *Lolium perene* | 9–10 |
| *Zea mays* | 8–11 |
| *Triticum aestivum* | 8 |
| *Hordeum vulgare* | 8 |
| *Oryza sativa* | 4–7 |

## Culture Medium

Generally, the mature endosperm of parasitic taxa shows optimal growth on a medium containing either only a cytokinin or a cytokinin+an auxin. Tamaoki and Ullstrup (1958) obtained satisfactory growth of maize endosperm on NM supplemented with yeast extract, whereas that of *Cucumis* (Nakajima, 1962) was grown on WM supplemented with 1,3-diphenylurea, yeast extract and/or IAA. Yeast extract appears to be essential for the growth of mature endosperm of many flowering plants. The immature endosperm also required complex substances like CH, YE or CM as well as cell division factors.

## Organogenesis in Endosperm Cultures

Organ formation from endosperm tissue was first demonstrated in *Exocarpus cupressifromis* (Santalaceae) by Johri and Bhojwani (1965). They observed that in "seed" cultures of *Exocarpus* on a medium supplemented with IAA, kinetin and CH, 10% cultures formed shoot buds all over the endosperm. In a single explant as many as eight buds developed.

In endosperm cultures of *Scurrula* and *Taxillus* a cytokinin alone brings about bud differentiation, but in the cultures of endosperm of *Dendrophthoe* and seedling tissues of *Lactuca* kinetin is effective only in the presence of a low level of IAA. Endosperm pieces of *Taxillus vestitus*, pre-soaked in 0.025% kinetin solution for 24 hrs, differentiated buds even on White's basal medium (lacking kinetin). In *Dendrophthoe* and *Leptomeria acida* casein hydrolysate had a promotive effect on shoot–bud differentiation (Nag and Johri,1971). Organogenic differentiation in endosperm cultures has been achieved in several autotrophic and parasitic taxa (Table 16.6).

## Applications of Endosperm Culture

Triploid plants of endosperm origin have implications in plant breeding programmes. Endosperm culture technique can be employed to raise superior triploid plants of many economically important plants whose triploids are presently in commercial use. These include several varieties of apple, banana, mulberry,

**TABLE 16.6**
**Species Reported to Form Shoots or Whole Plants from Endosperm Tissue**

| Families | Species |
|---|---|
| Apiaceae | *Pteroselinum hortense* |
| Euphorbiaceae | *Codiaeum variegatum, Putranjiva roxburghii* |
| Loranthaceae | *Dendrophthoe falcata, Scurrula pulverulenta, Taxillus vestitus* |
| Poaceae | *Oryza sativa* |
| Santalaceae | *Exocarpus cupressiformis, Santalum album* |

sugar beet, tea and watermelon. The cultured endosperm tissue has been shown to retain the ability to support the growth of young embryos. This feature of the cultured endosperm tissue may allow rearing of full plants from isolated very young hybrid embryos which normally abort prematurely.

## EMBRYO CULTURE

Embryo culture is the culture of isolated immature or mature embryos with an ultimate objective of obtaining viable and fertile plants. Embryos were the first plant tissue to be successfully cultured *in vitro* on artificial media. The first result in successful *in vitro* culture was by Hannig (1904) who grew under aseptic conditions relatively mature embryos of *Raphanus caudatus, R. landra, R. sativus* and *Cochlearia danica* (Brassicaceae) on Tollen's medium with sugars, amino acids and plant extracts and obtained transplantable seedlings.

Andronescu (1919) reported that maize embryos when cultured without their endosperm, gave rise to smaller plantlets than embryos cultured with their endosperm. Knudson (1922) succeeded in growing orchid embryos into plantlets by culturing them on agar medium containing sugar. These embryos were cultured without association of a symbiotic fungus which is an essential prerequisite for the germination of orchid seeds in nature. While working with embryos from a range of families and genera, Dietrich (1924) made two generalizations: (i) the embryos grown *in vitro* usually skipped a rest period and germinated; (ii) the embryos taken into culture at less than one-third of mature size were unable to produce viable plants capable of reaching maturity. Embryos from immature seeds tended to form malformed seedlings. Laibach (1925, 1929) demonstrated that in *Linum perenne* x *L. austriacum*, the seeds obtained were very light, shrunken and incapable of germination. He could raise the hybrid plants from the embryos.

Embryo culture can be broadly classified in two ways, namely on the basis of stage and histological origin of the isolated embryo explants. On the basis of stage of the isolated embryo explants, embryo culture is classified into mature embryo culture and immature embryo culture (embryo rescue). Based on the histological origin of the isolated embryo explants, embryo culture has been classified into zygotic and somatic embryo culture (Mondal et al., 2020).

## CULTURE REQUIREMENTS

### 1. Mineral Salts

Several formulations of mineral salts have been used for embryo culture. Umbeck and Norstog (1979) reported that $NH_4^+$, was essential for proper growth and differentiation of immature barley embryos. Embryos of jute showed an absolute requirement for nitrate nitrogen.

2. **Carbohydrate**

Carbohydrate is one of the most important ingredients required for the proper growth of cultured embryos. Sucrose is the most commonly used carbohydrate for embryo culture. Sucrose not only provides energy but also maintains a suitable osmolarity. 8–12% sucrose is generally adequate for the culture of proembryos (e.g., *Datura*, *Hordeum*). Raghavan and Torrey (1964) also noted a stimulatory effect of higher concentrations (12–18%) of sucrose on *in vitro* development of excised globular embryos of *Capsella.*

3. **Amino Acids**

The growth of the embryo may be stimulated by adding amino acids to the culture medium. Glutamine has been found to be the most effective amino acid for promoting the growth of excised embryos. Casein hydrolysate, an amino acid complex, is being widely used as an additive to embryo culture media.

4. **Natural Plant Extracts**

Addition of coconut milk (CM) in the medium has proved to be very beneficial for the development of the embryo. In barley, undifferentiated embryos (about 100 cells) were successfully cultured by supplementing the medium with 20% coconut milk. Extracts of dates and bananas, tomato juice and wheat glutene hydrolysate have also been found to promote the growth of excised embryos of barley.

5. **Growth Regulators**

An exogenous supply of growth regulators is essential for the normal development of excised embryos. Low concentrations of auxins promote growth of roots and shoots in pea embryos and also in *Avena sativa*, but at high concentrations, growth rate declines.

6. **pH**

Excised embryos grow well in media with pH 5.4–7.5. Rice embryos grow well at two pH values, i.e., 5 and 9.

## CULTURE TECHNIQUES

Raghavan and Torrey (1963) employed the following procedure for isolating embryos at different stages of development from the ovules of *Capsella bursa-pastoris.* Sterile capsules were kept in a few drops of the sterile culture medium. An incision was made in the placental region and the two halves were separated apart with the help of a forceps. Ten excised immature embryos and an ovule were removed from the placenta and was cut longitudinally; by careful teasing, the entire embryo, along with the attached suspensor, could be removed (Figure 16.10).

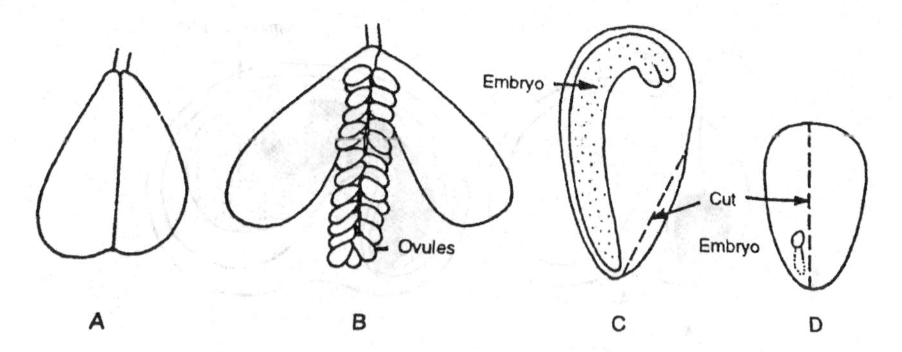


**FIGURE 16.10** Isolation of the embryos of *Capsella bursa-pastoris* (after Raghavan, 1966).

In *Manihot esculenta* (Lentini et al., 2020) individual carpels containing one ovule were cultured. This allowed the culture of immature zygotic embryos without physical injury. Embryos developed inside the ovule up to cotyledonary stage. Embryos were then dissected out from the ovules and were cultured on the medium. Fully grown plantlets were obtained using this technique.

## Embryo-Nurse Endosperm Transplant

Williams and De Leutour (1980) developed a technique of endosperm transplant for culture of hybrid embryos. The excised hybrid embryo was inserted into a cellular endosperm dissected out from a normally developing ovule (Figure 16.11). Endosperm with the transplanted embryo was then cultured on culture medium. Using this technique Williams and De Leutour produced several interspecific hybrids in *Trifolium*.

## Culture of Proembryos

Earlier workers observed that the immature embryos usually failed to survive on transfer to artificial media which supported growth of mature embryos. Young heart-shaped embryos of *Portulaca oleracea* cultured on a medium containing mineral salts, glucose and fibrin digest developed into nearly mature embryos. Young embryos of *Cuscuta reflexa*, when cultured on a medium containing IAA and CH, proliferated and produced numerous embryoids. Upon transfer to fresh medium, these embryoids formed normal shoots.

## Culture of Mature Embryos

Mature embryos require an extremely simple medium for their growth and differentiation. The mature embryos of *Amyema miquelii* and *A. pendula* cultured on a medium containing IAA and CH germinated in 3–5 weeks producing a massive holdfast and a pair or plumular leaves. The mature embryo (6–7 mm length) of

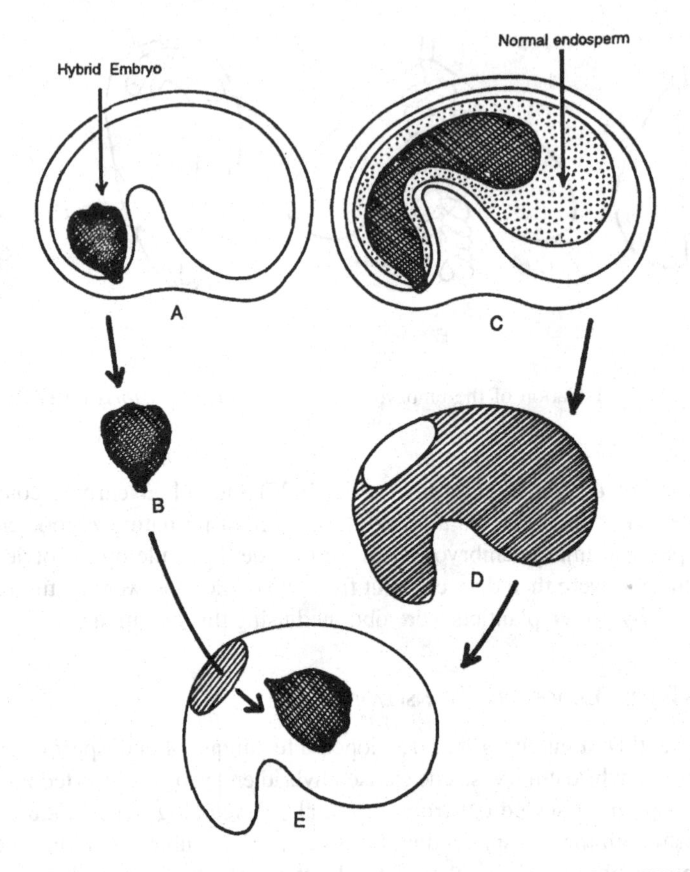


**FIGURE 16.11** Endosperm transplant for culture of hybrid embryos (after Williams and De Leutour, 1980).

*Allium cepa* when cultured on NM with 5% sucrose, germinated *in situ* after five days. The embryo of mature coconut (*Cocos nucifera*) grows well on MS medium.

## Culture of Embryonal Segments

The essentiality of cotyledons for normal growth of embryos is demonstrated in a large number of taxa. Muzik (1956) observed that it was necessary to culture the embryonal axis with at least two-thirds of the cotyledons for normal growth. The mature embryos of maize and other cereals cultured without the scutellum did not attain normal growth nor root and shoot primordia.

Rangaswamy and Rangan (1963, 1971) studied the effects of dicotyledon on embryos of *Cassytha filiformis* and reported that there is direct correlation between the partial or entire removal of the cotyledons and inhibition of morphogenesis of the shoot (Table 16.7). Kanta and Padmanabhan (1964) cultured embryonal segments of *Cajanus cajan* containing unorganized axillary bud

**TABLE 16.7**
**Growth Response of Decotylated Embryo Segments of *Cassytha filiformis* (Rangaswamy and Rangan, 1971)**

| Portions of Cotyledon(s) Severed | Response |
|---|---|
| 1. Plumular half of one cotyledon | Seedling well developed |
| 2. Radicular half of one cotyledon | Seedling well developed |
| 3. Plumular half of each cotyledon | Seedling well developed |
| 4. One entire cotyledon and plumular half of other cotyledon | Hypocotyl and plumule both inhibited |
| 5. One entire cotyledon and radicular Half of other cotyledon | Plumule quiscent |
| 6. Radicular half of each cotyledon | Plumule quiescent |
| 7. Radicular half of one cotyledon and plumular half of other cotyledon | Growth of hypocotyl limited, plumule inhibited |
| 8. Small portion from the equatorial region of each cotyledon | Seedling formation occasional |
| 9. Entire cotyledons retaining only a belt around the plumular pole of embryonal axis | Growth inhibited, plumule quiscent |
| 10. Both cotyledons leaving only the embryonal axis | Growth suppressed, plumule quiescent, rooting rare |

initials. By culturing separately, the plumule and two-axillary bud primordia of the cotyledonary node, they could obtain three seedlings from a single embryo.

## PRACTICAL APPLICATIONS

Embryo culture technique is used for culture where the embryo is incompletely developed (e.g., orchids), to overcome seed dormancy, for shortening the breeding cycle, determination of seed viability, microcloning of the source material and for rapid multiplication and conservation of rare, threatened and endangered species.

### Raising Rare Hybrids

In many intergeneric and interspecific crosses, the embryo shows normal development in early stages, but due to poor or abnormal endosperm, premature death of hybrid embryos take place. Hybrid plants can be raised by excising the embryos before the onset of abortion and culturing them on nutrient medium. The cross *Lycopersicon peruvianum* × *L. esculentum* is not successful due to fertilization barriers. Thomas and Pratt (1981) obtained hybrid plants of tomato using the embryo culture technique. Hybrid plants have been raised from the intergeneric crosses *Hordeum* x *Secale*, *Hordeum* x *Triticum* and *Hordeum* x *Agropyron* using embryo culture method (Kruse, 1974).

### Shortening of the Breeding Cycle

Long seed dormancy periods pose problems in breeding programmes. Embryo culture is a viable tool to plant breeders because this method can reduce the breeding cycle. By growing excised embryos on culture medium, the breeding cycle can be considerably reduced. For example, rose plants usually take one year to come to flower. Using embryo culture technique, it has become possible to produce two generations in a year. In *Iris* it has been possible to shorten the life cycle from seed to flowering by embryo culture.

### Determination of Seed Viability

Germination of excised embryos is regarded as a more reliable test for determining the seed viability than usual staining methods. In peach seeds and many cereals, long dormancy periods pose problems in testing the viability. In such cases, embryo culture technique can help in determining the seed viability.

### Propagation of Rare Plants

Some coconuts develop soft, fatty tissue in place of liquid endosperm as an abnormality. Such coconuts are termed "Makapuno". These nuts are very expensive and served only at special banquets in the Philippines. Under normal conditions the coconut seeds do not germinate. De Guzman and Del Rosario (1974) obtained plantlets from makapuno nuts through *in vitro* culture of embryos.

### Haploid Production

Embryo culture has been used for recovering haploid plants by the elimination of a set of chromosomes. Kasha and Kao (1970) obtained a high frequency of haploids in barley using embryo culture technique.

## SOMATIC EMBRYOGENESIS

Somatic embryogenesis is the process by which the somatic cells or tissues develop into differentiated embryos and each fully developed embryo is capable of developing into a plantlet. Haccius (1978) defined somatic embryogenesis as "a non-sexual developmental process which produces a bipolar embryo from somatic tissue".

Somatic embryogenesis was first demonstrated in 1958 simultaneously by Steward (USA) and Reinert (Germany). Since then, somatic embryogenesis has been reported in a large number of flowering plants. Plants belonging to families Solanaceae, Ranunculaceae, Apiaceae, Poaceae and Rutaceae respond better in producing somatic embryos in culture compared to other families.

Somatic embryogenesis is of two types: direct embryogenesis and indirect embryogenesis:

i. **Direct embryogenesis:** Refers to the development of an embryo directly from the original explant tissue (e.g., nucellar cells of polyembryonic varieties of *Citrus*, epidermal cells of hypocotyl in wild carrot, *Ranunculus sceleratus*, Figure 16.12).
ii. **Indirect embryogenesis:** Refers to the formation of embryos from callus or cell suspension or from cells or groups of cells or somatic embryo (e.g., secondary phloem of domestic carrot, inner hypocotyl tissues of wild carrot, leaf tissue explants of coffee). A comparative account of direct and indirect embryogenesis is given in Table 16.8.

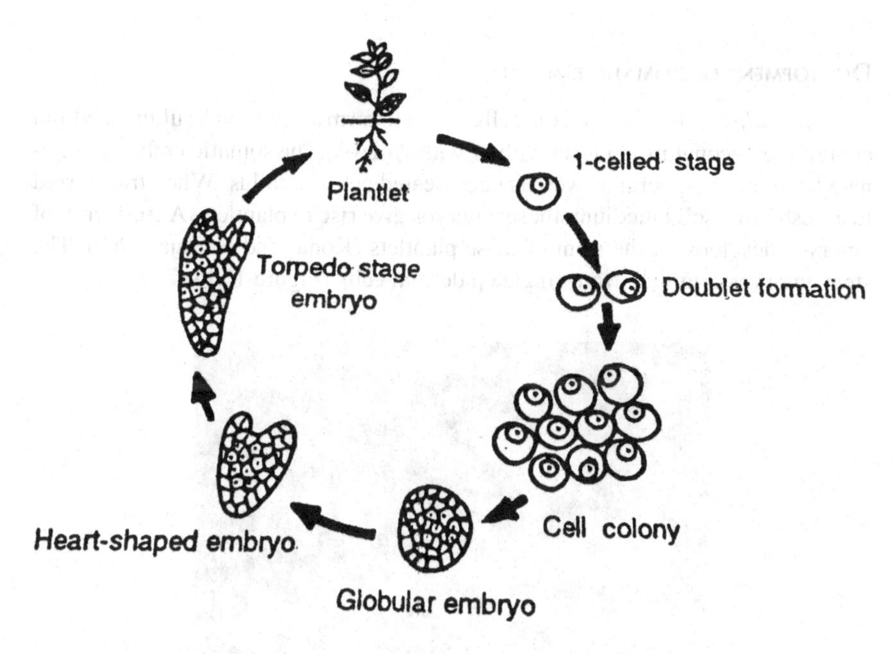


**FIGURE 16.12** Different stages of somatic embryogenesis

**TABLE 16.8**
**Comparison of General Patterns of Embryogenic Development**

| Direct Embryogenesis | Indirect Embryogenesis |
|---|---|
| 1. Development of embryo directly from explant tissue | 1. Callus proliferation occurs prior to embryo development. |
| 2. Direct embryogenesis proceeds from Pre-embryogenic determined cells (PEDC) | 2. Indirect embryogenesis requires the embryogenic determined cells (IEDC). |
| 3. PEDCs require synthesis of an inducer substance/removal of an inhibitory substance prior to resumption of mitotic exposure activity and embryo development. | 3. Cells undergoing IEDC differentiation require a mitogenic substance to re-enter the mitotic cell cycle and/or to specific concentration of growth regulators. |

The process of embryogenesis involves various stages of differentiation and development such as proembryo, globular, heart-shaped and torpedo stage embryo (Figures 16.12). Embryos formed in cultures have been referred to as embryoids, supernumerary embryos, adventive embryos and accessory embryos. Somatic embryogenesis is more or less similar in structure, function and biochemistry to that of zygotic embryogenesis (embryos derived from zygote). Somatic embryos share many characteristics with zygotic embryos. Both are bipolar structures, originate from a single cell and lack vascular differentiation (Figure 16.13).

## Development of Somatic Embryos

In *Ranunculus sceleratus* somatic cells divide to form a callus on a culture medium containing coconut milk (10%) with or without IAA. The somatic embryos originate from the peripheral as well as deep-seated cells of callus. When transferred to a fresh semi-solid medium, these embryos give rise to plantlets. A fresh crop of embryos develops on the stem of these plantlets (Konar and Natraja, 1969). The stem embryos originate from single epidermal cells (Figure 16.14).

**FIGURE 16.13** Different stages of somatic embryos from cotyledonary segment of *Melia azedarach* (courtesy: Dr C.R. Deb, Nagaland).

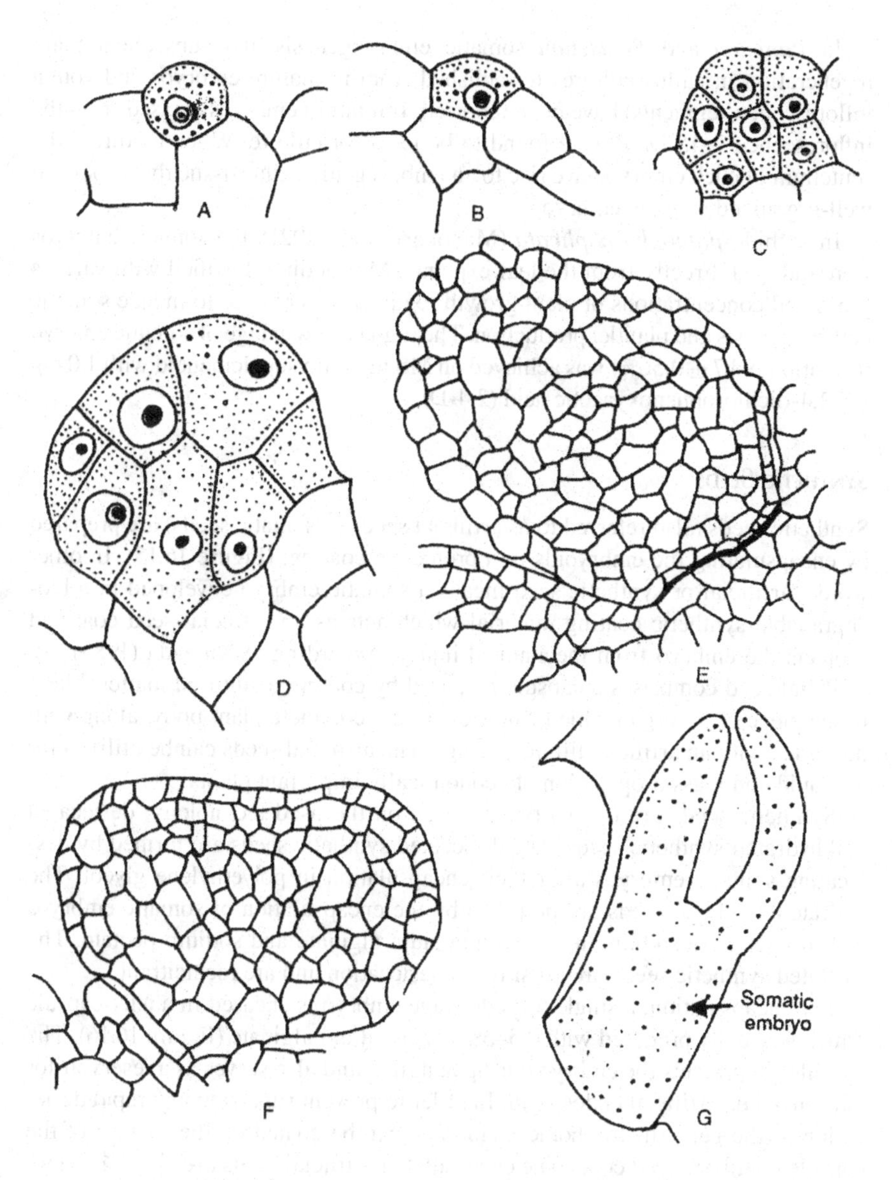


**FIGURE 16.14** Stages in embryo development from epidermal cells of *Ranunculus sceleratus*.

In *Trifolium repens* direct somatic embryogenesis occurred by the proliferation of hypocotyl epidermal cells of young cotyledonary (torpedo) embryos dissected 8–10 DAP on an EC6 medium containing 0.05 mg BAP (6-benzyl aminopurine). After three weeks, embryoids were separated from the parent embryo tissues and rooted on basal medium without BAP. Plantlets were transferred to soil after 1–2 weeks.

In *Panicum* and *Penisetum* somatic embryogenesis and subsequent plant regeneration in callus cultures (established from immature embryos and young inflorescence segments) have been reported. Immature embryos excised from the inflorescence 7–15 DAP were found to be ideal for culture. When cultured, the scutellum of each embryo gave rise to an embryogenic callus tissue that produced well-organized somatic embryos.

In orchid *Spathoglottis plicata* (Manokari et al., 2021), the somatic embryos were induced directly from the leaf explants. MS medium fortified with various types and concentrations of plant growth regulators were used to induce somatic embryogenesis and plantlet production. The highest percentage of somatic embryo formation ($93.7 \pm 0.56\%$) was achieved on MS medium supplemented with 1.0 mg $L^{-1}$ 2,4-dichlorophenoxyacetic acid (2,4-D).

## Synthetic Seeds

Synthetic seeds (also referred to as artificial seeds or somatic seeds) are prepared by encapsulating the embryoids in a protective coating. (Figure 16.15). In other words, artificial or synthetic seed means a somatic embryo enveloped in a biodegradable synthetic coating material which acts as an artificial seed coat and protects the embryo from mechanical injury. According to Kamada (1985) "an artificial seed comprises a capsule prepared by coating a cultured matter, like a tissue piece or an organ which can grow into a complete plant body, along with nutrients, with an artificial film/covering". The artificial seeds can be utilized for the rapid and mass propagation of economically important plants.

Synthetic seeds are of two types, based on the used technology: desiccated and hydrated synthetic seeds. The desiccated synthetic seeds are formed by desiccating somatic embryos after their encapsulation in polyethylene glycol. The hydrated synthetic seeds are prepared by the encapsulation of somatic embryos in hydrogels like sodium alginate, potassium alginate and sodium pectate. The hydrated synthetic seeds are sensitive to desiccation and are recalcitrant.

For encapsulation, a single torpedo stage embryo is encased in a jelly capsule which is usually prepared with 0.5–5% (w/v) sodium alginate (Figure 16.16). The capsule gel protects the embryo during handling and also serves as a reservoir for nutrients, i.e., artificial endosperm. In order to prevent rupturing and rapid desiccation of the gel, a hydrophobic membrane can be coated on the surface of the capsule (artificial seed coat). The encapsulated artificial seeds are about 4–6 mm in diameter. These seeds are spherical and transparent. They are stored in closed airtight containers to avoid desiccation of the capsules. Complete plants have been obtained through encapsulated artificial seeds in a number of plant species including *Zea mays*, *Oryza sativa*, *Lactuca sativa*, *Daucus carota*, *Brassica* species, *Gossypium* species, *Asparagus* species (Table 16.9). The recovery of alfalfa plants from encapsulated artificial seeds is quite encouraging because of high conversion frequency of 70–90% under culture conditions and about 20% under non-sterile soil conditions. Mature plants of alfalfa have been obtained through artificial seeds.

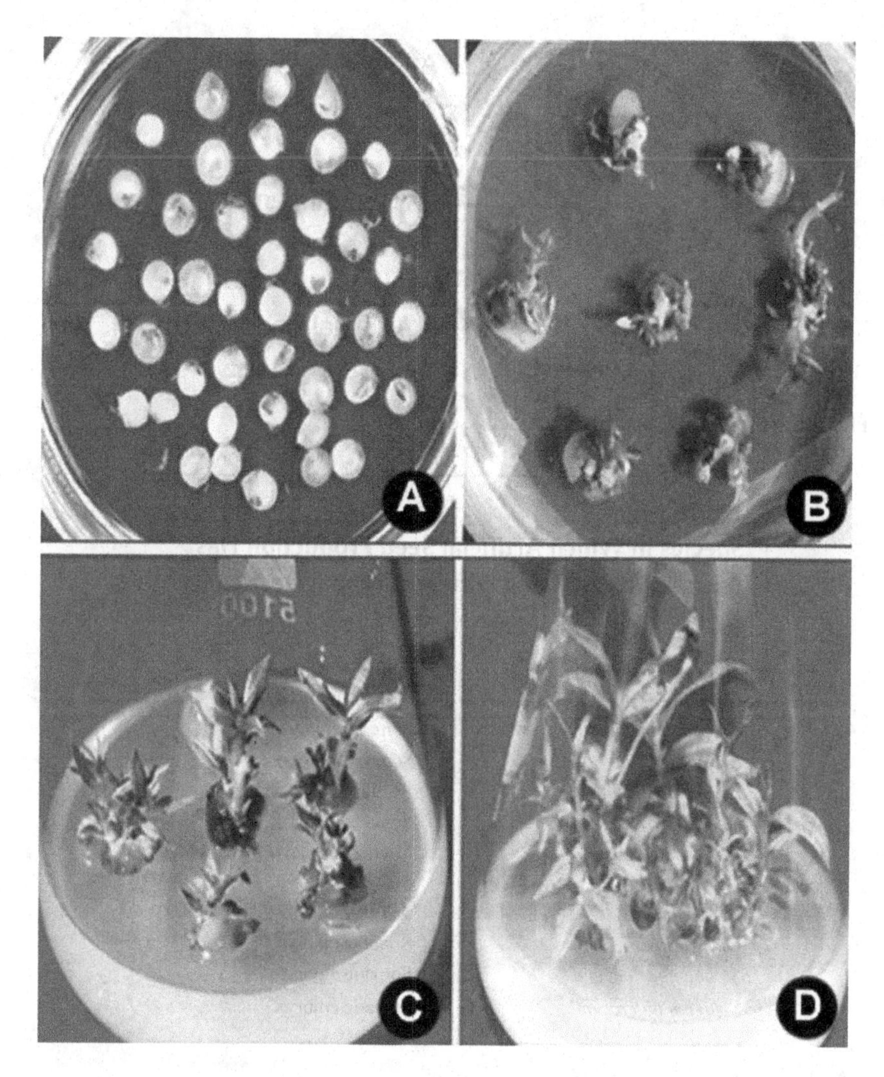


**FIGURE 16.15** Synthetic Seed Production and regeneration. A. Synthetic seeds; B. Germinating synthetic seeds; C, D. Plantlet formation (courtesy: Dr M. Anis, Aligarh).

## Encapsulation Methods

Redenbaugh et al. (1987) developed a method of encapsulation by gelation. In this method, hydrogel capsules are prepared by the dropping technique described below.

Somatic embryos are mixed with sodium alginate (2% w/v) and roped using a plastic pipette into a complexing salt such as calcium nitrate (100 mM). Normally, gelation of capsules takes about 20 minutes time for incubation. Somatic embryos may also be inserted into drops of sodium alginate as the drops fell from a separatory funnel into a solution of 100 mM calcium nitrate (Figure 16.17).

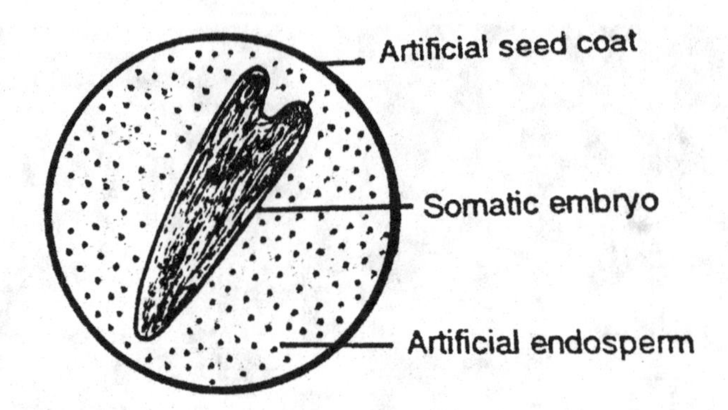


**FIGURE 16.16** The concept of artificial seed

**TABLE 16.9**
**Species in Which Artificial Seed Production Has Been Achieved Using Different Propagules**

| Species | Propagule Type |
|---|---|
| *Apium graveolens* | Somatic embryo |
| *Citrus* species | Somatic embryo |
| *Curcuma amada* | Somatic embryo |
| *Daucus carota* | Somatic embryo |
| *Dioscorea floribunda* | Axillary buds |
| *Gossypium hirsutum* | Somatic embryo |
| *Hordeum vulgare* | Microspore-developed embryo |
| *Lactuca sativa* | Somatic embryo |
| *Morus indica* | Axillary buds |
| *Oryza sativa* | Somatic embryo |
| *Solanum melongena* | Somatic embryo |
| *Solanum tuberosum* | Somatic embryo |

## Desiccated Coated or Non-coated Seeds

Kitto and Janick (1982) for the first time reported methodology for desiccated coated and non-coated artificial seeds. In this process, the somatic embryos are usually damaged which is essential for dehydration and rehydration protocols. The desiccated uncoated embryos may also be useful as they eliminate storage problems and do not require a membrane coating.

## Practical Applications of Artificial Seeds

1. Artificial seed technology can be useful for the propagation of crop plants especially in cases where true seeds are not used as planting

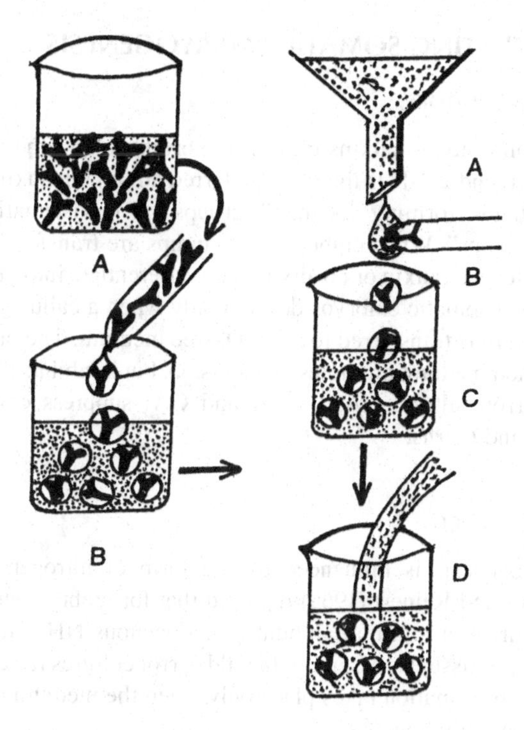


**FIGURE 16.17** Preparation of artificial seeds via encapsulation with calcium alginate beads using a dropping method.

material (e.g., potato) or true seeds are costly (e.g., cucumber) or hybrid plants (e.g., hybrid rice). In crops like sugarcane, sweet potato, grape, mango etc. which are vegetatively propagated and more prone to diseases, artificial seeds would be highly beneficial.

2. Artificial seeds would be useful for the multiplication of transgenic plants, somatic hybrids, cybrids and sterile and unstable genotypes. Artificial seeds would be useful material for preservation of desired genotypes. They can be directly delivered to the field thus avoiding transplantation and tissue handling steps.

## ADVANTAGES OF SYNTHETIC SEEDS

- Short-and long-term storage capacity
- Genetic uniformity
- Cost effective
- Easy transport to long distances
- Germplasm conservation

## FACTORS AFFECTING SOMATIC EMBRYOGENESIS

### 1. Growth regulators:

Of all the phytohormones, auxins play an important role in the development of somatic embryos, and 2,4-D is the most preferred auxin. On auxin-rich medium, callus differentiates forming localized groups of meristematic cells called "embryogenic clumps". When embryogenic clumps are transferred to a medium with a very low level of auxin or no auxin, they differentiate into mature embryos. In *Coffea arabica* somatic embryos develop only when a callus grown on 2,4-D containing medium is transferred to a 2,4-D–free medium. The presence of BAP in the proliferation medium promotes cell division, but it inhibits the embryogenic potential of carrot cultures. IAA, ABA and $GA_3$ suppress embryogenesis in *Daucus carota* and *Citrus.*

### 2. Nitrogen source:

*In vitro* embryogenesis is influenced by the form of nitrogen present in the medium. Tazawa and Reinert (1969) reported that for embryogenesis in the cultured cells of carrot, a minimal amount of endogenous $NH_4^+$, (around 5 mmol $kg^{-1}$ fresh weight of tissue) is essential. In wild carrot cultures raised from petiolar segments, embryo formation takes place only when the medium contained some amount of reduced nitrogen.

### 3. Potassium:

The presence of potassium in the medium affects embryogenesis. In domestic carrot, the increase in the level of potassium increases the number of somatic embryos formed. Brown et al. (1976) reported that high potassium (20 mmol $l^{-1}$) is necessary for embryogenesis in wild carrot.

### 4. Dissolved oxygen:

Dissolved oxygen below the critical level of 1.5 $mgl^{-1}$ favours embryogenesis, but higher levels of dissolved oxygen promote rooting.

## PRACTICAL APPLICATIONS OF SOMATIC EMBRYOGENESIS

1. Regeneration of plants through somatic embryogenesis pathway is of special significance in mutagenic studies. Unlike shoots, somatic embryos develop from single cells hence new strains of plants obtained through adventive embryogenesis in tissue cultures would be solid mutants.
2. Somatic embryogenesis can meet specific objectives by rapidly multiplying germplasm that is initially present as embryonic material, e.g.,

maternal embryos, haploid embryos and interspecific hybrid embryos that normally abort due to non-availability of endosperm tissue.

3. Disease-free plants can be raised by using nucellar embryos. In monoembryonate seedless varieties of *Citrus*, disease-free plants have been raised by culturing their nucelli and inducing somatic embryogenesis artificially.

## PROTOPLAST CULTURE AND SOMATIC HYBRIDIZATION

Hanstein (1880) first used the term protoplast (*protos*-means first, *plastos*-means being formed). Protoplasts are spherical naked plant cells obtained by removing the cell wall. Torrey and Landgren (1977) have defined higher plant protoplasts as "cells with their walls stripped-off and removed from the proximity of their neighbouring cells". According to Wu et al. (2009), a protoplast is a transitory state of a cell lacking its cell wall and can be obtained using pectocellulosic enzymes. Plant protoplasts are significant for plant cell culture, somatic cell fusion, genetics and breeding studies. Without cell walls, protoplasts can incorporate materials such as DNA and fuse. Somatic hybrids can be obtained when protoplasts of different species are fused.

## ISOLATION OF PROTOPLASTS

There are principally two methods of protoplast isolation: mechanical and enzymatic.

### MECHANICAL METHOD

Klercker (1892) was the first person to isolate protoplasts from higher plants using the mechanical method. This method involves isolation of protoplasts by cutting plasmolysed tissues with a sharp razor. The protoplasts are released from the cut ends of the cells. Using this procedure, protoplasts could be isolated from highly vacuolated cells of storage tissues, such as onion bulb scales, beet root tissue, etc. The mechanical method suffers from the limitations that only a small number of protoplasts can be obtained and it is not very useful for isolation of protoplasts from mature and meristematic cells.

### ENZYMATIC METHOD

Cocking (1960), for the first time, demonstrated the large-scale isolation of protoplasts from root tips of tomato (*Solanum lycopersicum*, syn. *Lycopersicon esculentum*). He used a concentrated solution of cellulase enzyme, obtained from cultures of the fungus *Myrothecium verrucaria* to degrade the cell walls. A list of commonly used cell wall digesting enzymes is given in Table 16.10.

Ahmed et al. (2021) obtained 88% viable protoplasts from callus tissues in *Jasminum* species. The optimized isolation protocol consisted of digestive callus

**TABLE 16.10**
**Cell Wall Digesting Enzymes**

| Hemicellulases | Pectinases | Cellulases |
|---|---|---|
| Hemicellulase | Pectolyase Y-23 | Meicelase R10 |
| Rhozyme | Macerozyme R-10 | Cellulase RS |
| | | Driselase |

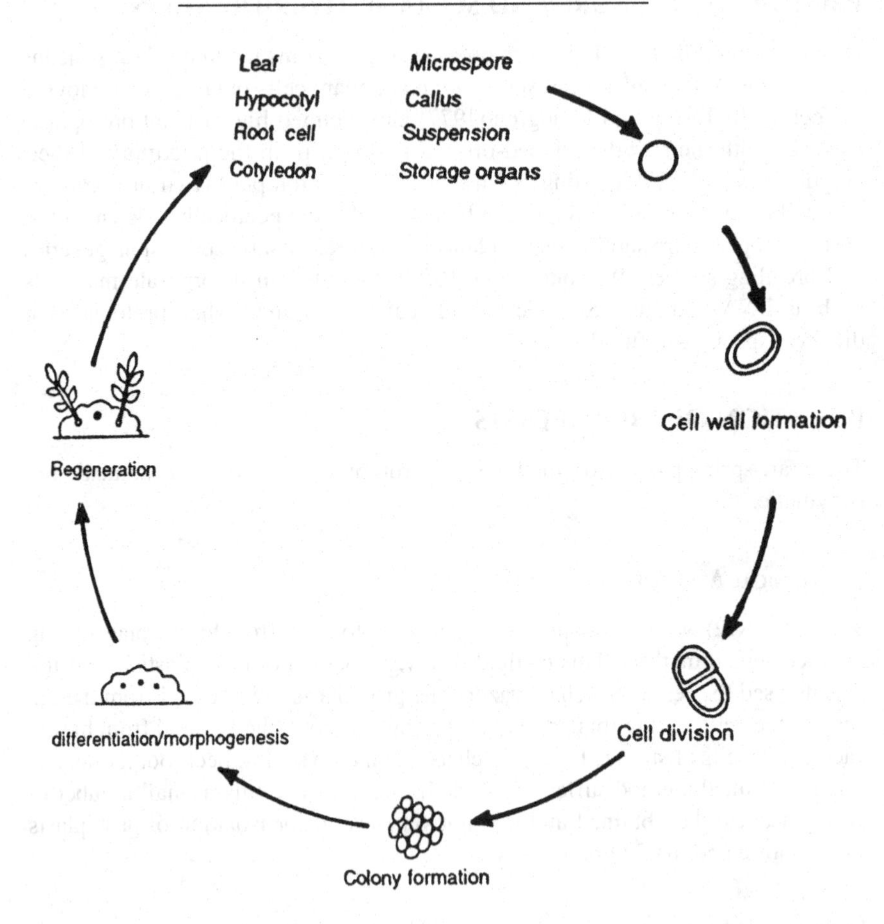


**FIGURE 16.18** Some of the events that occur when protoplasts are isolated and cultured.

in an enzyme solution containing 0.4 M mannitol, 0.2 M MES, 1 M CaCl2, 0.2 M KCL and 1 M NaH2PO4, 1.5% cellulases onozuka R-10, 0.4% Macerozyme R-10 and 0.8% Pectinase for 4 h at 26C in the dark.

Protoplast isolation has been reported from mesophyll cells, root tissues, shoot tips, coleoptile, tubers, petals, microspore mother cells, fruit tissues and cultured cells (Figure 16.18). Isolation of protoplasts through enzymatic method has certain advantages: (i) Large quantities of protoplasts can be isolated easily; (ii) osmotic shrinkage is very low; and (iii) cells are relatively intact and are not injured as observed in the mechanical method.

## ISOLATION PROCEDURE

For protoplast isolation, tissues should be pre-plasmolysed with enzymatic solutions followed by washing with Cell-Washing Protoplasts (CWP) and 13% mannitol. Protoplasts can be isolated from any part of the plant body, but it is easier to isolate protoplasts from the mesophyll of leaves. Since the mesophyll cells are loosely arranged, the enzymes have easy access to the cell wall. Takebe et al. (1968) developed a technique for the isolation of protoplasts from the mesophyll cells of *Nicotiana tabacum*. This technique involves four major steps: (i) Surface sterilization of the leaves, (ii) peeling off the lower epidermis, (iii) incubation of stripped leaf tissue in an enzyme solution and (iv) isolation of protoplasts.

The basic technique involved in the isolation of protoplasts from mesophyll cells of tobacco consists of following these steps. Fully expanded leaves from 7–8-week-old plants are selected. Surface sterilization of the leaves is done by immersing in 70% ethyl alcohol for 30 seconds. Leaves are then placed for 15–20 min. in 10% solution of calcium hypochlorite. The leaves are then repeatedly washed in sterile, distilled water. The lower epidermis of the leaves is carefully peeled with the help of forceps and the stripped leaves are cut into small pieces (4 $cm^2$). Peeled leaf pieces are placed on a thin layer of 600 $Mmol^{-1}$ mannitol–CPW (Cultured Protoplast Washing) solution. After 30 min. mannitol–CPW solution is replaced by a solution containing 4% cellulase SS, 0.4% macerozyme SS, 600 $mmol^{-1}$ mannitol and CPW salts. Petri plates containing the leaf pieces are sealed and kept in the dark at 24–26°C for 16–18 hrs.

Leaf pieces are then squeezed with the help of a Pasteur pipette in order to liberate the protoplasts. Large debris are removed by filtering, and the filtrate is poured in a centrifuge tube. The filtrate is centrifuged at 100 g for 3 min. The sediment is transferred to the top of 860 $mmoll^{-1}$ sucrose solution and again centrifuged at 100 g for 10 min. The green protoplasts are collected from the top of the sucrose pad and transferred to another centrifuge tube. Protoplast culture medium is added to suspend the protoplasts and again centrifuged at 100 g for 3 min. This washing is done thrice, and after the final washing, plating of protoplasts is done. Although the isolation of protoplasts can most easily be accomplished, the yield of protoplasts depends upon factors like physiological state of donor plant, the purity and composition of enzymes, selection of osmotic solution and pH. Optimal conditions needed for the isolation of protoplasts from cultured cells of tobacco are given in Table 16.11.

Gokhale (2015) isolated protoplast from leaves of *Oroxylum indicum* (Bignoniaceae) using protoplast releasing enzymes (cellulase, pectinase and hemicellulose) alone and in combination and obtained high yields of protoplasts.

**TABLE 16.11**
**Optimal Conditions for the Isolation of Tobacco Protoplast (after Uchimiya and Murashige, 1974)**

| Parameter | Cultured cells of tobacco |
|---|---|
| Plant material | 4–5-day–old subculture |
| Cellulase | 1% Onozuka R-10 |
| Macerozyme | 0.1–0.2% Onozuka R-10 |
| pH of enzyme solution | 4.7 or 5.7 |
| Volume of enzyme solution/fresh weight of tissue | 10 $mlg^1$ |
| Incubation period | 2–3 h |
| Incubation temperature | 22–37°C |
| Rate of agitation | 50 rpm |
| Osmoticum | 300–800 mmol $l^{-1}$ mannitol |

## CULTURE OF PROTOPLASTS

Protoplasts are totipotent. They have the capability to dedifferentiate, re-enter the cell cycle, go through repeated mitotic divisions and then proliferate or regenerate into various organs. Protoplasts may be cultured either by liquid drop method or plating method.

1. **Liquid drop method:** In this method, droplets of protoplast suspension are placed in liquid medium in a micro-chamber. Protoplasts may be transferred to a semisolid medium after a few divisions. Liquid drop method is generally preferred because (i) protoplasts of some species do not divide on agarified medium, (ii) it is possible to change the medium in case degenerating components of protoplasts produce certain toxic substances, which could kill the healthy cells; (iii) density of the cells can be reduced; and (iv) osmotic pressure of the medium can be effectively reduced after a few days of culture.
2. **Plating method**: A 2 ml aliquot of protoplasts is suspended in liquid culture medium, poured into falcon plastic petri dishes and mixed gently with an equal volume of the same medium containing 1% agar (Figure 16.19). Sealing of petri plates is done with paraffin. The dishes are then kept in an inverted position in a growth room at 28°C under continuous light (2300 lux). Plating method has certain advantages: (i) the entire sequence of development and differentiation can be easily observed, (ii) a higher plating efficiency can be achieved in protoplast cultures compared to single cells.

Generally, modified Murashige and Skoog (1962) or Nagata and Takebe's (1971) media are used for protoplast culture. Growth promoters like 2,4

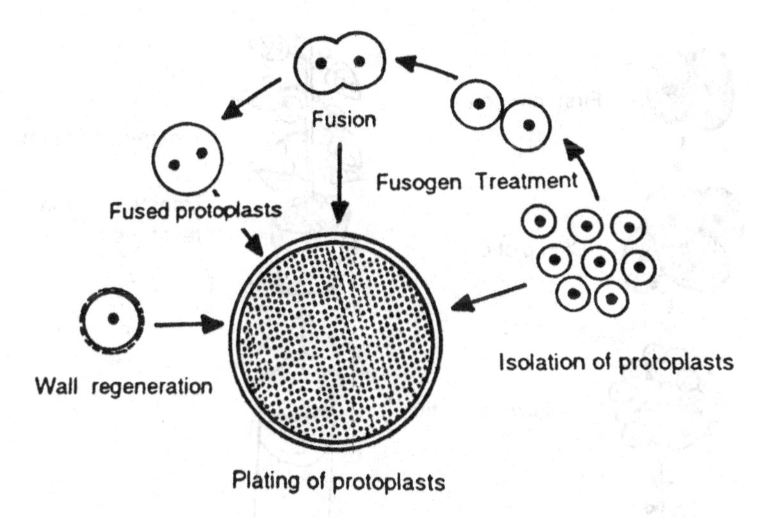


**FIGURE 16.19** Isolation, culture and fusion of protoplasts.

dichlorophenoxyacetic acid, 6-benzyl-aminopurine and kinetin induce division in protoplast cultures.

## Cell Wall Formation and Division

The synthesis of the cell wall begins after a few hours of culturing the protoplasts. The completion of the cell wall, however, takes two to several days. A newly formed cell wall is composed of loosely arranged microfibrils which later on organize to form a typical cell wall. In *Convolvulus* cell wall regeneration requires an exogenous supply of a readily metabolizable carbon source like sucrose. Protoplasts increase in size and numerous cytoplasmic strands develop during and after the early stages of cell wall regeneration. There is considerable increase in the number of cell organelles, respiration and synthesis of RNA, protein and polysaccharides. The majority of the cell organelles aggregate around the nucleus.

The protoplasts capable of dividing undergo the first division within 2–7 days. Multicellular colonies are formed after 2–3 weeks, which are then transferred on fresh medium (Figure 16.20).

## Regeneration and Organogenesis

Takebe et al. (1971) for the first time reported regeneration of plants from isolated protoplasts of *Nicotiana tabacum*. Since then, plant regeneration has been achieved from cultured protoplasts in a wide variety of plants (Table 16.12). Plants from the protoplasts may be regenerated either through embryoid formation or through the production of calli. Vasil and Vasil (1980) reported embryogenesis and plantlet formation in isolated protoplasts derived from suspension cultures of

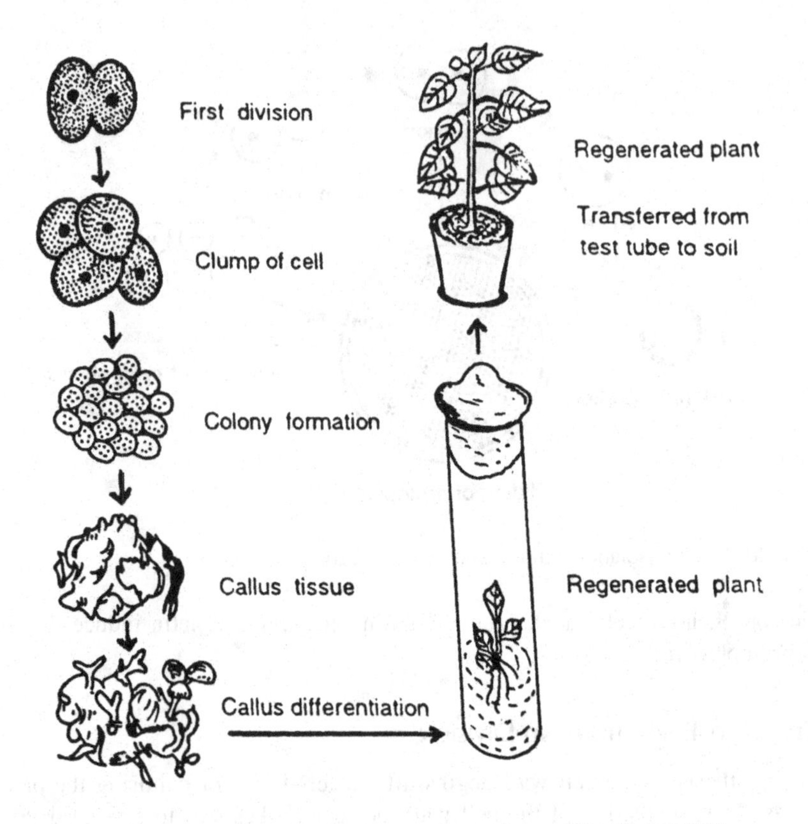


**FIGURE 16.20** Plant regeneration from protoplast cultures (after Bajaj, 1974).

**TABLE 16.12**
**Species in Which Plant Regeneration Has Been Achieved from Cultured Protoplasts**

| Family | Species |
|---|---|
| Apiaceae | *Apium graveolens, Daucus carota* |
| Asteraceae | *Cichorium endiva, Helianthus annuus, Senecio vulgaris* |
| Brassicaceae | *Arabidopsis thaliana, Brassica napus, B. oleracea* |
| Euphorbiaceae | *Mannihot esculenta* |
| Fabaceae | *Medicago sativa, Trifolium repens* |
| Poaceae | *Saccharum officinarum, Triticum aestivum, Zea mays* |
| Rutaceae | *Citrus sinensis* |
| Solanaceae | *Atropa belladonna, Solanum lycopersicum, Nicotiana tabacum, Petunia hybrida, Solanum melongena* |
| Zingiberaceae | *Zingiber officinale* |

immature embryos of *Pennisetum americanum* (pearl millet). In *Medicago sativa* (Kao and Michayluk, 1980) and *Trifolium repens* (Gresshoff, 1980), complete plants have been raised from isolated protoplasts.

Protoplasts obtained from *Asparagus officinalis* cladodes when cultured on the medium containing NAA, Zeatin and L-glutamine gave rise to small colonies. When protoplast colonies were transferred to a medium containing a reduced level of glutamine, root differentiation started. Shoot formation began when zeatin was substituted for BAP and adenine was added to the medium. Variations in the levels of hormones resulted in the regeneration of embryoids from protoplasts. When these embryoids were transferred on semi-liquid medium containing IAA and zeatin, they developed into whole plants.

Protoplasts isolated from haploid tobacco (*Nicotiana tabacum*) and *Petunia hybrida* have been successfully regenerated into entire plantlets. Callus formation and plant regeneration has also been reported from mesophyll protoplasts of *Brassica napus* (rape plant).

## SOMATIC HYBRIDIZATION

The technique of hybrid production through the fusion of protoplasts from different genetic backgrounds is known as somatic hybridization or parasexual hybridization.

Somatic hybridization by protoplast fusion is a promising technique for breeding ornamental species and requires reliable *in vitro* protocols. Somatic hybridization can fuse two complete genomes, which is an alternative to sexual reproduction (Wu et al., 2009). This technique was successfully used to breed citrus, sunflower, brassica and wheat (Davey et al., 2005).

Somatic hybridization offers the following possible genomic manipulations: (i) overcoming sexual incompatibility, (ii) producing amphyploids, (iii) transferring part of one species genome to another (cybrids); (iv) transferring cytoplasmatic DNA to produce male-sterile plants; and (v) producing plants resistant to environmental stresses, pests and diseases (Wu et al., 2009). Several somatic hybrids between sexually compatible and incompatible parents have been raised (Table 16.13). Production of somatic hybrids involves a number of steps (Figure 16.21)

## PROTOPLAST FUSION

The fusion of protoplasts may be of two types: (i) spontaneous fusion and (ii) induced fusion.

i. **Spontaneous fusion:** Protoplasts may fuse at the time of enzymatic degradation of cell walls. In somatic hybridization, spontaneous fusion is of no significance, and only induced fusion has received popularity.
ii. **Induced fusion:** This type of fusion is achieved by using suitable agents called fusogens. These include sodium nitrate, polyethylene glycol (PEG),

**TABLE 16.13**
**Hybrid Plants Raised Through Protoplast Fusion**

| SPECIFIC HYBRIDIZATION |
|---|
| **Sexually Compatible Combinations** |
| *Daucus carota* (2n=18)+*D. capillifolius* (2n=18) |
| *Nicotiana tabacum* (2n=48)+*N. glauca* (2n=24) |
| *Nicotiana tabacum* (2n=24)+*N. sylvestris* (2n=24) |
| *Petunia parodii* (2n=14)+*P. hybrida* (2n=14) |
| *Solanum tuberosum* (2n=24)+*S. chcoense* (2n=24) |
| **Sexually Incompatible Combinations** |
| *Nicotiana tabacum* (2n=24)+*N. nesophila* (2n=48) |
| *Datura innoxia* (2n=24)+*D. candida* (2n=24) |
| *Petunia parodii* (2n=14)+*P. parviflora* (2n=18) |
| **INTERGENERIC HYBRIDIZATION** |
| *Arabidopsis* (2n=40)+*Brassica campestris* (2n=20) |
| *Datura innoxia* (2n=24)+*Atropa belladonna* (2n=24) |

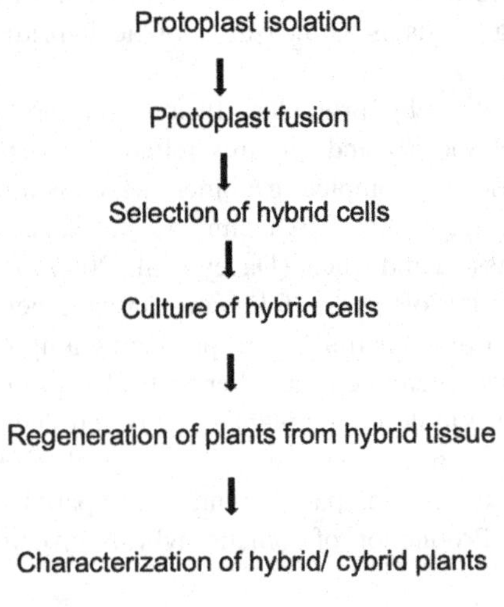

Protoplast isolation
↓
Protoplast fusion
↓
Selection of hybrid cells
↓
Culture of hybrid cells
↓
Regeneration of plants from hybrid tissue
↓
Characterization of hybrid/ cybrid plants

**FIGURE 16.21** Sequential steps involved in somatic hybridization.

polyvinyl alcohol or dextran. PEG is the most preferred fusogen because of the reproducible high frequency of heterokaryon formation and low toxicity. PEG-induced fusion is enhanced by enriching the PEG solution with $Ca^{2+}$ ions. The PEG-induced fusion is non-specific and has been successfully used to fuse protoplasts from entirely unrelated taxa like soybean–maize, soybean–tobacco and soybean–barley (Kao et al., 1974).

Another method of protoplast fusion is by using electric fields. Oliveres-Fuster et al. (2005) developed the electrochemical protoplast fusion method that combines

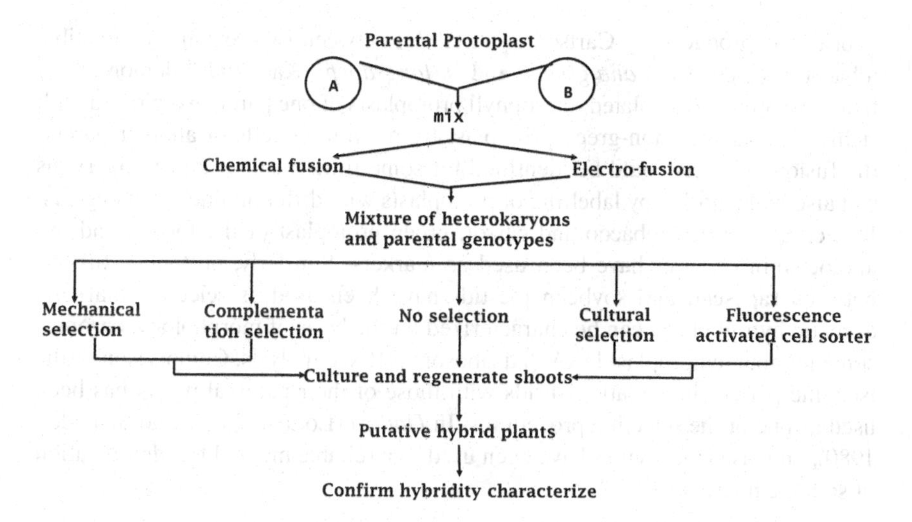


**FIGURE 16.22** General scheme followed for the production of somatic hybrids between different protoplast populations, A and B.

the advantages of the two methods. It is based on chemically induced protoplast aggregation and direct current pulse promoted membrane fusion. When protoplasts are introduced into a high frequency alternating electric field, they act as dipoles and are attracted to each other. Once the protoplasts come in contact, a high frequency of fusion may be achieved by disrupting the protoplast membranes with a short direct current pulse which, unlike chemical fusogens, have relatively little effect on protoplast viability. This method has now been demonstrated to provide high frequency of fusion. Zimmerman and Scheurich (1981) demonstrated that an electric field pulse of very short duration but of high intensity can trigger the fusion process between mesophyll cell protoplasts of *Avena sativa*.

Protoplast fusion consists of three main stages: agglutination, membrane fusion and rounding-off of fused protoplasts. During agglutination, the plasma membrane of two or more protoplasts are brought into close proximity. Agglutination is followed by membrane fusion at localized places resulting in the formation of cytoplasmic channels between protoplasts. These channels gradually expand, and the fused protoplasts become spherical forming either heterokaryons or homokaryons. Fusion between the isolated protoplasts of same species gives homokaryons, whereas fusion between different species results in heterokaryons. The general scheme followed for the production of somatic hybrids between different protoplasts is given in Figure 16.22.

## SELECTION PROCEDURE

In somatic hybridization, the most important aspect is to use certain selective markers for the identification of heteroplasmic fusion products. The first somatic hybrid was produced by Carlson et al. (1972) between two sexually compatible tobacco species, *Nicotiana glauca* and *N. longsdorfii*. Kao (1977) demonstrated that when green, vacuolated, mesophyll protoplasts of one parent were fused with richly cytoplasmic, non-green protoplast from cultured cells of another parent, the fusion products could be identified for some time in culture. Heterokaryons can also be identified by labelling of protoplasts with different fluorescent agents. In a cross between tobacco and carrot, green protoplast of the former and red anthocyanin of latter have been used as markers. Similarly, protoplast fusions between rapeseed and soybean plastids have been used as selective markers. Somatic hybrid plants can be characterized on the basis of morphology, chromosome number, chloroplast DNA and ribosomal RNA analysis. Comparison of the isozyme pattern in somatic hybrids with those of their parental plants has been used as one of the selection procedures. In *Datura* (Lonnendonker and Schieder, 1980), amylase isoenzymes have been used as a reliable method for identification of somatic hybrids.

## CYBRIDS

Cybrids (cytoplasmic hybrids) may be defined as a "plant or cell which is a cytoplasmic hybrid produced by fusion of a protoplast and a cytoplast". By combining the cytoplasm of a donor of one species with the nucleus of an acceptor belonging to another species, interspecies nuclear and organelle genomic configurations are created. These cybrids occasionally display cytoplasmic male sterility (CMS) as a result of the incongruence of nuclear and cytoplasmic DNA expression.

Power et al. (1975) demonstrated that following protoplast fusion and culture, cell lines can be isolated which carry the nucleus of one of the parents and the cytoplasm of both. When mesophyll protoplast of *Petunia* were fused with cultured cell protoplasts of the crown gall of *Parthenocissus*, the resultant line contained the chromosomes of only *Parthenocissus* but exhibited some of the characteristics of *Petunia* for some time.

In cultures, cybrids may arise by one of the following means: (i) removal of one of the nuclei after the formation of heterokaryon, (ii) fusion between a normal protoplast with an enucleate protoplast, (iii) fusion between normal protoplast having a non-viable nucleus and (iv) selective loss of genome of one of the nuclei.

Maliga et al. (1982) fused enucleated protoplasts with nucleated protoplasts in *Nicotiana*. Protoplasts of *Nicotiana tabacum* (SRI) carrying a maternally inherited streptomycin-resistant mutation were enucleated. The resulting cytoplasts were then fused with streptomycin sensitive protoplasts of *Nicotiana plumbaginifolia*. The SRI cytoplasts, having no nuclei, could not form callus. All resistant clones recovered after fusion were supposed to be derived from interspecific cytoplast protoplast fusion. Plants regenerated from eight of the resistant clones were resistant to streptomycin and inherited the resistance maternally.

Somatic hybrids and cybrids are identified on the basis of morphological characters and the karyology of the $F_1$ plants (Figure 16.23). Cybrids could be

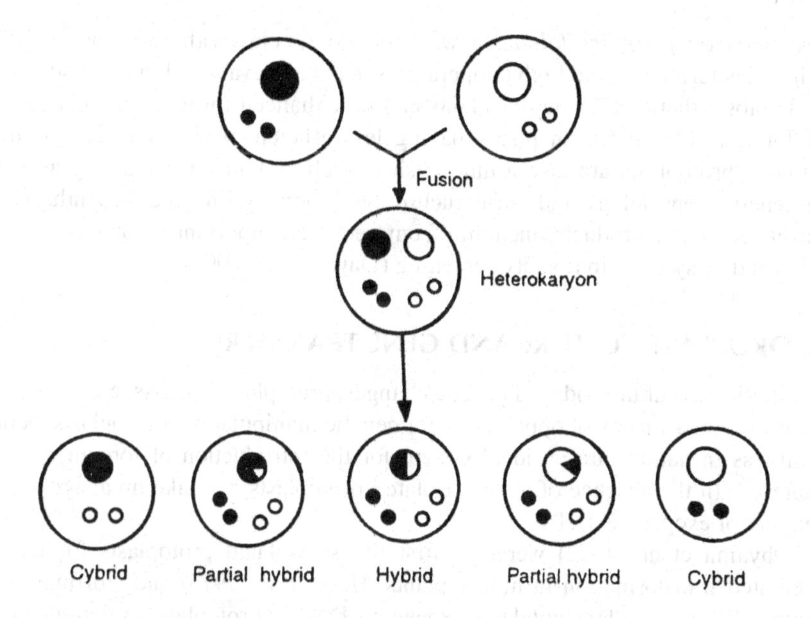


**FIGURE 16.23** Fusion of two genetically different protoplasts can result in complete hybrids, partial hybrids and cybrids.

exploited for transferring male sterility from one species to another, especially where the sexual crosses are incomplete.

## Applications of Somatic Hybridization

The most promising potential application of the protoplast culture is in inducing protoplasts from genetically different backgrounds to fuse and regenerate to form a somatic hybrid. Cell fusion techniques thus allow the transfer or combination of identifiable agronomically useful traits without the need for detailed molecular information and also enables transfer of polygenic factors that may underline such traits. Heterozygous lines can be developed within a single species which are usually propagated by means of vegetative methods. Cytoplasmic male sterile lines can be developed through fusion of an 'elite' cultivar with the x-ray-irradiated protoplasts of the CMS parent. Somatic hybrids have been obtained by fusion of mesophyll protoplasts of *Solanum lycopersicum* var. *cerasiforme* and protoplasts obtained from callus culture of originally dihaploid *Solanum tuberosum*.

Disease resistance can be induced by incorporating selective genome from the disease resistant to susceptible crop plants. The wild potato species *Solanum brevidens* is resistant to potato leaf roll virus (PLRV) and potato virus Y (PVY), two important viral diseases of commercial potato (*Solanum tuberosum*). *Solanum brevidens* cannot be sexually crossed directly with *Solanum tuberosum*. Somatic hybrids, produced by protoplast fusion, exhibit resistance to PLRV and PVY from the wild parent. In Japan somatic hybrids of *Solanum tuberosum* and

*Lycopersicon pimpinellifolium*, a wild tomato species with resistance to late blight, bacterial wilt and high temperatures, has been evaluated in the field.

Photosynthetic efficiency of plants can be enhanced through transplantation of foreign chloroplasts in plants having less efficient photosynthetic systems. Isolated protoplasts are also a unique single-cell system for evaluating aspects of genetics, physiology and ultrastructure, with potential for the biosynthesis of novel secondary products, including commercially important recombinant proteins and as systems in toxicity screening (Davey et al., 2005).

## PROROPLAST CULTURE AND GENE TRANSFERS

In addition to culture-induced genetic changes, protoplasts and tissue culture systems present a variety of opportunity for genetic manipulation. Protoplasts, being wall-less in nature, are an ideal system for the introduction of foreign genetic material. In the absence of a wall, isolated protoplasts can take up a significant amount of exogenous DNA.

Ohyama et al. (1972) were the first to use isolated protoplasts for DNA-mediated transformation in higher plants. Hess et al. (1973) and Hoffman and Hess (1973) reported the uptake of exogenous DNA in protoplasts of *Petunia*. The transformation of higher plants by introducing DNA into their cells involves a highly complex process. This includes uptake, integration, replication, expression and transmission of the introduced DNA in the alien environment of the recipient cells. Direct DNA uptake by protoplasts has been successfully demonstrated in *Triticum monococcum* and *Lolium multiflorum*.

Cocking (1966) and Cocking and Pojnar (1969) reported that tomato fruit protoplasts take up tobacco mosaic virus particles by pinocytosis, and demonstrated the initial stages of infection by electron microscopic studies. Davey and Cocking (1972) reported that *Rhizobium japonicum* can be introduced into pea leaf protoplasts during enzymatic digestion of the cell wall. Davey et al. (1973) also successfully isolated legume root nodule protoplasts containing packets of bacteria. The *Agrobacterium*-induced transformation can be achieved by culturing *Agrobacterium* with isolated protoplasts. The advantages of this method are: (i) A large number of protoplasts can be treated in each co-cultivation to produce large numbers of independent transformation events. (ii) The cells can be exposed uniformly to the selective agent used. However, the disadvantage is that protoplast co-culture is limited to those taxa which can readily be regenerated back to plants.

In tobacco, high protoplast transformation frequencies have been achieved using either chemical methods (usually based on polyethylene glycol treatment) or electric methods (electroporation) or a combination of physical and chemical treatments. It has been demonstrated that direct gene transfer could be applied equally to protoplasts of graminaceous species, such as *Triticum monococcum*, *Zea mays* and *Lolium multiflorum*.

Protoplast fusion overcomes pre- and post-zygotic sexual incompatibility barriers and generates novel germplasm through new nuclear-cytoplasmic combinations.

## PARTHENOCARPY

The term parthenocarpy was first introduced by Noll (1902) who defined parthenocarpy as the "development of fruits without pollination or any other stimulus". Since then, the definition of parthenocarpy has undergone a slight modification. According to Nitsch (1965), parthenocarpy refers to the formation of fruits without fertilization. Parthenocarpic fruits are seedless. The fruit development without pollination is termed vegetative parthenocarpy and with pollination, stimulative parthenocarpy. In pear, both stimulative and autonomous parthenocarpy have been identified, and some commercial cultivars show parthenocarpic fruit development (Nyeki et al., 1998).

Plants which are cultivated by man and propagated for long periods of time like navel orange, banana and grape variety Thompson bear seedless fruits. Many of the plants cultivated for their fruits show seeded as well as parthenocarpic varieties. Parthenocarpy has been reported in over 100 flowering plants. About half of the parthenocarpic species are polyploid or show instances of polyploidy or evidence of hybrid origin (Picorella and Mozzucato, 2019). The most common groups are Rosideae, Anacardiaceae, Rutaceae, Solanaceae, etc.

Nitsch (1963) recognized three types of parthenocarpy: (i) genetical, (ii) environmental and (iii) chemically induced parthenocarpy.

### 1. Genetical Parthenocarpy

Parthenocarpy can be natural or genetic. In genetic parthenocarpy the expression of the parthenocarpic trait is not influenced by external factors. Rather, mutations and hybridizations are the basic reasons for this condition. In cucumber, PP gives rise to non-parthenocarpic fruits, and heterozygous Pp produces parthenocarpic fruits. Genetic parthenocarpy occurs in *Citrus*, *Cucurbita*, *Punica*, *Vitis* and *Eugenia*.

Plantain and banana (*Musa* spp.) cultivation comprises of polyploid (mainly triploid) and highly sterile plants which develop parthenocarpic fruits. At least three complimentary dominant genes control parthenocarpy. In tomato (*Solanum lycopersicum*, formerly *Lycopersicon esculentum*), mutant lines carrying non-allelic genes causing parthenocarpic fruit development have been identified.

### 2. Environmental Parthenocarpy

A number of environmental factors, such as frost, fog, low temperatures etc., greatly influence parthenocarpic development of fruits. Parthenocarpy has been induced in certain varieties of *Cucurbita pepo*, by preventing the pollination factor and subjecting the plant to 15°C at night. In pears, parthenocarpic fruits can be obtained by exposing the flowers to freezing temperatures. In *Capsicum* if plants are shifted from temperature of 32–38°C to 10–16°C at the time of anthesis, parthenocarpic fruits can be obtained.

### 3. **Chemically Induced Parthenocarpy**

Application of auxin and gibberellins can induce seedless fruits in a wide variety of plants. Indolbutyric acid (IBA) has been found to be one of the most effective auxins, and the fruits resulting from its application are as large as those produced after natural pollination.

Gibberellic acid has been reported to induce parthenocarpy in a number of plants (e.g., figs, grapes, tomato). In strawberry, 2-naphthoxy acetic acid at the concentration of 50 mg/l sprayed after two weeks of pollination is known to increase production of seedless fruits by 30%.

### 4. **Genetically Engineered Parthenocarpy**

Rotino et al. (1997) induced parthenocarpy in tobacco and eggplant by inserting a chimeric gene *DefH9-iaam*. Transgenic tobacco and eggplants expressing the coding region of the *iaaM* gene from *Pseudomonas syringae pv. savastanoi*, under the control of the regulatory sequences of the ovule-specific *DefH9* gene from *Antirrhinum majus*, showed parthenocarpic fruit development. The transgenic plants of eggplant carrying the gene-produced fruits with viable seeds when pollinated and produced parthenocarpic fruits in the absence of pollination.

Ficcadenti et al. (1999) engineered parthenocarpy in two genotypes of tomato by using DefH9-iaaM chimeric gene. Seedless parthenocarpic fruits were obtained from emasculated flowers and fruits with seeds from pollinated flowers.

### Significance of Parthenocarpy

Parthenocarpy offers the possibility of improving fruit quality and productivity in many crop plants grown for their fruits. Parthenocarpy is considered as the most efficient way to produce fruits under environmental conditions adverse for pollination and / or fertilization. Parthenocarpy allows early fruit production and harvest. In horticultural crops, seedlessness of parthenocarpic fruits is the most desirable trait for their consumption as fresh fruit and for their use in the juice, jelly and jam industries.

## SUMMARY


- Experimental plant reproductive biology is concerned with an imitation and a modification of the course of nature, with a view to understanding various processes underlying the development and differentiation of the embryo.
- Cells, tissues and organs can only be cultured *in vitro* when all the required nutrients are available in the medium.
- Callus is an organized mass of loosely arranged parenchymatous cells. Callus initiation begins when segments of plant organs (explants) are cultured on solidified culture media.

- In anther cultures, homozygous lines can be produced after doubling the pollen-derived haploid plants.
- Flower, ovary, ovule, nucellus, endosperm and embryo have been successfully cultured on artificial media.
- Somatic embryogenesis is the process by which the somatic cells or tissues develop into differentiated embryos.
- Artificial seeds (also referred to as synthetic or somatic seeds) are prepared by encapsulating the embryoids in a protective covering.
- The technique of hybrid production through the fusion of protoplasts from different genetic backgrounds is known as somatic hybridization.
- Cybrids (cytoplasmic hybrids) are plants or cells in which a cytoplasmic hybrid is produced by the fusion of a protoplast and a cytoplast.
- Development of fruit without pollination is known as parthenocarpy.

## SUGGESTED READING


Ahmed, A.A.A., M. Miao, E.D. Pratsinakis, H. Zhang, W. Wang, Y. Yuan, M. Lyu, J. Ifftikha, A.F. Yousef, P. Madesis and B. Wu. 2021. Protoplast isolation, fusion, culture and transformation in the woody plant *Jasminum* spp. *Agriculture* 11 (8): 699.

Ali, J., K.L.C. Nocolas, S. Akhtar, et al. 2021. Improved anther culture media for enhanced callus formation and plant regeneration in rice (*Oryza sativa* L.). *Plants*. https://doi.org/10.839/plants10050839.

Andronescu, D.I. 1919. Germination and further development of the embryo of *Zea mays* separated from the endosperm. *Am. J. Bot.* 6: 443–452.

Bajaj, Y.P.S. 1964. Development of ovule of *Abelmoschus esculentus* L. var. Pusa Sawani in vitro. *Proc. Natl. Inst. Sci. India* 30: 175–185.

Bajaj, Y.P.S. 1974. Isolation and culture studies on pollen tetrad and pollen mother cell protoplasts. *Plant Sci. Lett.* 3: 93–99.

Bansal, Y.K. and S. Bansal. 2012. Protoplast culture, fusion and somatic hybridization. In *Plant Tissue Culture: Totipotency to Transgenics*, ed. H.P. Sharma, 216–240. Jodhpur: Agrobios (India).

Beasley, C.A. 1971. In vitro culture of fertilized cotton ovules. *Bioscience* 21: 906–907.

Beasley, C.A. 1973. Hormonal regulation of growth in unfertilized cotton ovules. *Science (Wash, D.C.)* 179: 1003–1005.

Bhojwani, S.S. and M.K. Rajdan. 1983. *Plant Tissue Culture: Theory and Practice*. Amsterdam: Elsevier.

Bourgin, J.P. and J.P. Nitsch, 1967. Obtention de *Nicotiana* haploids à partir d'étamines cultivées *in vitro*. *Ann. Physiol. Veg.* 9: 377–382.

Brown, S., D.F. Wetherell and D.K. Dougall. 1976. The potassium requirement for growth and embryogenesis in wild carrot suspension cultures. *Physiol. Plant.* 37: 73.

Brulfert, J. and D. Fontaine. 1967. Utilisation de la culture in vitro pour une étude de développement floral chez *Anagallis arvensis* ssp. *phoenicea* Scop. *Biol. Plant.* 9: 439–446.

Button, J. and C.H. Bornman. 1971. Development of nucellar plants from unpollinated and unfertilized ovules of the Washington Navel Orange in vitro. *J. S. Afr. Bot.* 37: 127–133.

Button, J. and J. Kochba. 1977. Tissue culture in the citrus industry. In *Applied and Fundamental Aspects of Plant Cell, Tissue and Organ Culture*, eds. J. Reinert and Y.P.S. Bajaj. Berlin: Springer-Verlag.

Carlson, P.S., H.H. Smith and R.D. Dearing. 1972. Parasexual interspecific plant hybridization. *Proc. Nat. Acad. Sci. USA* 69: 2292–2294.

Chopra, R.N. 1962. Effect of some growth substances and calyx on fruit and seed development of *Althaea rosea* Cav. In *Plant Embryology - A Symposium*, 170–181. New Delhi: CSIR.

Cocking, E.C. 1960. A method for the isolation of plant protoplasts and vacuoles. *Nature (London)* 187: 927–929.

Cocking, E.C. and E. Pojnar. 1969. An electron microscopy study of the infection of isolated tomato fruits protoplasts by tobacco mosaic virus. *J. Gen. Virol.* 4: 305–312.

Cocking, E.C. 1979. Parasexual reproduction in flowering plants. *New Zealand. J. Bot.* 17: 665–671.

Cocking, E.C. 1985. Protoplast fusion and its application in agriculture. In *Biotechnology and International Agriculture Research. Proc. Inter-Cener Sem. Intl. Agr. Res. Cent.* Biotech., International Rice Research Institute, Manila.

Dieterich, K. 1924. Über Kultur von Embryonen ausserhalb des Samens. *Flora* 117: 379–417.

Eeckhaut, T., P.S. Lakshmanan, D. Deryckere, et al. 2013. Progress in plant protoplast research. *Planta*. https://doi.org/10.1007/s00425-013-1936-7.

Galun, E., Y. Jung and A. Lang. 1962. Culture and sex modification of male cucumber buds in vitro. *Nature (London)* 194: 596–598.

Galun, E., Y. Jung and A. Lang. 1963. Morphogenesis of floral buds of cucumber cultured in vitro. *Dev. Biol.* 6: 370–387.

Gamborg, O.L., R.A. Miller and K. Ojima. 1968. Nutrient requirements of suspension cultures of soybean Roct cells. *Exp. Cell Res.* 50: 151–158.

Guha, S. and S.C. Maheshwari 1964. In vitro production of embryos from anthers of Datura. *Nature* 204: 497.

Guha, S. and S.C. Maheshwari 1966. Cell division and differentiation of embryos in the pollen grains of *Datura in vitro. Nature* 212: 97–98.

Guha, S. and B.M. Johri. 1966. *In vitro* development of ovary and ovule of *Allium cepa* L. *Phytomorphology* 16: 353–364.

Haccius, B. 1978. Question of unicellular origin of non-zygotic embryos in callus cultures. *Phytomorphology* 28: 74–81.

Haccius, B. and N.N. Bhandari. 1975. Zur Frage der 'Befestigung' junger Pollen-Embryonen von Nicotiana tabacum aus der Antheren Wand. Beitr. *Biol. Pflanz.* 51: 53–56.

Hanning, E. 1904. Zur Physiologie pflanzlicher embryonen. I. Uber die culture von crucifever embryonen ausserhalb des embryosac. *Bot. Ztg.* 62: 45–80.

Javed, S.B., A.A. Alatar, M. Anis and M. Faisal. 2017. Synthetic seeds production and germination studies for short term storage and long distance transport of *Erythrina variegata* L: A multipurpose tree legume. *Ind. Crops Prod.* 105: 41–46.

Johri, B.M. and S.S. Bhojwani. 1965. Growth responses of mature endosperm in culture. *Nature (London)* 208: 1345–1347.

Joshi, P.C. and B.M. Johri. 1972. In vitro growth of ovules of *Gossypium hirsutum. Phytomorphology* 22: 195–209.

Kang, H.H., A.H. Naing and C.K. Kim. 2020. Protoplast isolation and shoot regeneration from protoplast derived callus of *Petunia hybrida* cv. Morage Rose. *Biology* 9 (8): 228. https://doi.org/10.3390/biology9080228.

Kanta, K. and D. Padmanabhan. 1964. In vitro culture of embryo segments of *Cajanus cajan* (L.) Millsp. *Current Science* 33 (23): 704–706.

Kao, K.N., F. Constabel, M.R. Michayluk and O.L. Gamborg. 1974. Plant protoplast fusion and growth of intergeneric hybrid cells. *Planta* 120 (3): 215–227.

Kao, K.N. 1977. Chromosomal behaviour in somatic hybrids of soyabean, *Nicotiana glauca. Mol. Gen. Genet.* 105: 225–230.

Kasha, K.J. and K.N. Kao. 1970. High frequency haploid production in barley (*Hordeum vulgare* L.). *Nature* 225: 874–875.

Kitto, S.K. and J. Janick. 1982. Polyox as an artificial seed coat for asexual embryos. *Hort. Science* 17: 488.

Knudson, L. 1922. Nonsymbiotic germination of orchid seeds. *Bot. Gaz.* 73: 1–25.

Konar, R.N. and K. Natraja. 1969. Morphogenesis of isolated floral buds of *Ranunculus sceleratus* L. in vitro. *Acta Bot. Neerl.* 18: 680–699.

Kruse, A. 1974. An *in vivo/vitro* embryo culture technique. *Hereditas* 77: 219–224.

Kumar, P.P. and C.S. Loh. 2012. Plant tissue culture for biotechnology. *Plant Biotechnology and Agriculture*, 131–138. https://doi.org/10.1016/b978-0-12-381466-1.00009-2.

Laibach, F. 1925. Das Taubwerden der Bastardsamen und die kunstliche Aufzucht fruh absterbender Bastardembryonen. *Ztschr. F. Bot.* 17: 417–459.

Laibach, F. 1929. Ectogenesis in plants: Methods and genetic possibilities of propagating embryos otherwise dying in the seed. *J. Hered.* 20: 201–208.

Lampe, L. and C.O. Mills. 1933. Growth and development of isolated endosperm and embryo of maize. *Abstr. Pap. Bot. Soc. Boston.*

La Rue, C.D. 1936. The growth of plant embryos in culture. *Bull. Torrey Bot. Club* 63: 365–382.

La Rue, C.D. 1942. The rooting of flower in sterile culture. *Bull. Torrey Bot. Club* 69: 332–341.

LaRue, C.D. 1949. Cultures of the endosperm of maize. *Am. J. Bot.* 36: 798.

Lentini, Z., G. Restrepo, M.E. Buitrago and E. Tabares. 2020. Protocol for rescuing young cassava embryos. *Front. Plant Sci.* 11: 522. https://doi.org/10.3389/fpls.2020.00522.

Li, F., Y. Cheng, X. Zhao, R. Yu, H. Li, L. Wang, S. Li and Q. Shan. 2020. Haploid induction via unpollinated ovule culture in *Gerbera hybrida. Scientific Reports* 10: 1702 https://doi.org/10.1038/s41598-020-58552-z.

Maheshwari, N. 1958. In vitro culture of excised ovules of *Papaver somniferum. Science, N.Y.* 127: 342.

Majumdar, S.K. 1970. In vitro culture of flower buds of Haworthia and Astroloba. *Фyton (Argentina)* 27: 31–34.

Manokary, M., S. Priyadharshini and M. Shekhawat. 2021. Direct somatic embryogenesis using leaf explants and short term storage of synseeds in *Spathoglottis plicata* Blume. *Plant Cell Tissue Organ Cult.* 145 (3): 1–11.

Maliga P., H. Larz, G. Lazer and F. Nagy. 1982. Cytoplast-protoplast fusion for interspecific chloroplast transfer in Nicotiana. *Mol. Zen. Genet.* 185: 211–215.

Massart, J. 1902. Sur la pollination sans fecundation. Bul. Jard. Bot. *Brussels de P'Etat* 1: 85–95.

Mitra, G.C. and H.C. Chaturvedi. 1972. Embryoids and complete plants from unpollinated ovaries and from ovules of in vivo-grown emasculated flower buds from *Citrus* spp. *Bull Torrey Bot Club* 9: 184–189.

Mondal, T.K. 2020. Somatic embryogenesis and alternative in vitro techniques. In *Tea: Genome and Genetics*, pp. 85–126. Singapore: Springer.

Mullins, M.G. and C. Srinivasan. 1976. Somatic embryos and plantlets from an ancient clone of the grapevine (cv. Cabernet-Sauvignon) by apomixis in vitro. *J. Exp. Bot.* 27 (5): 1022–1030.

Murashige, T. and F. Skoog. 1962. A revised medium for rapid growth and bioassays with tobacco tissue cultures. *Physiol. Plant.* 15: 473–497.

Muzik, T.J. 1956. Studies on the development of the embryo and seed of *Hevea brasiliensis* in culture. *Lloydia* 19: 86–91.

Nag, K.K. and B.M. Johri. 1971. Morphogenic studies on endosperm of some parasitic angiosperms. *Phytomorphology* 21: 202–218.
Nagata, T. and I. Takabe. 1971. Plating of isolated tobacco mesophyll protoplasts on agar medium. *Planta* 99: 12–20.
Nakajima, T. 1962. Physiological studies of seed development, especially embryonic growth and endosperm development. *Univ. Osaka Prefect. Ser. B* 13: 13–48.
Nakajima, T., Y. Doyama and H. Matsumoto. 1969. In vitro culture of excised ovules of white clover, *Trifolium repens* L. *Jpn. J. Breed.* 19 (5): 373–378.
Nakamura, A., T. Yamada, N. Kadotani, R. Itagaki and M. Oka, M. 1974. Studies on the haploid method of breeding in tobacco. *SABRAO J.* 6: 107–131.
Nitsch, P. 1963. The *in vitro* culture of flowers and fruits. In *Plant Tissue and Organ Culture - A Symposium*, eds. Maheshwari, P. and Rangaswamy, N.S., 198–214. International Society of Plant Morphologists. Delhi: University of Delhi.
Nitsch, L.P. 1969. Experimental androgenesis in Nicotiana. *Phytomorphology* 19: 389–404.
Nitsch, C. 1974. La culture de pollen isole sur milieu synthetic. *C.R. Acad. Sci. Paris* 278: 1031–1034.
Nitseh, C. and B. Norreel. 1973. Effeet d'un ehoe thermique sur le pouvoir embryogene du pollen de *Datura innoxia* eultive dans I'anthere on isole de I'anthere. *C.R. Acad. Sci. Paris* 276: 303–306.
Ohyama, K. and J.P. Nitsch. 1972. Flowering haploid plants obtained from protoplasts of tobacco leaves. *Plant and Cell Physiol.* 13: 229–236.
Olivares-Fuster, O., N. Duran-Vila and L. Navarro. 2005. Electrochemical protoplast fusion in citrus. *Plant Cell Rep.* 24 (2): 112–119.
Poddubnaya-Arnoldi, V.A. 1959. Study of fertilization and embryogenesis in certain angiosperms using living material. *The American Naturalist* 93 (870). https://doi .org/10.1086/282071.
Poddubnaya-Arnoldi, V.A. 1960. Study on fertilization in the living material of some angiosperms. *Phytomorphology* 10: 185–198.
Power, J.B., E.M. Frearson, C. Hayward and E.C. Cocking. 1975. Some consequences of the fusion and collective culture of Petunia and Parthenocissus protoplast. *Plant Sci. Lett.* 5: 197–207.
Picarella, M.E. and A. Mazzucato. 2019. The occurrence of seedlessness in higher plants; Insights on roles and mechanisms of parthenocarpy. *Front. Plant Sci.* https://doi.org /10.3389/fpls.2018.01997.
Pundir, N.S. 1972. Experimental embryology of *Gossypium arboreum* L. and *G. Htrsutum* L. and their reciprocal crosses. *Bot. Gaz.* 133: 7–26.
Raghavan, V. and J.G. Torrey. 1963. Growth and morphogenesis of globular and older embryos of Capsella in culture. *Am. J. Bot.* 50: 540–551.
Raghavan, V. 1966. Nutrition, growth and morphogenesis of plant embryos. *Biol. Rev.* 41: 1–58.
Rangan, T.S., T. Murashige and W.P. Bitters. 1969. In vitro studies of zygotic and nucellar embryogenesis in citrus. In *Proceedings of the International Citrus Symposium*, ed. H.D. Chapman, 225–229. Riverside: University of California.
Rangan, T.S. 1982. Ovary, ovule, and nucellus culture. In *Experimental Embryology of Vascular Plants*, ed. Johri, B.M., 105–129. Berlin, Heidelberg, and New York: Springer.
Rangaswamy, N.S. 1958. In vitro culture of nucellus and embryos of Citrus. In *Proc. Symp. Modern Developments in Plant Physiology*, ed. P. Maheshwari, 104–106. Univ. Delhi, Delhi.

Rangaswamy, N.S. 1961. Experimental studies on female reproductive structures of *Citrus microcarpa* Bunge. *Phytomorphology* 11: 109–127.

Rangaswamy, N.S. and T.S. Rangan. 1963. In vitro culture of embryos of *Cassytha filiformis* L. *Phytomorphology* 13: 445–449.

Rangaswamy, N.S. and P.S. Rao. 1963. Experimental studies on *Santalum album*: Establishment of tissue culture of endosperm. *Phytomorphology* 13: 450–454.

Rangaswamy, N.S. and T.S. Rangan. 1971. Morphogenic investigations on parasitic angiosperms. IV. Morphogenesis in decotylated embryos of *Cassytha filiformis* L. Lauraceae. *Bot. Gazette* 132 (2): 113–119.

Redenbaugh, K., D. Slade, P. Viss and J. Fujii. 1987. Encapsulation of somatic embryos in synthetic seed coats. *HortScience* 22: 803–809.

Rotino, G.L., E. Perri, M. Zottini, H. Sommer and A. Spena. 1997. Genetic engineering of parthenocarpic plants. *Nat. Biotechnol.* 15: 1998–1401.

Sabharwal, P.S. 1963. In vitro culture of ovules, nucelli, and embryos of *Citrus retieulata* Blanco var. Nagpuri. In *Plant Tissue and Organ Culture - A Symposium*, eds. P. Maheshwari and N.S. Rangaswamy, 265–274. Intl. Soc. Plant Morphologists, Univ. Delhi, Delhi.

Sachar, R.C. and K. Kanta. 1958. Influence of growth substances on artificially cultured ovaries of *Tropaeolum majus* L. *Phytomorphology* 8: 202–218.

Sachar, R.C. and M. Kapoor. 1959. *In vitro* culture of ovules of *Zephyranthes*. *Phytomorphology* 9: 147–156.

Sangwan, R.S. and B.S. Sangwan-Norreel. 1987. Ultrastructural cytology of plastids in pollen grain of certain androgenic and nonandrogenic plants. *Protoplasma* 138: 11–22.

Sharma, V., B. Kamal, N. Srivastava, A.K. Dobriyal and V.S. Jadon. 2014. In vitro flower induction from shoots regenerated from cultured axillary buds of endangered medicinal herb *Swertia chirayita* H. Karst. *Biotechnol. Res. Int.* 2014: 264690. https://doi.org/10.1155/2014/264690.

Sharp, W.R., R.S. Ruskin and H.E. Sommer 1972. The use of nurse culture in the development of haploid clones in tomato. *Planta* (Berl) 104: 357–361.

Shmykova, N., E. Domblides, T. Vjurtts and A. Domblides. 2021. Haploid embryogenesis in isolated microspore culture of carrots (*Daucus carota* L.). *Life* 11: 20. https://doi.org/10.3390/life11010020.

Spena, A. and G.L. Rotino. 2001. Parthenocarpy. In *Current Trends in the Embryology of Angiosperms*, eds. Bhojwani, S.S. and Soh, W.Y. Dordrecht: Springer. https://doi.org/10.1007/978-94-017-1203-3_17.

Stringam, G.R., V.K. Bansal, M.R. Thiagarajah, D.F. Degenhardt and J.P. Tewari. 1995. Development of an agronomically superior blackleg resistant cultivar in *Brassica napus* L. using doubled haploidy. *Can. J. Plant Sci.* 75: 437–439.

Sunderland, N. and M. Roberts. 1977. New approach to pollen culture. *Nature* 270: 236–238.

Takebe, I., G. Lebib and G. Melchers. 1971. Regeneration of whole plants from isolated mesophyll protoplasts of tobacco. *Naturwissenschaften* 58: 318–320.

Takebe, I., Y. Otsuki and S. Aoki. 1968. Isolation of tobacco mesophyll cells in intact and active state. *Plant Cell Physiol. Tokyo* 9: 115–124.

Tamaoki, T. and A.J. UIlstrup. 1958. Cultivation in vitro of excised endosperm and meristem tissues of com. *Bull. Torrey Bot. Club* 85: 260–272.

Tazawa, M. and J. Reinert. 1969. Extracellular and intracellular chemical environments in relation to embryogenesis *in vitro*. *Protoplasma* 68: 157–173.

Tepfer, S.S., R.I. Greyson, W.R. Craig and J.L. Hindman. 1963. *In vitro* culture of floral buds of *Aquilegia*. *Am. J. Bot* 50: 1035–1045.

Tepfer, S.S., A.J. Karpoff and R.I. Greyson. 1966. Effect of growth substances on excised floral buds of Aquilegia. *Am. J. Bot.* 53: 148–157.

Thomas, B.R. and D. Pratt. 1981. Efficient hybridization between *Lycopersicon esculentum* and *L. peruvianum* via embryo callus. *Theor. Appl. Genet.* 59: 215–219.

Uchimiya, H., T. Kameya and N. Takanashi. 1971. In vitro culture of unfertilized ovules in *Solanum melongena* and *Zea mays*. *Jpn. J. Breed.* 21: 247–250.

Umbeck, P.F. and K. Norstog. 1979. Effects of abscisic acid and ammonium ion on morphogenesis of cultured barley embryos. *Bull. Torrey Bot. Club* 106: 110–116.

Vasil, I.K. 1984. *Cell Culture and Somatic Cell Genetics of Plants*. Florida: Academic Press.

Wang, Y., T. Yan, L. YueFei, Z. Yun, Z, Jun, F. JianYun, F. Hui 2011. High frequency plant regeneration from microspore-derived embryos of ornamental kale (*Brassica oleracea* L. var. *acephala*). *Sci. Hortic.* 130: 296–302.

Wenzel, G. 1980. Anther culture and its role in plant breeding. In *Proc. Symp. Plant Tissue Culture: Genetic Manipulation and Somatic Hybridization of Plant Cells*, eds. P.S. Rao, M.R. Heble and M.S. Chadha, 68–78. Bombay: Bhabha Atomic Research Centre.

White, P.R. 1932. Plant tissue cultures. A preliminary report of results obtained in the culturing of certain plant meristems. *Arch. Exp. Zellforsch* 12: 602–620.

Williams, E.G. and G. de Lautour. 1980. The use of embryo culture with transplanted nurse endosperm for the production of interspecific hybrids in pasture legumes. *Int. J. Plant Sci.* 141 (3). https://doi.org/10.1086/337152.

Yang, H.-Y. and C. Zhou. 1992. Experimental plant reproductive biology and reproductive cell manipulation in higher plants. *Am. J. Bot.* 79: 354–363.

Zimmermann, U. and P. Scheurich. 1981. High frequency fusion of plant protoplasts by electric fields. *Plants* 151: 26–32.

Zenkteler, M. 1990. In vitro fertilization and wide hybridization in higher plants. *Critical Rev. Plant Sci.* 9: 267–279.

# Glossary

**Adventitious:** Developing from any plant part other than the normal one, e.g., adventitious roots develop from stems or leaves not from the embryonic root (radicle).

**Adventitious embryony:** Development of the embryo directly from a somatic cell of the ovule without formation of a gametophyte.

**Agamospermy, nonrecurrent:** Development of sporophyte from reduced or unreduced megasporangial cells. Development pollination dependent. Embryo haploid.

**Agamospermy, recurrent:** Pollination not necessary. Embryo diploid.

**Albuminous seed:** A seed that contains relatively large amounts of endosperm at maturity.

**Amphimixis:** Sexual reproduction Amphimixis is the process of reproduction where megaspore mother cells undergo meiosis to form a reduced embryo sac and then form embryo after double fertilization to give rise to seeds with both maternal and paternal contribution.

**Anatropous ovule:** Ovule that is inverted and fused to the funiculus so that the micropyle is situated next to the funiculus.

**Aneuploid:** Cell or organism possessing a chromosome number other than the haploid number or an exact multiple of it.

**Androecium (pl.: androecia)** A collective term referring to all the stamens of one flower.

**Androgenesis:** Development of plants from male gametophyte.

**Anther:** The pollen-bearing portion of the stamen, borne at the top of filament.

**Anther culture:** Culture of anthers (or pollen grains) on a suitable medium for production of callus or haploid plants.

**Anthesis:** The process of dehiscence of the anthers; the period of pollen distribution.

**Antipodals:** Generally, a unit of three cells present at the chalazal end of the organized embryo sac.

**Aperture of pollen:** Thin, variously shaped region of the pollen wall through which the pollen tube emerges during germination.

**Apetalous:** Lacking petals.

**Apomixis:** Apomixis (asexual seed formation) is a phenomenon in which a plant bypasses meiosis and fertilization to form a viable seed.

**Apospory:** It is the type of gametophytic apomixis where the unreduced embryo sac arises from a somatic nucellar cell which acquires the developmental program of a functional megaspore.

**Aril:** An integument-like outgrowth from the funicular region of a fertilized ovule. It may cover the seed partially or completely.

**Arrilode:** An aril-like outgrowth but growing downwards from the tip of the outer integument.

**Asymmetrical flower:** Flower lacking a plane of symmetry, i.e., neither radial nor bilateral.

**Auxins:** A group of hormones of plants which are synthesized by growing tips of stem and roots and regulate many aspects of plant growth e.g., IAA.

**Axile placentation:** Ovules attached to the central axis of an ovary with two or more locules.

**Basal placentation:** Ovule or ovules attached at the base of the ovary.

**Bisexual flower:** Flower with both androecium (stamens) and gynoecium (carpels).

**Batch culture:** A suspension culture in which cells grow in a finit volume of nutrient medium and follow a sigmoid pattern.

**Callose:** A carbohydrate that forms an enveloping layer around both the microspore and megaspore mother cell. Callose is also present on the walls of pollen tubes and endothelial cell.

**Callus:** A tissue consisting of dedifferentiated cells generally produced as a result of tissue wounding or of culturing tissues in the presence of auxins in particular.

**Campylotropous ovule:** Ovule that is curved so the micropyle is positioned near the funiculus.

**Carpel:** Organ of a flower that contains ovules and is involved in the production of megaspores, seeds and fruits.

**Carpellate flower:** Flower with a gynoecium (carpel or carpels) but without functional androecium (stamens).

**Caruncle:** A spongy tissue mass produced by the rim part of the integument on the funicular side of the seed or all around the micro Pi.

**Cell culture:** Culture of single or groups of similar cells.

**Cell differentiation:** The process whereby descendants of common parental cell achieve and maintain specialization of structure and function.

**Cellular endosperm:** Division of the primary endosperm nucleus, and all subsequent nuclear divisions, immediately followed by formation of cell walls.

**Chalaza:** The basal part of the ovule that is in contact with funicle.

**Chalazogamy:** Entry of pollen tube through chalaza.

**Cleistogamy:** Pollination and fertilization in an unopened flower bud.

**Cleistogamous:** Condition of a flower that never opens and is self-pollinated and self- fertilized.

**Clone:** Individuals obtained from a single plant through a sexual reproduction.

**Cloning:** Multiplication methods via clones.

**Coenomegaspore:** A cell containing four megaspore nuclei.

**Compound pollen grain:** A condition where the individual microspores derived from a microspore mother cell remain in contact with one another and never become separated.

**Complete flowers:** Flower having all usual parts (sepals, petals stamens and pistils).

**Cross pollination:** is the transfer of pollen from an anther on one plant to a stigma in a flower on a different plant.

**Cultivar:** A category of plants of the same species but with a characteristic phenotype, generated through cultivation by man.

**Culture medium:** Nutrient material that supports cell growth.

**Cybrid:** Plant or cell which is a cytoplasmic hybrid produced by fusion of a protoplast and cytoplasm.

**Cytokinins:** A group of hormones with stimulatory effect divisions of plant cells, e.g., kinetic and zeatin.

**Cytomixis:** Migration of nuclear material from cell to cell.

**Cytoplast:** Enucleated protoplast.

**Diaspores or disseminules:** Units of dispersal.

**Dictyosome:** A stack of thin vesicles held together in a flat or cup-shaped array; dictyosomes receive vesicles from endoplasmic reticulum along their forming face, then modify the material in the vesicle lumen physical or synthesize new material. Vehicles swell and are released from the measuring face.

**Differentiation:** Modification in structure and function of the parts of an organism, owing to increase in specialization.

**Dioecious:** Having staminate and pistillate flowers on different plants of the same species.

**Diploid:** Refers to two full sets of chromosomes in each nucleus, as typically found in sporophytes and zygotes.

**Diplospory:** It is the type of gametophytic apomixis where the unreduced embryo sac arises from a megaspore mother cell with suppressed or modified meiosis. Parthenogenesis: It is the process of spontaneous fertilization-independent development of the embryo from reduced or unreduced egg cell.

**Double fertilization:** The process unique to angiosperms in which one sperm fertilizes the egg (forming a zygote) and the other sperm fertilizes the polar nuclei (forming the primary endosperm nucleus).

**Egg:** Particular cell of egg apparatus destined to be fertilized by a male gamete.

**Egg apparatus:** A group of generally three cells (one egg and two synergids) present towards the micropylar end of the embryo sac.

**Elaiosome:** Hard, oily outgrowth (aril) of a seed that attracts ants.

**Emasculate:** To remove the anthers from a bud or flower before pollen is shed. Emasculation is a normal preliminary step in crossing to prevent self-pollination.

**Embryo:** The rudimentary plant in a seed. The embryo develops from the zygote.

**Embryogenesis:** A pathway of differentiation which is characterised by the formation of organized structures that resemble zygotic embryos.

**Embryoids:** Embryo-like structures produced as a consequence of differentiation processes such as embryogenesis and androgenesis.

**Embryo sac:** Typically, an 8-nulceate female gametophyte. The embryo sac arises from the megaspore by successive mitotic divisions.

**Endoplasmic reticulum (ER):** A system of narrow tubes and sheets of membrane that form a network throughout the cytoplasm. If ribosomes are attached, it is rough ER (RER) and is involved in protein synthesis. If no ribosomes are attached, it is smooth ER (SER) and is involved in lipid synthesis.

**Endosperm:** Triploid tissue which arises from the triple fusion of a sperm nuclei with the polar nuclei of the embryo sac. In seeds of certain species, the endosperm persists as a storage tissue and is used in the growth of the embryo and by the seedling during germination.

**Endothecium:** Hypodermal layer of anther wall, which at maturity, shows fibrous thickenings.

**Exalbuminous seed:** A seed with little or no endosperm at maturity.

**Exaptive evolution:** A character not selected by natural selection for its current utility.

**Explant:** A plant organ or piece of tissue used to initiate a tissue culture.

**Fertilization:** Union of an egg and a sperm (gamete) to form a zygote.

**Filament:** Stalk of a stamen.

**Filiform apparatus:** Apical wall of synergids form finger-like projections.

**Gamete:** A haploid sex cell, such as an egg or sperm.

**Gametoclonal variation:** Variation observed among plants regenerated from gametic culture.

**Gametophytic apomixis:** Those apomictic pathways where a diploid cell divides mitotically and differentiates to form an unreduced embryo sac which eventually gives rise to the maternal embryo.

**Generative cell:** In the pollen grains of seed plants, the cell that gives rise directly to the sperm cells.

**Genetic engineering:** The experimental manipulation of DNA (or RNA) of different species producing recombinant DNA, which includes some genes from both species.

**Genome:** Genome consists of all the chromosomes of a diploid species that are distinct from each other with respect to their gene content, and often morphology. Members of a genome do not pair.

**Genotype:** Genetic constitution of an organism.

**Geographic parthenogenesis:** It is the term used for higher-latitude geographic distribution of asexuals compared to their sexual counterparts.

**Germplasm:** Total genetic variability available to species.

**Gibberellins:** A class of hormones involved in stem elongation, seed germination and other processes.

**Gynoecium (pl.: gynoecia)** A collective term referring to all the carpels of a flower.

**Haploid:** Having a single set (genome) of chromosomes in a cell or an individual; the reduced number (n), as in a gamete.

**Heterokaryon:** A cell in which two or more nuclei of unlike genetic makeup are present. Intron sequence of DNA interrupting the coding sequence of gene.

**Hilum (pl. hyla):** Scar produced when a seed breaks from the funiculus.

**Homokaryon:** Protoplast or cell with two, generally identical nuclei, a product of somatic hybridization.

**Hybrid:** (1) The first-generation offspring of a cross between two individuals differing in one or more genes; (2) the progeny of a cross between species of the same genus or of different genera.

**Hypocotyl:** The portion of an embryo axis located between the cotyledons and the radicle.

**Imperfect flower:** A flower lacking either stamens or pistils.

**Incompatibility:** Failure to obtain fertilization and seed formation after self-pollination.

**Incomplete flower:** A flower lacking one or more of the four essential flower parts.

**Integument:** In flowers, the covering layer over the nucellus of an ovule.

**Intine:** Inner layer of pollen grain below the exine.

***In vitro*:** (Latin, "in glass") Any process carried out in sterile cultures,

***In vivo*:** (Latin, "in life") Any process occurring in whole organism.

**Karyogamy:** Fusion of the nuclei of two gametes after protoplasmic fusion (plasmogamy).

**Karyokinesis (pl.: Karyokineses):** Division of a nucleus, as opposed to the cell division, cytokinesis. The two types of karyokinesis are mitosis and meiosis.

**Male sterility:** A condition in which pollen is absent or non-functional in flowering plants.

**Megagametophyte:** Female gametophyte.

**Megaspore:** One of the four haploid spores originating from the meiotic divisions of the diploid megaspore mother cell in the ovary and which gives rise to the megagametophyte.

**Megaspore mother cell:** Diploid cell in ovary which gives rise, through meiosis, to four haploid megaspores.

**Meiosis:** Two successive nuclear divisions, in the course of which the diploid chromosome number is reduced to the haploid.

**Meiosis I:** The first division of meiosis, during which the chromosome number per nucleus is reduced.

**Meiosis II:** The second division of meiosis, during which centromeres divide and the two chromatids of one chromosome become an independent chromosome.

**Meristem culture:** Culture of apical meristems, particularly shoot apical meristem, for production of shoots and plantlets,

**Mentor pollen:** Compatible pollen treated in several ways to block its fertilizing power but retains its ability to stimulate incompatible pollen to effect fertilization.

**Meristem:** A group of cells specialized for the production of new cells. **Apical meristem:** Located at the farthest point of the tissue or organ produced. **Basal meristem:** Located at the base. **Intercalary meristem:** Located between the apex and the base. **Lateral meristem:** Located along the side.

**Microsporangium (pl.: microsporangia):** A structure that produces microspores.

**Microspore:** One of the four haploid spores originating from the meiotic division of the microspore mother cell in the anther which gives rise to the pollen grain.

**Microspore mother cell:** Diploid cell in the anther which gives rise, through meiosis, to four haploid microspores. Synonym: microsporocyte.

**Microspore tetrad:** Aggregate of four microspores.

**Microtubules:** A skeletal element in eukaryotic cells, composed of alpha and beta tubulin. Microtubules constitute the mitotic spindle, phragmoplast, and axial component of flagella.

**Mitosis:** A process of nuclear division in which the chromosomes are duplicated longitudinally, forming two daughter nuclei each having a chromosome complement equal to that of the original nucleus.

**Monad:** A single pollen grain.

**Monoecious:** Having staminate and pistillate flowers on the same plant.

**Monocolpate:** Pollen grain with a single, long, grooved aperture.

**Monoporate:** Pollen grain with a single, pore-like aperture.

**Monosulcate:** Pollen grain with a single, long and grooved aperture located at the pole.

**Morphogenesis:** Developmental pathways in differentiation which result in the formation of recognized tissues.

**Morphogenic (morphogenetic) response:** A response in which the quality of the plant changes such as conversion from a vegetative state to a floral state.

**Nectary:** A gland that secretes a sugary solution that typically attracts pollinators.

**Nexine:** Inner portion of the exine of pollen composed of the endexine and foot layer.

**Nucellus:** In flowering plants, another name for the sporangium wall within the ovule.

**Nuclear endosperm:** Successive divisions of the primary endosperm nucleus not followed by cell wall formation.

**Ontogeny:** Synonym for development, morphogenesis.

**Orthotropus ovule:** Straight ovule with the micropyle at the apex and the funiculus at the base.

**Ovary:** The enlarged basal portion of the pistil, in which the seeds are borne.

**Ovule:** The structure which bears the female gamete and becomes a seed after fertilization.

**Palynology:** Study of the form and structure of pollen and spores.

**Pantoporate:** Pollen grain with numerous pore-like apertures.

**Parthenocarpy:** The production of fruits without fertilization.

**Parthenogenesis:** Development of an embryo from an unfertilized egg.

**Perfect flower:** Flower possessing both stamens and pistils.

**Perianth:** Collective term for calyx and corolla, or all tepals (when calyx and corolla are not distinguished), of a flower.

**Pericarp:** Technical term for the fruit wall, composed of one or more of the following exocarp, mesocarp, endocarp.

**Perisperm:** A nutritive tissue in seeds of the dicot order Caryophyllales, formed as nucellus cells proliferation.

**Phenotype:** (1) Physical or external appearance of an organism as contrasted with its genetic constitution (genotype); (2) a group of organisms with similar physical or external make up.

**Phragmoplast:** That part of the spindle fibres which is directly or indirectly involved in the organization of cell plate during cell division.

**Pistil:** The seed-bearing organ in the flower, composed of the ovary, the style, and the stigma.

**Pistillate flower:** A flower bearing pistils but no stamens.

**Pistillode:** Sterile pistil.

**Placenta (pl.: placentae or placentas):** Tissue in the ovary of a carpel to which the ovules are attached.

**Plantlet:** A small rooted shoot or germinated embryo.

**Plumule:** Synonym for epicotyl.

**Polar nuclei:** Two centrally located nuclei in the embryo sac which unite with the second sperm in triple fusion. In certain seeds the product of this triple fusion develops into the endosperm.

**Pollen grain:** The male gametophyte, originating from a microspore.

**Pollen mother cell:** See microspore mother cell.

**Pollination:** Transfer of pollen from the anther to the stigma of the same individual or a different individual.

**Pollination syndromes:** Floral characteristics associated with pollination by various biotic or abiotic means, e.g., bee pollination, bird pollination, wind pollination, etc.)

**Pollen sac:** Chamber in the anther that contains the pollen grain.

**Pollen tube:** Tube formed by the germinating pollen grain that carries the sperm to ovule.

**Pollinarium (pl. pollinaria):** Mass of pollen grains transported as a unit, as in many Orchidaceae and Apocynaceae (subfamily Asclepiadioideae).

**Polyad:** Small cluster of pollen grains.

**Polycolpate:** Pollen grain with several long, grooved apertures.

**Polycolporate:** Pollen grain with several long, grooved apertures, each with a central pore.

**Polyphenic origin:** When multiple forms or phenotypes can arise from a single genotype by differing external conditions.

**Polyphyletic origin:** When a phenotype is derived independently from more than one common evolutionary ancestor.

**Polyploid:** An organism with more than two sets (genomes) of chromosomes in its body cells.

**Polyporate:** Pollen grain with many pore-like apertures.

**Polyspory:** Polysporic species include both bisporic (meiosis produces only two megaspores, each with two haploid nuclei, and one of them forms functional megaspore; therefore, two haploid nuclei are involved in the formation of embryo-sac) and tetrasporic species (meiosis produces one four-nucleate megaspore, and all haploid nuclei are involved in the formation of embryo sac).

**Porate aperture:** Round and pore-like aperture.

**Proembryo:** An early stage of embryo development, usually considered to encompass the stages between the zygote and the initiation of the cotyledon primordia.

**Protoandry:** Maturation of anthers, and shedding of pollen before stigmas become receptive (in a bisexual flower or monoecious individual).

**Protoclonal variation:** Variation observed among plants regenerated from protoplast cultures.

**Protogyny:** Maturation of stigmas before pollen is shed (in a bisexual flower or monoecious individual).

**Protoplast:** A cell which lacks a wall.

**Raphe:** In seeds, a ridge caused by the fusion of the funiculus to the side of the ovule.

**Receptacle:** The stem (axis) of a flower, to which all the other parts are attached.

**Recombinant DNA (rDNA):** Hybrid DNA produced by joining pieces of DNA from different organisms together *in vivo*.

**Ruminate endosperm:** Endosperm with regular to irregular ingrowths of the seed coat.

**$S_0$:** Symbol used to designate the original selfed plant.

**$S_1$, $S_2$, etc:** Symbols for designating first selfed generation (progeny of $S_0$ plant), second selfed generation (progeny of $S_1$ plant) etc.

**Scutellum (pl. scutella):** Shield-shaped and specialized cotyledon of the embryo of members of Poaceae (grass family), to which the remaining portion of the embryo is laterally attached.

**Seed:** A mature ovule with its normal coverings. A seed consists of the seed coat, embryo, and in certain plants, an endosperm.

**Seed coat:** Protective outer layer or layers of the seed, developed from the integument or integuments.

**Self-compatibility:** Ability of an individual's own pollen to effect self-fertilization.

**Self-fertile:** Capable of fertilization and setting seed after self-pollination.

**Self-fertilization:** Union of a sperm and egg from the same individual.

**Self-incompatibility:** Failure of an individual's own pollen to effect self-fertilization, typically due to its failure to germinate on the stigma (sporophytic incompatibility) or failure to grow through the style (gametophytic incompatibility).

**Self-pollination:** The transfer of pollen from an anther to the stigma of the same flower or another flower on the same plant, or within a clone.

**Self-sterility:** Failure to complete fertilization and setting seed after self-pollination.

**Sexine:** Outer portion of the exine of a pollen grain.

**Sexual reproduction:** Reproduction involving germ cells and union of gametes.

**Somaclonal variation:** Variation observed among plants regenerated from somatic cultures.

**Somatic:** Referring to diploid body cells, normally with one set of chromosomes coming from the male parent and one set from the female parent.

**Somatic embryogenesis:** Development of embryos from somatic cells in culture whose structure is similar to zygotic embryos found in seeds and with analogous embryonic organs such as cotyledons or cotyledonary leaves.

**Somatic hybridization:** A technique of fusing protoplasts from two contrasting genotypes for production of hybrids or cybrids which contain various mixtures of nuclear and/or cytoplasmic genomes, respectively.

**Sperm:** A male gamete.

**Sporophytic apomixis or adventitious embryony:** This variation of apomixis involves the formation of a maternal embryo from one or more somatic cells of the ovule.

**Sporopollenin:** Polymer with saturated and unsaturated hydrocarbons and phenolics that are cross-linked to each other, the biopolymer most resistant to biodegradation.

**Stamen:** The pollen-bearing organ in the flower; composed of an anther and a filament (stalk).

**Staminate flower:** A flower bearing stamens but not a functional gynoecium (carpel or carpels).

**Staminode:** Sterile stamen.

**Sterility:** Failure to complete fertilization and obtain seed as a result of defective pollen, ovules or other aberrations.

**Stigma:** The portion of the carpel (several fused carpels) that receives and facilitates the germination of the pollen.

**Style:** The stalk connecting the ovary and the stigma,

**Sulcate:** Pollen grain with a sulcus.

**Sulcus:** Long, grooved aperture positioned at the pole of a pollen grain.

**Suspensor:** In seed plant embryos, the stalk of cells that pushes the embryo into the endosperm.

**Sympatric:** Two species are said to be sympatric if they occupy in the same or overlapping geographic areas.

**Synergids:** In the egg apparatus of an angiosperm megagametophyte. There is an egg and one or two adjacent cells, synergids; the pollen tube enters into one of the synergids.

**Syngamy:** The act of fusion of sperm and egg nucleus.

**Syngenesious stamens:** Stamens fused by their anthers and filaments free.

**Synthetic seeds:** Embryoids preserved in a thin layer of calcium alginate.

**Syntopic:** Two species are said to be syntopic if they can coexist without interference with each other.

**Taxon:** A term that refers to any taxonomic group such as species, genus, family and so on.

**Tectate exine:** Exine with the outer layer, or tectum, supported by numerous small columns (columellae).

**Tegmen (pl. tegmina):** Inner part of the seed coat, developing from the inner integument.

**Testa (pl. testae):** Synonym for seed coat. Outer part of the seed coat, developing from the outer integument.

**Tetrad:** Group of four; usually applied to pollen grains united in a group of four that have not separated after meiosis.

**Tetradynamous stamen:** With four long and two short stamens.

**Thrum:** Short-styled floral form in a heterostylous species.

**Tissue culture:** Culture of plant cells and tissues in vitro on artificial media.

**Totipotency:** Potentiality or property of a cell to produce a whole organism.

**Transgenic organisms:** A genetically manipulated organism containing in its genome one or more inserted genes of another species.

**Tricolpate:** Pollen grain with three, long, grooved apertures, each with a central pore.

**Tricolporate:** Pollen grain with three, long, grooved apertures, each with a central pore.

**Triploid:** A nucleus that contains three sets of chromosomes.

**Triporate:** Pollen grains with three equatorial, pore-like apertures.

**Tubule:** Little tube, as on the anther of species of Vaccinium (blueberries) and relatives (in the family Ericaceae).

**Unisexual flower:** Flower lacking either an androecium (stamens) or a gynoecium (carpels).

**Unitegmic:** With one integument.

**Vacuole:** A membrane-bounded (tonoplast) space larger than a vesicle which stores material either dissolved in water or by a crystalline or flocculent mass.

**Vegetative cell:** In the pollen grain of seed plants, the cell or cells that do not give rise to sperm cells; the cell that is not the generative cell.

**Versatile:** Structure, such as an anther, that is attached at its midpoint.

**Viscin:** Elastic/ or somewhat sticky material often covering pollen grains.

**Xenia:** The immediate effect of pollen on the character of the endosperm.

**Zygote:** The cell resulting from the fusion of the gametes.

# Subject and Plant Index

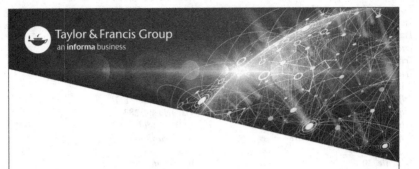

Printed in the United States
by Baker & Taylor Publisher Services